W9-AWR-181

ETHICS IN INFORMATION TECHNOLOGY

Second Edition

ETHICS IN INFORMATION TECHNOLOGY

Second Edition

George W. Reynolds

THOMSON
™
COURSE TECHNOLOGY

Australia • Canada • Mexico • Singapore • Spain • United Kingdom • United States

Ethics in Information Technology, Second Edition

by George W. Reynolds

Senior Product Manager:
Eunice Yeates-Fogle

Acquisitions Editor:
Maureen Martin

Development Editor:
Dan Seiter

Production Editor:
Jill Klaffky

Cover Designer:
Laura Rickenbach

Composition/Prepress:
GEX Publishing Services

Senior Manufacturing Coordinator:
Justin Palmeiro

Copy Editor:
Mark Goodin

Proofreader:
Kim Kosmatka

Indexer:
Rich Carlson

BRIEF CONTENTS

Preface   xiii

Chapter 1
An Overview of Ethics   1

Chapter 2
Ethics for IT Professionals and IT Users   33

Chapter 3
Computer and Internet Crime   67

Chapter 4
Privacy   105

Chapter 5
Freedom of Expression   145

Chapter 6
Intellectual Property   171

Chapter 7
Software Development   207

Chapter 8
Employer/Employee Issues   237

Chapter 9
The Impact of Information Technology on the Quality of Life   267

Appendix A
A Brief Introduction to Morality   293

Appendix B
*Association for Computing Machinery (ACM) Code of Ethics and
Professional Conduct*   309

Appendix C
Association of Information Technology Professionals (AITP) Code of Ethics   317

Appendix D
Software Engineering Code of Ethics and Professional Practice   321

Appendix E
PMI Member Ethical Standards and Member Code of Ethics   329

Answers to Self-Assessment Questions   331

Glossary   333

Index   341

TABLE OF CONTENTS

Preface   xiii

Chapter 1   *An Overview of Ethics*   1
  Vignette   1
    Parent Company of Philip Morris Strives for Integrity   1
  What Is Ethics?   3
    Definition of Ethics   3
    The Importance of Integrity   4
  Ethics in the Business World   5
    Why Fostering Good Business Ethics Is Important   8
    Improving Corporate Ethics   10
    When Good Ethics Result in Short-Term Losses   15
    Creating an Ethical Work Environment   15
    Ethical Decision Making   17
  Ethics in Information Technology   20
    An Overview of This Text   21
  Summary   22
  Self-Assessment Questions   22
  Review Questions   23
  Discussion Questions   24
  What Would You Do?   24
  Cases   25
  End Notes   30

Chapter 2   *Ethics for IT Professionals and IT Users*   33
  Vignette   33
    Cleveland State University Accuses PeopleSoft of Selling Vaporware   33
  IT Professionals   35
    Are IT Workers Professionals?   35
    Professional Relationships That Must Be Managed   36
  The Ethical Behavior of IT Professionals   44
    Professional Codes of Ethics   44
    Professional Organizations   45
    Certification   46
    Government Licensing   48
  IT Professional Malpractice   50
  IT Users   51
    Common Ethical Issues for IT Users   51
    Supporting the Ethical Practices of IT Users   52
  Summary   54
  Self-Assessment Questions   55
  Review Questions   56
  Discussion Questions   57
  What Would You Do?   57
  Cases   59
  End Notes   64

Chapter 3   *Computer and Internet Crime*   67
  Vignette   67
    Treatment of Sasser Worm Author Sends Wrong Message   67
  IT Security Incidents: A Worsening Problem   68
    Increasing Complexity Increases Vulnerability   69
    Higher Computer User Expectations   69
    Expanding and Changing Systems Introduce New Risks   70
    Increased Reliance on Commercial Software with Known Vulnerabilities   70
    Types of Attacks   72
    Perpetrators   75
  Reducing Vulnerabilities   81
    Risk Assessment   82
    Establishing a Security Policy   82
    Educating Employees, Contractors, and Part-Time Workers   83
    Prevention   84
    Detection   88
    Response   89
  Summary   94
  Self-Assessment Questions   95
  Review Questions   96
  Discussion Questions   96
  What Would You Do?   97
  Cases   98
  End Notes   102

Chapter 4   *Privacy*   105
  Vignette   105
    Have You Been Swiped?   105
  Privacy Protection and the Law   107
    The Right of Privacy   108
    Recent History of Privacy Protection   108
  Key Privacy and Anonymity Issues   113
    Governmental Electronic Surveillance   113
    Data Encryption   118
    Identity Theft   121
    Consumer Profiling   123
    Treating Consumer Data Responsibly   126
    Workplace Monitoring   127
    Spamming   128
    Advanced Surveillance Technology   130
  Summary   132
  Self-Assessment Questions   134
  Review Questions   135
  Discussion Questions   135
  What Would You Do?   136
  Cases   137
  End Notes   141

Chapter 5   *Freedom of Expression*   145
  Vignette   145
    China Stifles Online Dissent   145
  First Amendment Rights   146
    Obscene Speech   147
    Defamation   148

Freedom of Expression: Key Issues   148
    Controlling Access to Information on the Internet   148
    Anonymity   153
    National Security Letters   156
    Defamation and Hate Speech   157
    Pornography   158
Summary   161
Self-Assessment Questions   162
Review Questions   163
Discussion Questions   164
What Would You Do?   164
Cases   165
End Notes   168

Chapter 6   *Intellectual Property*   171
Vignette   171
    Apple Trade Secrets Revealed?   171
What Is Intellectual Property?   172
    Copyrights   173
    Patents   176
    Trade Secret Laws   180
Key Intellectual Property Issues   184
    Plagiarism   184
    Reverse Engineering   186
    Open Source Code   188
    Competitive Intelligence   189
    Cybersquatting   192
Summary   194
Self-Assessment Questions   195
Review Questions   196
Discussion Questions   197
What Would You Do?   198
Cases   198
End Notes   204

Chapter 7   *Software Development*   207
Vignette   207
    Antivirus Bug Brings Computers to a Standstill   207
Strategies to Engineer Quality Software   209
    The Importance of Software Quality   211
    Software Development Process   215
    Capability Maturity Model Integration for Software   216
Key Issues in Software Development   217
    Development of Safety-Critical Systems   218
    Quality Management Standards   221
Summary   224
Self-Assessment Questions   225
Review Questions   226
Discussion Questions   227
What Would You Do?   227
Cases   229
End Notes   233

Chapter 8   *Employer/Employee Issues*   237
  Vignette   237
    Microsoft and Tata Promote Chinese Outsourcing   237
  Use of Nontraditional Workers   239
    Contingent Workers   240
    H-1B Workers   243
    Offshore Outsourcing   246
  Whistle-Blowing   251
    Protection for Whistle-Blowers   252
    Dealing with a Whistle-Blowing Situation   254
  Summary   257
  Self-Assessment Questions   258
  Review Questions   259
  Discussion Questions   259
  What Would You Do?   259
  Cases   261
  End Notes   265

Chapter 9   *The Impact of Information Technology on the Quality of Life*   267
  Vignette   267
    Technological Advances Create Digital Divide in Healthcare   267
  The Impact of IT on the Standard of Living and Productivity   269
    The Digital Divide   272
  The Impact of IT on Healthcare Costs   275
    Electronic Health Records   276
    Use of Mobile and Wireless Technology   278
    Telemedicine   278
    Medical Information Web Sites for Lay People   279
  Summary   281
  Self-Assessment Questions   282
  Review Questions   283
  Discussion Questions   284
  What Would You Do?   284
  Cases   285
  End Notes   289

Appendix A   *A Brief Introduction to Morality*   293
  Introduction   293
  The Knotty Question of Goodness   294
  Relativism: Why "Common Sense" Won't Work   295
    Egoism vs. Altruism   297
  Deontology, or the Ethics of Logical Consistency and Duty   298
  Happy Consequences, or Utilitarianism   300
    Promises and Contracts   303
  A Return to the Greeks: The Good Life of Virtue   304
    Feminism and the Ethics of Care   305
  Pluralism   306
  Summary   307

Appendix B   *Association for Computing Machinery (ACM) Code of Ethics and Professional Conduct*   309
    Preamble   309
    1. General Moral Imperatives   310
    2. More Specific Professional Responsibilities   312
    3. Organizational Leadership Imperatives   314
    4. Compliance with the Code   316

Appendix C   *Association of Information Technology Professionals (AITP) Code of Ethics*   317
    AITP Standards of Conduct   317

Appendix D   *Software Engineering Code of Ethics and Professional Practice*   321
    Preamble   321
    Principles   322

Appendix E   *PMI Member Ethical Standards and Member Code of Ethics*   329
    Preamble   329
    Member Code of Ethics   329

Answers to Self-Assessment Questions   331

Glossary   333

Index   341

PREFACE

We are excited to publish the second edition of *Ethics in Information Technology*. This new edition builds on the success of the first and meets the need for a book that helps readers understand many of the legal, ethical, and societal issues associated with IT. We have responded to the feedback from first edition adopters, students, and other reviewers to create an improved text. We think you will be pleased with the results.

Ethics in Information Technology, Second Edition fills a void of practical business information for business managers and IT professionals. The typical introductory Information Systems book devotes one chapter to ethics and IT but provides little practical advice. Such limited coverage does not meet the needs of business managers and IT professionals—the people who are primarily responsible for addressing ethical issues in the workplace. What is missing is an examination of the different ethical situations that arise in IT and practical advice for addressing these issues.

Ethics in Information Technology, Second Edition has enough substance for an instructor to use it in a full-semester course in computer ethics. You can also use the book as a reading supplement for such courses as Introduction to Management Information Systems, Principles of Information Technology, Managerial Perspective of Information Technology, Computer Security, E-Commerce, and so on.

ORGANIZATION

The nine chapters in this book address different aspects of ethics in information technology.

- Chapter 1, "An Overview of Ethics," provides an introduction to ethics, ethics in business, and the relevance of discussing ethics in IT. The chapter also discusses some philosophical approaches to ethical decision making and suggests a model for ethical decision making.

- Chapter 2, "Ethics for IT Professionals and IT Users," explains the importance of ethics to IT professionals in their business relationships, and the role that certification and licensing have (or don't have) in legitimizing professional standards. The chapter also emphasizes the significance of IT professional organizations and their codes of ethics.

- Chapter 3, "Computer and Internet Crime," describes the types of ethical decisions IT professionals must make and the business needs they must balance when dealing with security issues. It provides a useful classification of computer crimes and their perpetrators. In addition, the chapter explains how to manage security vulnerabilities and respond to specific security incidents to fix problems quickly and improve ongoing security measures.

- Chapter 4, "Privacy," covers the issue of privacy and how the use of IT affects privacy rights. Although legislation has often addressed privacy rights over the years, little of it has specifically affected private businesses. Personal information that businesses gather using IT can be useful to obtain or keep customers (or to monitor employees), but privacy advocates are concerned about how much information can be gathered, with whom it can be shared, how the information is gathered in the first place, and how it is used. These concerns also extend to law enforcement and government.

- Chapter 5, "Freedom of Expression," addresses issues raised by the growing use of the Internet as a means for freedom of expression. The First Amendment of the U.S. Constitution provides extensive protection of free speech, even if it is offensive or annoying. However, the ease and anonymity with which Internet users communicate can pose problems for people who might be adversely affected by such communication. For example, the chapter discusses attempts to control access to Internet content that is unsuitable for children or unnecessary in a business environment; in both cases, the solutions have had only limited success.

- Chapter 6, "Intellectual Property," defines intellectual property and explains how technology is protected and not protected by copyright, patent, and trade secret laws. The chapter discusses several key issues that are relevant to ethics in IT, including reverse engineering of software, software license legislation, competitive intelligence gathering, and cybersquatting.

- Chapter 7, "Software Development," provides a thorough discussion of the software development process and the importance of software quality. The chapter covers issues that software manufacturers must consider when deciding "how good is good enough?" with regard to their software products, particularly when the software is safety-critical and its failure can cause loss of human life. These issues include deciding what risks are being taken, the extent of quality assurance to use, and standards that may be adopted to achieve quality goals.

- Chapter 8, "Employer/Employee Issues," addresses two significant topics regarding employee and employer relationships. The first topic addresses the ethics of using contingent workers, contractors, foreign workers, and employees of offshore outsourcing companies. The second topic addresses the risks and protections of whistle blowing, the ethical decisions that whistle blowers must make, and a process they might use after deciding they must take action against their own company. Both topics have ethical, social, and economic implications that should be of great interest to students.

- Chapter 9, "The Impact of Information Technology on the Quality of Life," examines IT's effect on major issues of the early 21st century, including the standard of living and productivity, the so-called digital divide, and healthcare costs.

- Appendix A provides an in-depth discussion of how ethics and moral codes developed through time. Appendices B through E include codes of ethics for several important IT professional organizations.

PEDAGOGY

Ethics in Information Technology, Second Edition employs a variety of pedagogical features to enrich the learning experience and provide interest for the instructor and student.

- **Opening Quotation.** Each chapter begins with a quotation to stimulate interest in the chapter material.

- **Vignette.** At the beginning of each chapter, a brief real-world example illustrates the issues to be discussed and piques the reader's interest.

- **Focus Questions.** Carefully crafted focus questions follow the vignette to further highlight topics that are covered in the chapter.

- **Key Terms.** Key terms appear in bold in the text and are defined in the glossary at the end of the book.

Additional Features

Two additional features appear in most chapters to maintain the reader's interest and motivation, support the goals and themes of the chapter, and generate class discussion.

- **Legal Overview.** This overview summarizes an important aspect of the law as it applies to ethics and IT.

- **Manager's Checklist.** The checklist provides a practical and useful list of questions to consider when making a business decision.

End-of-Chapter Material

To help students retain key concepts and expand their understanding of important IT concepts and relationships, the following sections are included at the end of every chapter:

- **Summary.** Each chapter includes a summary of the key issues raised. These items relate to the Focus Questions posed at the start of the chapter.

- **Self-Assessment Questions.** These questions help students review and test their understanding of key chapter concepts. This is a new feature for the second edition. The answers to the self-assessment questions are listed at the back of the book.

- **Review Questions.** These questions are directly linked to the text; they reinforce the key concepts and ideas in each chapter.

- **Discussion Questions.** These more open-ended questions help instructors generate class discussion to move students deeper into the concepts and explore the numerous aspects of ethics in IT.

- **What Would You Do?** These exercises present realistic dilemmas that encourage students to think critically about the ethical principles presented in the text. The number of these exercises has been increased for the second edition.

- **Case Projects.** The number of real-world cases has been increased to three per chapter for this edition. These cases reinforce important ethical principles and

IT concepts and show how real companies have addressed ethical issues associated with IT. Questions after each case focus students on its key issues and ask them to apply the concepts presented in the chapter.

ABOUT THE AUTHOR

George Reynolds brings a wealth of computer and industrial experience to this project. He has more than 30 years of experience in government and commercial IT organizations. He has authored 16 IT-related college texts and has taught as an Adjunct Assistant Professor at Xavier University (Ohio), College of Mount St. Joseph, and Miami University (Ohio). He is currently an Assistant Professor at the University of Cincinnati in the Information Systems department.

TEACHING TOOLS

The following supplemental materials are available when this book is used in a classroom setting. All of these tools are provided to the instructor on a single CD-ROM. You can also find some of these materials on the Thomson Course Technology Web site at *www.course.com*.

- **Electronic Instructor's Manual**. The Instructor's Manual that accompanies this textbook includes additional instructional material to assist in class preparation, including suggestions for lecture topics. It also includes solutions to all end-of-chapter exercises.

- **ExamView®**. This textbook is accompanied by ExamView, a powerful testing software package that allows instructors to create and administer printed, computer (LAN-based), and Internet exams. ExamView includes hundreds of questions that correspond to the topics covered in this text, enabling students to generate detailed study guides that include page references for further review. The computer-based and Internet testing components allow students to take exams at their computers, and save the instructor time by grading each exam automatically.

- **PowerPoint Presentations**. This book comes with Microsoft PowerPoint slides for each chapter. The slides can be included as a teaching aid for classroom presentation, made available to students on the network for chapter review, or printed for classroom distribution. Instructors can add their own slides for additional topics they introduce to the class.

- **Distance Learning**. Thomson Course Technology is proud to present online courses in WebCT and Blackboard. For more information on how to bring distance learning to your course, contact your local Thomson sales representative.

ACKNOWLEDGMENTS

I want to thank a number of people who helped greatly in the creation of this book: Jill Klaffky, Production Editor, for guiding the book through the production process; Maureen Martin, Acquisitions Editor, for her encouragement and support; Eunice Yeates-Fogle, Senior Product Manager, for overseeing and directing this effort; Dan Seiter, Development Editor, for his many useful suggestions and helpful edits; and my many students at Miami University and the University of Cincinnati, who provided excellent ideas and constructive feedback on the text. In addition, I want to thank a marvelous set of reviewers who offered many useful suggestions:

- Karen Williams, University of Texas-San Antonio

- Catharine Kuchar, Southern Institute for Business and Professional Ethics, Decatur, GA

- Cherie Ann Sherman, Ramapo College of NJ

- Jeff Stewart, Ph.D., Macon State College

Last of all, thanks to my family for all their support and for giving me the time to do this.

—George Reynolds

CHAPTER **1**

AN OVERVIEW OF ETHICS

VIGNETTE

Parent Company of Philip Morris Strives for Integrity

"Nothing is more important than our commitment to integrity—no financial objective, no marketing target, no effort to outdo the competition. Our commitment to integrity must always come first," said Louis C. Camilleri. Coming from the CEO of Altria Group, the parent company of Philip Morris of tobacco fame, such a statement might give pause to many Americans. Yet, Camilleri was referring to a massive initiative to establish a corporate code of conduct among Altria enterprises.

Led by Camilleri, the board of directors, and senior officers, the Altria Compliance and Integrity program has searched for ways to ensure that the code filters down to its employees at all levels. Employees receive customized handbooks that are translated into their native languages and cover information they need to know. Managers receive much more information about the code than their workers. Online courses focus on specific risks and on-the-job scenarios. Employees also watch skits staged with professional actors, answer questions, and engage in discussions with managers and other workers.

At the same time, Altria encourages employees to report violations of the code. Not only can employees approach the Human Resources Department and other departments, they can call a 24-hour Integrity Helpline and voice concerns anonymously. The helpline provides translation services for more than 100 languages. Altria's policy states that employees can be disciplined and even fired for retaliating against anyone who has made a complaint in good faith.

Altria conducts annual audits to determine risk and review best practices. Compliance and Integrity managers have also met with more than 150 employees to help shape the content of their code, which includes such goals as producing accurate records, protecting company assets, and dealing with customers honestly. Altria appears committed to implementing this code. Yet one goal—responding to society's expectation of Altria—may prove increasingly challenging to a company whose products include a carcinogen known to cause approximately 87 percent of all lung cancer deaths in the United States.[1,2,3]

LEARNING OBJECTIVES

As you read this chapter, consider the following questions:

1. What is ethics, and why is it important to act according to a code of principles?
2. Why is business ethics becoming increasingly important?
3. What are corporations doing to improve business ethics?
4. Why are corporations interested in fostering good business ethics?
5. What approach can you take to ensure ethical decision making?
6. What trends have increased the risk of using information technology unethically?

WHAT IS ETHICS?

Each society forms a set of rules that establishes the boundaries of generally accepted behavior. These rules are often expressed in statements about how people should behave, and they fit together to form the **moral code** by which a society lives. Unfortunately, the different rules often have contradictions, and you can be uncertain about which rule to follow. For instance, if you witness a friend copy someone else's answers while taking an exam, you might be caught in a conflict between loyalty to your friend and the value of telling the truth. Sometimes, the rules do not seem to cover new situations, and you must determine how to apply the existing rules or develop new ones. You may strongly support personal privacy, but in a time when employers track employee e-mail and Internet usage, what rules do you think are acceptable to govern the appropriate use of company resources?

The term **morality** refers to social conventions about right and wrong that are so widely shared that they become the basis for an established consensus. However, one's view of what is moral may vary by age, cultural group, ethnic background, religion, and gender. There is widespread agreement on the immorality of murder, theft, and arson, but other behaviors that are accepted in one culture might be unacceptable in another. For example, in the United States it is perfectly acceptable to place one's elderly parents in a managed care facility in their declining years. In most Middle Eastern countries, however, elderly parents would never be placed in such a facility; they remain at home and are cared for by other family members.

Another example concerns attitudes toward the illegal copying of software (piracy), which range from strong opposition to acceptance as a standard approach to business. In 2003, 36 percent of all software in circulation worldwide was pirated, at a cost of $29 billion to software vendors. The highest piracy rates were in Vietnam and China, where 92 percent of the software was pirated. In the United States, the piracy rate was 22 percent.[4]

Even within the same society, people can have strong disagreements over important moral issues—in the United States, for example, issues such as abortion, the death penalty, and gun control are continuously debated, and both sides feel their arguments are on solid moral ground.

Definition of Ethics

Ethics is a set of beliefs about right and wrong behavior. Ethical behavior conforms to generally accepted social norms, many of which are almost universal. However, although nearly everyone would agree that lying and cheating are unethical, what constitutes ethical behavior on many other issues is a matter of opinion. For example, most people would not steal an umbrella from someone's home, but a person who finds an umbrella in a theater might be tempted to keep it. A person's opinion of what represents ethical behavior is strongly influenced by a combination of family influences, life experiences, education, religious beliefs, personal values, and peer influences.

As children grow, they learn complicated tasks—walking, riding a bike, writing the alphabet—that they perform out of habit for the rest of their lives. People also develop habits that make it easier to choose between what society considers good or bad. **Virtues** are habits that incline people to do what is acceptable, and **vices** are habits of unacceptable behavior. Fairness, generosity, honesty, and loyalty are examples of virtues, while vanity, greed, envy, and anger are considered vices. People's virtues and vices help define their **value system**, the complex scheme of moral values by which they live.

The Importance of Integrity

Your moral principles are statements of what you believe to be rules of right conduct. As a child, you may have been taught not to lie, cheat, or steal or have anything to do with those who do. As an adult who makes more complex decisions, you often reflect on your principles when you consider what to do in different situations: Is it okay to lie to protect someone's feelings? Can you keep the extra $10 you received when the cashier mistook your $10 bill for a $20 bill? Should you intervene with a coworker who seems to have an alcohol or chemical dependency problem? Is it okay to exaggerate your work experience on a résumé? Can you cut some corners on a project to meet a tight deadline?

A person who acts with **integrity** acts in accordance with a personal code of principles—integrity is one of the cornerstones of ethical behavior. One approach to acting with integrity is to extend to all people the same respect and consideration that you desire. Unfortunately, this consistency can be difficult to achieve, particularly when you are in a situation that conflicts with your moral standards. For example, you might believe it is important to do as your employer requests and that you should be fairly compensated for your work. However, if your employer insists that you not report recent overtime hours due to budget constraints, a moral conflict arises. You can do as your employer requests or you can insist on being fairly compensated, but you cannot do both. In this situation, you may be forced to compromise one of your principles and act with an apparent lack of integrity.

Another form of inconsistency emerges if you apply moral standards differently according to the situation or the people involved. To be consistent and act with integrity, you must apply the same moral standards in all situations. For example, you might consider it morally acceptable to tell a "little white lie" to spare a friend some pain or embarrassment, but would you lie to a work colleague or customer about a business issue to avoid unpleasantness? Clearly, many ethical dilemmas are not about right versus wrong but involve choices between right versus right. For example, it is right to protect the Alaskan wildlife from being spoiled, and it is right to find new sources of oil to maintain U.S. reserves, but how do you balance these two concerns?

The remainder of this chapter provides an introduction to ethics in the business world. It discusses the importance of ethics in business, outlines what businesses can do to improve their ethics, points out that good ethics is not always good business, provides advice for creating an ethical work environment, and suggests a model for ethical decision making. The chapter concludes with a discussion of ethics as it relates to information technology (IT) and provides a brief overview of the remainder of the text.

Risk is the product of multiplying the likelihood of an event by the impact of its occurrence. Thus, if the likelihood of an event is high and its potential negative impact is large, the risk is considered great. Ethics has risen to the top of business agendas because the risks associated with inappropriate behavior have increased, both in their likelihood and their potential negative impact.

Several corporate trends have increased the likelihood of unethical behavior. First, greater globalization has created a much more complex work environment that spans diverse societies and cultures, making it much more difficult to apply principles and codes of ethics consistently. For example, numerous U.S. companies have garnered negative publicity for moving operations to third-world countries where employees work in conditions that would not be acceptable in most developed parts of the world.

Employees, shareholders, and regulatory agencies are increasingly sensitive to violations of accounting standards, failures to disclose substantial changes in business conditions to investors, nonconformance with required health and safety practices, and production of unsafe or substandard products. Such heightened vigilance raises the risk of financial loss for businesses that do not foster ethical practices or run afoul of required standards. For example, Enron's accounting practices hid the real value of the firm, and in late 2001 the energy company was forced to file for bankruptcy. The case was notorious, but many other recent scandals have occurred in IT companies in spite of safeguards that were enacted as a result of the Enron debacle:

- The U.S. Securities and Exchange Commission (SEC) filed fraud charges against WorldCom in June 2002 for inflating its earnings by $11 billion. The stock peaked at $64.50 in June 1999 and dropped to less than $1 per share three years later. WorldCom eventually became the largest bankruptcy in U.S. history after Chairman Bernard Ebbers resigned and the extent of the fraud was revealed. (It since has emerged from bankruptcy under the name MCI Inc. to become the object of an acquisition war between Qwest and Verizon Communications, which Verizon eventually won.) Under a judgment released in May 2003 by a U.S. District Court in New York, MCI was required to pay a civil penalty of $1.51 billion to defrauded shareholders and bondholders, although the amount was reduced to $500 million under terms of a bankruptcy settlement.[5] Ebbers was convicted in March 2005 of helping to orchestrate the massive accounting fraud.[6] In addition, 10 members of the WorldCom board of directors avoided trial by agreeing to pay a total of $18 million on top of the $250 million they lost when the stock collapsed.[7]
- Qwest Communications International Inc., the primary local phone provider in 14 western states, had been under investigation by the SEC since 2002. On the same day that Ebbers was convicted, the SEC charged former Qwest CEO Joseph Nacchio and six other executives with orchestrating massive financial fraud from April 1999 to March 2002. In 2000, the company allegedly misstated that $3 billion from a one-time sale was a recurring revenue in order to ensure a merger with US West. Qwest executives allegedly reaped tens of millions of dollars in profit while hiding the scheme from investors and the public.[8]

- In 2004, Adelphia Communications Corp. founder John Rigas and his son Timothy were convicted in federal court on charges of conspiracy, bank fraud, and securities fraud. They were charged with hiding $2.3 billion in debt at the cable company, deceiving investors, and stealing company cash to line their own pockets. Although most of the fraud took the form of hidden debt, the trial provided examples of excessive extravagance that has marked other white-collar trials. The prosecutor alleged that Rigas ordered 17 company cars and the company purchase of 3600 acres of timberland for $26 million to preserve the pristine view outside his Coudersport home.[9]

- Several former Computer Associates (CA) executives pleaded guilty to civil and criminal fraud and obstruction of justice for systematically recording sales revenue before contracts were finalized, inflating CA's financial results by about $2.2 billion during 2000 and 2001. The scandal eventually led to the resignation of Sanjay Kumar, the company's CEO, in April 2004.[10] In May 2005, the company said that a continuing review of its accounting practices turned up more improperly recorded transactions from 1998 through 2001. This required CA to restate its financial reports again, reducing revenue in prior periods by up to $110 million in aggregate.[11] (For more on the CA scandal, see the Cases section at the end of the chapter.)

These cases have led to an increased focus on business ethics. Read the following Legal Overview to find out more about one attempt by the U.S. Congress to improve business ethics.

LEGAL OVERVIEW

The Sarbanes-Oxley Act

The U.S. Public Company Accounting Reform and Investor Protection Act of 2002—better known as the Sarbanes-Oxley Act or simply SOX—was enacted in response to public outrage over several major accounting scandals, including those at Enron, WorldCom, Tyco, Adelphia, Global Crossing, and Qwest, plus numerous restated financial reports that clearly demonstrated a lack of oversight within corporate America. The sponsors, Senator Paul Sarbanes (D-Maryland) and Representative Michael Oxley (R-Ohio), wanted to renew investors' trust in corporate executives and their financial reports.

Sarbanes-Oxley Act Section 404, Management Report on Internal Control over Financial Reporting, states that annual reports must contain a signed statement by the CEO and CFO attesting that the information in any SEC filing is accurate. The company must also submit to an audit to prove that it has controls in place to ensure accurate information. The penalties for false attestation can include up to 20 years in jail and significant monetary fines for senior executives. As a result, CEOs and CFOs, their staff, and others spend significant time and energy to document and test internal control processes. The average company spends an estimated 0.1 percent of annual revenues to conform to SOX; for example, a company with annual sales of $10 billion will spend about $10 million, including employee time, fees for outside resources, and software.

continued

A key provision of the act was the creation of the Public Company Accounting Oversight Board (PCAOB). The PCAOB provides oversight for auditors of public companies, including establishing quality control standards for company audits and inspecting the quality controls at audit firms under its oversight. The PCAOB is made up of five full-time appointed members and is overseen by the SEC.

The act attempts to ensure that internal controls or rules are in place to govern the creation and documentation of financial statements. However, it is not specific; for example, it does not define a required set of internal control practices or specify how a business must store records. It simply describes which records to store and for how long. The legislation not only affects the financial side of corporations, it affects IT departments that must store a corporation's electronic records. The act specifies that a firm must have adequate controls, but not necessarily automated controls. Thus, a company can decide whether to make a significant investment in technology that automates its manual processes or make a smaller investment in additional people to double-check everything manually. The SEC set a deadline of November 15, 2004, for the rule to take effect, but problems didn't surface until companies began to file their annual reports and 10K reports in the first quarter of 2005. On March 2, the SEC extended the deadline to July 15, 2006, for small and midsized companies and foreign firms.

In November 2004, SunTrust Banks Inc. became one of the first companies to report an accounting problem that made it impossible to meet its SOX reporting requirements. The firm said an internal audit had found numerous errors in the loan loss allowance calculations for its first and second quarters that were not immediately investigated and corrected. Additional problems revealed inadequate internal control procedures, insufficient validation and testing, and a failure to detect errors in the allowance calculations. The bank had to restate its first- and second-quarter 2004 financial results, and three employees were fired.

In March 2005, more than a dozen companies reported deficiencies with their internal accounting controls, forcing them to delay the filing of annual reports to regulators. The share prices of most of these companies dropped substantially after they disclosed that they needed more time to file audited financial reports with the SEC. These companies joined some 500 others that told shareholders they couldn't ensure their financial reports were accurate and reliable under the new rules.

Because IT systems are used to generate, change, store, and disseminate data, the IT organization must build controls that ensure the information stands up to audit scrutiny. Critical elements include controls that ensure the overall performance and integrity of financial systems, business process applications, and other applications. The audit emphasizes segregation of duties to avoid potential fraud—for example, the same person cannot generate a purchase order and approve its payment—and limiting the authorization to perform critical functions to a few people; for example, only people in quality control should be able to release products that are placed on hold, pending the results of quality tests. The audit also requires companies to document who performs specific roles within the system. In the future, the SEC might revise its interpretation of SOX, creating additional work for IT organizations.[12,13,14,15,16,17,18]

Why Fostering Good Business Ethics Is Important

Corporations have at least five reasons for promoting a work environment in which they encourage employees to act ethically when making business decisions:

1. To gain the goodwill of the community
2. To create an organization that operates consistently
3. To produce good business
4. To protect the organization and its employees from legal action
5. To avoid unfavorable publicity

Gaining the Goodwill of the Community

Although organizations exist primarily to earn profits or provide services to customers, they also have some basic responsibilities to society. Many corporations recognize these responsibilities and make a serious effort to fulfill them. Often, they declare these responsibilities in a formal statement of their company's principles or beliefs. Their socially responsible activities include making contributions to charitable organizations and nonprofit institutions, providing benefits for employees in excess of any legal requirements, and choosing economic opportunities that might be more socially desirable than profitable.

The goodwill that socially responsible activities create can make it easier for corporations to conduct their business. For example, a company known for treating its employees well will find it easier to compete for the best job candidates. On the other hand, companies viewed as harmful to their community may suffer a disadvantage. For example, a corporation that pollutes the air and water (see Figure 1-1) may find that adverse publicity reduces sales, impedes relationships with some business partners, and attracts unwanted government attention.

FIGURE 1-1   Companies that harm a community can harm themselves

Creating an Organization That Operates Consistently

Organizations develop and abide by values to create a consistent approach that meets the needs of their stakeholders—shareholders, employees, customers, suppliers, and the community. They need to emphasize workplace issues that affect their corporate strengths, weaknesses, opportunities, and threats. Although each company's value system is different, many share the following values:

- Operate with honesty and integrity, staying true to corporate principles.
- Operate according to standards of ethical conduct, in words and action.
- Treat colleagues, customers, and consumers with respect.
- Strive to be the best at what matters most to the company.
- Accept personal responsibility for actions.
- Value diversity.
- Make decisions based on facts and principles.

Good Ethics Can Mean Good Business

In many cases, good ethics can mean good business and improved profits. Companies that produce safe and effective products avoid costly recalls and lawsuits. Companies that provide excellent service maintain their customers instead of losing them to competitors. Companies that develop and maintain strong employee relations suffer less turnover and enjoy better employee morale. Suppliers and other business partners often prefer to work with companies that operate in a fair and ethical manner.

Likewise, bad ethics can lead to bad business results. For example, many employees can develop negative attitudes if they perceive a difference between their own values and the values stated or implied by an organization's actions. In such environments, employees often act to defend themselves against anticipated punishment or retaliate against poor treatment. A bad ethical environment destroys employee commitment to organizational goals and objectives, creates low morale, fosters poor performance, erodes employee involvement in corporate improvement initiatives, and builds indifference to the organization's needs.

Protecting the Corporation and Its Employees from Legal Actions

In 1991, the U.S. Justice Department published sentencing guidelines that suggested more lenient treatment for convicted executives if their companies had ethics programs. Fines for criminal violations can be lowered by up to 80 percent if the organization has implemented an ethics management program and cooperates with authorities. These measures are covered in Chapter Eight of the Federal Sentencing Guidelines for Organizations.[19]

The following list briefly describes the key features an organization must implement to show it has an effective program of compliance and ethics.

- Identify its core beliefs, which need to include a commitment to complying with the letter and spirit of the law and ethical conduct.
- Understand the strengths and weaknesses of its culture and organizational capacities.

- Scan its business environment, presumably on an enterprise-wide basis, to determine what pressures the organization faces, especially the risk of criminal conduct and violating other applicable laws, and more broadly, to gather benchmarking data to compare to industry standards and best practices.
- Determine, relative to its goals and objectives and baseline data of its prior performance, what outcomes should be expected of the program.
- Identify targets and measurable indicators of expected program outcomes.
- Design, implement, and enforce a program that will "exercise due diligence to prevent, detect, and report criminal conduct and otherwise promote an organizational culture that encourages ethical conduct and a commitment to compliance with all applicable law."
- Regularly evaluate its program to determine if it is effective, and capture what the organization learns along the way.[20]

Avoiding Unfavorable Publicity

The public reputation of a company strongly influences the value of its stock, how consumers regard its products and services, the degree of oversight it receives from government agencies, and the amount of support and cooperation it receives from business partners. Thus, some companies are motivated to build a strong ethics program to avoid negative publicity. If an organization is perceived as operating ethically, customers, business partners, shareholders, consumer advocates, financial institutions, and regulatory bodies will regard it more favorably.

Companies that operate unethically often suffer negative consequences and bad publicity. A recent example involves the Federal National Mortgage Association (Fannie Mae), which helps low-income and middle-income Americans finance home mortgages. The Office of Federal Housing Enterprise Oversight (OFHEO) began investigating Fannie Mae in September 2004 and found serious accounting problems, including earnings manipulation and poor internal controls. The SEC ordered Fannie Mae in December 2004 to restate its earnings back to 2001, a correction estimated at $9 billion. In February 2005, the OFHEO reported additional problems with accounting for securities and loans and improper practices to spread the impact of income and expenses over time. As a result, several Fannie Mae executives were forced out by the board of directors and changes were made in its staff of 1400 information systems and services workers.[21] Fannie Mae stock dropped below $58 per share on the New York Stock Exchange (NYSE) by March 2005, their lowest level in more than four years and 30 percent below a high of nearly $80 in March 2004.[22]

Improving Corporate Ethics

The risks of unethical behavior are increasing, so the improvement of business ethics is becoming more important. The following sections explain some of the actions corporations can take to improve business ethics.

Appointing a Corporate Ethics Officer

Corporate ethics can be broadly defined to include ethical conduct, legal compliance, and corporate social responsibility. The primary functions of a corporate ethics policy include setting standards, building awareness, and handling internal reports—tasks that are either not consolidated or handled well in many organizations. As a result, many organizations pull these functions together under a corporate officer to ensure that they receive sufficient emphasis and cohesive treatment.

The **corporate ethics officer** is a senior-level manager who provides vision and direction in the area of business conduct. Ethics officers come from diverse backgrounds such as legal staff, human resources, finance, auditing, security, or line operations.[23] Their role includes "integrating their organization's ethics and values initiatives, compliance activities, and business conduct practices into the decision-making processes at all levels of the organization."[24] Typically, the ethics officer tries to establish an environment that encourages ethical decision making through the actions described in this chapter. Specific responsibilities include "complete oversight of the ethics function, collecting and analyzing data, developing and interpreting ethics policy, developing and administering ethics education and training, and overseeing ethics investigations."[25]

The presence of a corporate ethics officer has become increasingly common. There is even a professional association, the Ethics Officer Association (EOA), for managers of ethics, compliance, and business conduct programs. "The EOA provides ethics officers with training and a variety of conferences and meetings for exchanging best practices in a frank, candid manner."[26] As of March 2005, there were more than 700 EOA member companies, including the well-known IT companies shown in the following list.

Adelphia Communications Group	Intel
AOL Time Warner	MCI
AT&T	Microsoft
BAE Systems	Oracle
British Telecom	QUALCOMM
Cingular Wireless	Qwest
Computer Associates, International	Sprint
Dell Computer	Sun Microsystems
Hewlett-Packard	Texas Instruments

However, simply naming a corporate ethics officer does not automatically improve ethics; hard work and effort are required to establish and provide ongoing support for an organizational ethics program.

Ethical Standards Set by Board of Directors

The board of directors is responsible for the careful and responsible management of an organization. In a for-profit corporation, the board's primary objective is to oversee the organization's business activities and management for the benefit of all stakeholders, including shareholders, customers, suppliers, and the community. In a nonprofit corporation, the board reports to a different set of stakeholders, particularly the local communities that the nonprofit serves.

The board fulfills some of its responsibilities directly and assigns others to various committees. The board is not normally responsible for day-to-day management and operations; these responsibilities are delegated to the organization's management team. However, the board is responsible for supervising the management team.

Directors of the company are expected to conduct themselves according to the highest standards of personal and professional integrity. Directors are also expected to set the standard for company-wide ethical conduct and ensure compliance with laws and regulations.

As you learned earlier in this chapter, the passage of Sarbanes-Oxley led to significant reforms in the content and preparation of disclosure documents by public companies. Section 406 of the act requires public companies to disclose whether they have codes of ethics and to disclose any waivers of those codes for certain members of senior management. The SEC also approved significant reforms by the NYSE and NASDAQ that, among other things, require companies listed on these exchanges to have codes of ethics that apply to all employees, senior management, and directors.

Establishing a Corporate Code of Ethics

A **code of ethics** highlights an organization's key ethical issues and identifies the overarching values and principles that are important to the organization and its decision making. The code frequently includes a set of formal, written statements about the purpose of the organization, its values, and the principles that guide its employees' actions. An organization's code of ethics applies to its directors, officers, and employees. The code of ethics should focus employees on areas of ethical risk relating to their role in the organization, provide guidance to help them recognize and deal with ethical issues, and provide mechanisms for reporting unethical conduct and fostering a culture of honesty and accountability in an organization. The code of ethics helps ensure that employees abide by the law, follow necessary regulations, and behave in an ethical manner.

A code of ethics cannot gain company-wide acceptance unless it is developed with employee participation and fully endorsed by the organization's leadership. It must also be easily accessible by employees, shareholders, business partners, and the public. The code of ethics must continually be applied to a company's decision making and emphasized as an important part of its culture. Breaches in the code of ethics must be identified and treated appropriately so that its relevance is not undermined.

Establishing a code of ethics is an important step for any company, and a growing number have done so. Figure 1-2 shows that almost 80 percent of surveyed companies have developed a code of ethics and are very satisfied with it.

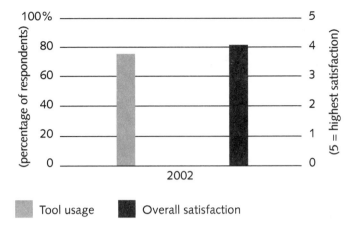

Source: Rigby, Darrell, "Management Tools," www.bain.com/management_tools/tools_ethics.asp?
groupcode=2, March 17, 2005.

FIGURE 1-2 Corporate satisfaction with their codes of ethics

In March 2005, *Business Ethics* magazine rated U.S.-based, publicly held companies using a statistical analysis of corporate service to seven stakeholder groups—employees, customers, community, minorities and women, shareholders, the environment, and non-U.S. stakeholders. The top IT company, based on performance between 2000 and 2004, was Intel Corporation, the world's largest computer chip maker. Intel's code of ethics is summarized in the following paragraph.

> Honest and ethical conduct, including the ethical handling of actual or perceived conflicts of interest between personal and business relationships, is our rule every day and for all that we do. If we are to maximize the value we create for our stockholders, it is the personal responsibility of each of us to: (1) comply with applicable laws (including statutes, controlling case law, agency regulations and orders, and other administrative directives) and Intel guidelines, (2) employ technical excellence and integrity to do the best we can to provide timely, accurate and understandable reporting of actual and forward-looking financial information, (3) employ our business processes and guidelines to do the best we can to base our business decisions on sound economic analysis (including a prudent consideration of risks), and (4) safeguard and utilize our physical, financial, and intellectual property assets to the best and most prudent effect.[27]

A more detailed version of Intel's code of ethics is spelled out in a 22-page document that offers employees guidelines designed to deter wrongdoing, promote honest and ethical conduct, and comply with applicable laws and regulations. Intel's code of ethics also expresses its policies regarding the environment, health and safety, diversity, nondiscrimination, supplier expectations, privacy, and business continuity.

Conducting Social Audits

An increasing number of companies conduct social audits of their policies and practices. In a **social audit**, companies identify ethical lapses they committed in the past and set directives for avoiding similar missteps in the future. For example, each year Intel sets social responsibility goals and tracks results against those goals. Intel's annual report on its social responsibility efforts shares the information with employees, shareholders, investors, analysts, customers, suppliers, government officials, and the communities in which Intel operates. Here are a few highlights from its 2003 report:

- Intel contributed more than $100 million in cash gifts worldwide.
- Global waste recycling teams exceeded their goals by recycling more than 66 percent of chemical waste and 74 percent of solid waste worldwide. These totals represented 40,000 tons of materials recycled.
- Intel has trained more than 1.5 million teachers in 33 countries to use technology effectively and improve student learning.
- Intel remained at world-class levels of health and safety performance.[28]

Requiring Employees to Take Ethics Training

The ancient Greek philosophers believed that personal convictions about right and wrong behavior could be improved through education. Today, most psychologists agree with them. Lawrence Kohlberg, the late Harvard psychologist, found that many factors stimulate a person's moral development, but one of the most crucial is education. Other researchers have repeatedly supported these findings—people can continue their moral development through further education that involves critical thinking and examining contemporary issues.

Thus, a company's code of ethics must be promoted and continually communicated within the organization, from top to bottom. Organizations should show employees examples of how to apply the code of ethics in real life. One approach is through a comprehensive ethics education program that encourages employees to act responsibly and ethically. Such programs are often presented in small workshop formats in which employees apply the organization's code of ethics to hypothetical but realistic case studies. For example, Procter & Gamble requires all its employees to take a workshop on the topic of **principle-based decision making**, based on principles in the corporate code of ethics. Workshop participants must decide how to best respond to real-life ethical problems, such as giving honest and constructive feedback to an employee who is not meeting expectations. Employees are also given examples of recent company decisions made using principle-based decision making. Not only do these courses make employees more aware of a company's code of ethics and how to apply it, the courses demonstrate that the company intends to operate in an ethical manner. The existence of formal training programs can also reduce a company's liability in the event of legal action.

Including Ethical Criteria in Employee Appraisals

Employees are increasingly evaluated on their demonstration of qualities and characteristics that are stated in the corporate code of ethics. For example, many companies base a portion of their employee performance evaluations on treating others fairly and with respect, operating effectively in a multicultural environment, accepting personal accountability to meet business needs, continually developing themselves and others, and operating openly and honestly with suppliers, customers, and other employees. These factors are considered along with more traditional criteria used in performance appraisals, such as an employee's overall contribution to moving the business ahead, successful completion of projects, and maintenance of good customer relations.

When Good Ethics Result in Short-Term Losses

Operating ethically does not always guarantee business success. Many organizations that operate outside the United States have found that the "business as usual" climate in some foreign countries can place them at a significant competitive disadvantage.

For example, a major global telecommunications company faced significant competitive disadvantages by consistently applying its corporate values to its South American business. Although the organization's code of ethics prohibited the practice of financially "influencing" decision makers on project bids, its competition did not play by the same rules. As a result, the company lost many projects and millions of dollars in revenues. Senior management argued in favor of integrity and the consistent application of corporate ethics, reasoning that situational ethics was wrong and that the practice could be hard to stop once it was started. Their hope was that good ethics would prove to be good business in the long term.

Creating an Ethical Work Environment

Most employees want to perform their jobs successfully and ethically, but good employees sometimes make bad ethical choices. Employees in highly competitive workplaces often feel pressures from aggressive competitors, cutthroat suppliers, unrealistic budgets, minimum quotas, tight deadlines, and bonus incentives for meeting performance goals. Employees may also be encouraged to do "whatever it takes" to get the job done. Such environments can make some employees feel pressure to engage in unethical conduct to meet management's expectations, especially if there are no corporate codes of conduct and no strong examples of senior management practicing ethical behavior. Table 1-1 shows how management's behavior can result in unethical employee behavior, and Table 1-2 provides a manager's checklist for establishing an ethical workplace; to each question in the latter table, the preferred answer is *yes*.

Employees must have a knowledgeable and potent resource with whom they can discuss perceived unethical practices. For example, Intel expects employees to report suspected violations of its code of ethics to a manager, the Legal or Internal Audit Departments, or a business unit's legal counsel. Employees may also report violations anonymously through an internal Web site dedicated to ethics. Senior management at Intel has made it clear that any employee can report suspected violations of corporate business principles without fear of reprisal or retaliation.

TABLE 1-1 How management can affect employees' ethical behavior

Managerial behavior that can encourage unethical behavior	Possible employee reaction
Set and hold people accountable for meeting "stretch" goals, quotas, and budgets	"My boss wants results, not excuses, so I have to cut corners to meet the goals my boss has set."
Fail to provide a corporate code of ethics and operating principles to guide decision making	"Because there are no guidelines, I don't think my conduct is really wrong or illegal."
Fail to act in an ethical manner and set a poor example for others to follow	"I have seen other successful people take unethical actions and not suffer negative repercussions."

TABLE 1-1 How management can affect employees' ethical behavior (continued)

Managerial behavior that can encourage unethical behavior	Possible employee reaction
Fail to hold people accountable for unethical actions	"No one will ever know the difference, and if they do, so what?"
When employees are hired, put a 3-inch binder titled "Corporate Business Ethics, Policies, and Procedures" on their desks. Tell them to "read it when you have time and sign the attached form that says you read and understand the corporate policy."	"This is overwhelming. Can't they just give me the essentials? I can never absorb all this."

TABLE 1-2 Manager's checklist for establishing an ethical work environment

Questions	Yes	No
Does your company have a corporate code of ethics?	___	___
Was the corporate code of ethics developed with broad input from employees at all levels within the organization, and does it have their support?	___	___
Is the corporate code of ethics concise and easy to understand, and does it identify the values you need to operate consistently and meet the needs of your stakeholders?	___	___
Do all employees have easy access to a copy of the corporate code of conduct, and have they all signed a document stating that they have read and understood it?	___	___
Do employees participate in annual training to reinforce the values and principles that make up the corporate code of ethics?	___	___
Do you set an example by communicating the corporate code of ethics and actively using it in your decision making?	___	___
Do you evaluate and provide feedback to employees on how they operate with respect to the values and principles in your corporate code of ethics?	___	___
Do you seek feedback from your employees to ensure that their work environment does not create conflicts with the corporate code of ethics?	___	___
Do employees believe that you are fair, and do they seek your advice when they see coworkers violating the company's code of ethics?	___	___
Do employees have an avenue, such as an anonymous hotline, for reporting infractions of the code of ethics?	___	___
Are employees aware of sanctions for breaching the code of ethics?	___	___

Ethical Decision Making

Often in business, the ethically correct course of action is clear and easy to follow. Exceptions occur, however, when ethical considerations come into conflict with the practical demands of business. Dealing with these situations is challenging and can even be risky to one's career. How, exactly, should you think through an ethical issue? What questions should you ask, and what factors should you consider? This section lays out a seven-step approach that can help guide your ethical decision making; however, the process is not a simple, linear activity. Keep in mind that information you gain or a decision you make in one step may cause you to go back and revisit previous steps.

The seven steps are summarized in the following list and explained in the following sections:

- Get the facts.
- Identify stakeholders and their positions.
- Consider the consequences of your decision.
- Weigh various guidelines and principles.
- Develop and evaluate options.
- Review your decision.
- Evaluate the results of your decision.

Getting the Facts

Innocent situations can often become unnecessary controversies because no one bothers to check the facts. For example, you might see your boss receive what appears to be an employment application from a job applicant and then throw the application in the trash after the applicant leaves. This would violate your company's policy to treat each applicant with respect and to maintain a record of all applications for one year. You could report your boss for failure to follow the policy or you could take a moment to speak directly to your boss. You might be pleasantly surprised to find out that the situation was not as it appeared. Perhaps the "applicant" was actually a salesperson promoting a product for which your company had no use, and the "application" was marketing literature.

Identifying the Stakeholders and Their Positions

A **stakeholder** is someone who stands to gain or lose from how a situation is resolved. Stakeholders often include others besides people who are directly involved in an issue. Identifying the stakeholders helps you better understand the impact of your decision and could help you make a better decision. Unfortunately, it may also cause you to lose sleep from wondering how you may affect the lives of others. You may recognize the need to involve stakeholders in the decision and thus gain their support for the recommended course of action. What is at stake for each stakeholder? What does each stakeholder value, and what outcome does the stakeholder want? Do some stakeholders have a greater stake because they have special needs or because the company has special obligations to them? To what degree should they be involved in the decision?

Considering the Consequences of Your Decision

You can view the consequences of a decision from several perspectives. Often, your decision directly affects you, although you must guard against thinking too narrowly and focusing on what is best for you. Another perspective is to consider the harmful and beneficial effects your decision might have on the stakeholders. A third perspective is to ask whether your decision will help the organization meet its goals and objectives. Finally, you should consider the decision's impact on the broader community of other organizations and institutions, the public, and the environment. As you view problems and proposed solutions from each of these perspectives, you may gain additional insights that affect your decision.

Weighing Various Guidelines and Principles

Do any laws apply to your decision? You certainly don't want to violate a law that can lead to a fine or imprisonment for yourself or others. If the decision does not have legal implications, what corporate policies or guidelines apply? What guidance does the corporate code of ethics offer? Will any of your personal principles affect your decision?

Philosophers have developed many approaches to deal with moral issues. Four of the most common approaches, which are summarized in Table 1-3 and discussed later in this section, provide a framework for decision makers to reflect on the acceptability of their actions and evaluate moral judgments. People must find the appropriate balance between all applicable laws, corporate principles, and moral guidelines to help them make decisions. (For a more in-depth discussion of ethics and moral codes, see Appendix A.)

TABLE 1-3 Philosophical theories for ethical decision making

Approach to dealing with moral issues	Principle
Virtue ethics approach	The ethical choice best reflects moral virtues in yourself and your community
Utilitarian approach	The ethical choice produces the greatest excess of benefits over harm
Fairness approach	The ethical choice treats everyone the same and shows no favoritism or discrimination
Common good approach	The ethical choice advances the common good

Virtue ethics approach.   Virtue ethics focuses on how you should behave and think about relationships if you are concerned with your daily life in a community. It does not define a formula for ethical decision making, but suggests that when faced with a complex ethical dilemma, people either do what they are most comfortable doing or what they think a person they admire would do. The assumption is that people are guided by their virtues to reach the "right" decision. A proponent of virtue ethics believes that a disposition to do the right thing is more effective than following a set of principles and rules and that people should perform moral acts out of habit, not introspection.

Virtue ethics can be applied to the business world by equating the virtues of a good businessperson with those of a good person. However, businesspeople face situations that are peculiar to business, so they may need to tailor their ethics accordingly. For example, honesty and openness when dealing with others is generally considered virtuous; however, a corporate purchasing manager who negotiates a multimillion dollar deal might need to be vague in discussions with competing suppliers.

A problem with the virtue ethics approach is that it doesn't provide much of a guide for action. The definition of virtue cannot be worked out objectively; it depends on the circumstances—you work it out as you go. For example, bravery is a great virtue in many circumstances, but in others it may be foolish. The right thing to do in a situation depends on which culture you're in and what the cultural norm dictates.

Utilitarian approach.   This approach to ethical decision making states that you should choose the action or policy that has the best overall consequences for all people who are directly or indirectly affected. The goal is to find the single greatest good by balancing the interests of all affected parties.

Utilitarianism fits easily with the concept of value in economics and the use of cost-benefit analysis in business. Business managers, legislators, and scientists weigh the benefits and harm of policies when deciding whether to invest resources in building a new plant in a foreign country, to enact a new law, or to approve a new prescription drug, respectively.

A complication of this approach is that measuring and comparing the values of certain benefits and costs is often difficult, if not impossible. How do you assign a value to human life or to a pristine wildlife environment? It can also be difficult to predict the full benefits and harm that result from a decision.

Fairness approach.   This approach focuses on how fairly actions and policies distribute benefits and burdens among people affected by the decision. The guiding principle of this approach is to treat all people the same. However, decisions made with this approach can be influenced by personal biases toward a particular group, and the decision makers may not even realize their bias. If the intended goal of an action or policy is to provide benefits to a target group, other affected groups may consider the decision unfair.

Common good approach.   This approach to decision making is based on a vision of society as a community whose members work together to achieve a common set of values and goals. Decisions and policies that use this approach attempt to implement social systems, institutions, and environments that everyone depends on and that benefit all people. Examples include an effective education system, a safe and efficient transportation system, and accessible and affordable health care.

As with the other approaches to ethical decision making, there are complications. People clearly have different ideas about what constitutes the common good, which makes consensus difficult. In addition, maintaining the common good often requires some groups to bear greater costs than others—for instance, homeowners pay property taxes to support public schools, but apartment dwellers do not.

Developing and Evaluating Options

In many cases, you can identify several answers to a complex ethical question. By listing the key principles that apply to the decision, you can usually focus on the two or three best options. What benefits and harm will each course of action produce, and which alternative will lead to the best overall consequences? The option you choose should be ethically defensible and should meet the legitimate needs of economic performance and the company's legal obligations.

Reviewing Your Decision

Is the decision consistent with your personal values as well as those of the organization? How would coworkers, stakeholders, business partners, friends, and family regard your decision if they knew the facts of the situation and the basis for your decision? Would they see it as right, fair, and good? If you belonged to any of the other stakeholder groups, would you be able to accept the decision as fair?

Evaluating the Results of Your Decision

After the organization implements the decision, monitor the results to see if it achieved the desired effect and observe its impact on employees and other affected parties. This evaluation will allow you to adjust and improve the process for future decisions.

ETHICS IN INFORMATION TECHNOLOGY

The growth of the Internet, the ability to capture and store vast amounts of personal data online, and greater reliance on information systems in all aspects of life have increased the risk of using information technology unethically. In the midst of the many IT breakthroughs in recent years, the importance of ethics and human values has been underemphasized—with a range of consequences. Here are some examples that raise public concern about the ethical use of information technology:

- Today's workers might have their e-mail and Internet access monitored while at work, as employers struggle to balance their need to manage important company assets and work time with employees' desire for privacy and self-direction.
- Millions of people have used peer-to-peer networks to download music and movies at no charge and in apparent violation of copyright laws.
- Organizations contact millions of people worldwide through unsolicited e-mail (spam) at an extremely low cost.
- Hackers break into databases of financial institutions and steal customer information, then use it to commit identity theft, opening new accounts and charging purchases to unsuspecting victims.
- Students around the world have been caught downloading material from the Internet and plagiarizing content for their term papers.
- Web sites plant cookies or spyware on visitors' hard drives to track their Internet activity.

This book is based on two fundamental tenets. First, the general public has not realized the critical importance of ethics as they apply to IT; too much emphasis has been placed on the technical issues. However, unlike most conventional tools, IT has a profound effect on society. IT professionals need to recognize this fact when they formulate policies that will affect the well-being of millions of consumers and have legal ramifications.

Second, in the corporate world, important technical decisions are often left to the technical experts. General business managers must assume greater responsibility for these decisions, but to do so they must be able to make broad-minded, objective, ethical decisions based on technical savvy, business know-how, and a sense of ethics. They must also try to create a working environment in which ethical dilemmas can be discussed openly, objectively, and constructively.

Thus, the goals of this text are to educate people about the tremendous impact of ethical issues in the successful and secure use of information technology; to motivate people to recognize these issues when making business decisions; and to provide tools, approaches, and useful insights for making ethical decisions.

An Overview of This Text

The remaining chapters in this book cover a range of topics that relate to ethics in information technology:

- Chapter 2 discusses how ethics is important to IT professionals and IT users.
- Chapter 3 addresses computer crime, an area of growing concern to networked IT users.
- Chapter 4 covers the important issues of personal data privacy and employee monitoring.
- Chapter 5 deals with the ethical issues raised by Internet communications and freedom of expression.
- Chapter 6 covers the protection of intellectual property rights through patents, copyrights, and trade secrets.
- Chapter 7 addresses ethical issues raised by the software development process.
- Chapter 8 covers the use of nontraditional employees, the ethical dilemmas it can cause, and the implications of whistle-blowing.
- Chapter 9 addresses the impact of IT on society.

Summary

1. What is ethics, and why is it important to act according to a code of principles?

   Ethics is a set of beliefs about right and wrong behavior. A person who acts with integrity acts in accordance with a personal code of principles. Integrity is one of the cornerstones of ethical behavior.

2. Why is business ethics becoming increasingly important?

   Ethics in business is becoming more important because the risks associated with inappropriate behavior have grown in number, complexity, likelihood, and significance.

3. What are corporations doing to improve business ethics?

   Corporations can appoint a corporate ethics officer, set ethical standards at a high organizational level, establish a corporate code of ethics, conduct social audits, require employees to take ethics training, and include ethical criteria in employee appraisals.

4. Why are corporations interested in fostering good business ethics?

   Corporations want to protect themselves and their employees from legal action, to create an organization that operates consistently (because good ethics can be good business), to avoid negative publicity, and to gain the goodwill of the community. Being ethical, however, does not always guarantee business success.

5. What approach can you take to ensure ethical decision making?

   One approach involves seven steps: get the facts of the issue, identify the stakeholders and their positions, consider the consequences of the decision, weigh various guidelines and principles, develop and evaluate various options, review the decision, and evaluate the results. This is not a linear process; some backtracking and repeating of previous steps may be required.

6. What trends have increased the risk of using information technology unethically?

   The growth of the Internet, the ability to capture and store vast amounts of personal data online, and greater reliance on information systems in all aspects of life have increased the risk of using information technology unethically. In the midst of the many IT breakthroughs in recent years, the importance of ethics and human values has been underemphasized—with a range of consequences.

Self-Assessment Questions

NOTE

The answers to the self-assessment questions are listed at the back of the book.

1. Habits that incline people to do what is acceptable are called _____ .

2. Integrity is a cornerstone of ethical behavior and is the practice of acting in accordance with one's own code of principles. True or False?

3. Increased corporate globalization is one of several trends that have made it more difficult to apply principles and codes of ethics consistently. True or False?

4. Which of the following companies has *not* been charged with serious accounting irregularities?

   a. Qwest

   b. MCI

   c. Adelphia

   d. Intel

5. Corporations need to operate consistently so that they can promote an ethical work environment by encouraging employees to act ethically when making business decisions and by supporting them when they do. True or False?

6. The sentencing of organizations for violating federal law changes periodically and is now covered in Chapter Eight of the Federal Sentencing Guidelines for Organizations, which provides for lesser punishment if the organization can demonstrate that it had an "effective compliance and ethics program." True or False?

7. The _____ highlights an organization's key ethical issues and identifies the overarching values and principles that are important to the organization and its decision making.

8. In a _____ , companies set goals for social responsibility, define improvement programs, and track progress toward meeting their goals.

Review Questions

1. Define the word *ethics* and the term *value system*.

2. What trends have increased the need for organizations to foster an ethical environment?

3. In what ways do good ethics engender good business?

4. Briefly summarize the key provisions of Section 404 of the Sarbanes-Oxley Act. How might it affect the accounting practices of an organization?

5. What are the key reasons that corporations need to promote an ethical work environment?

6. The goodwill that socially responsible activities create can make it easier for corporations to conduct their business. Explain what this means, and provide an example.

7. Identify specific actions that corporations can take to improve business ethics.

8. What is the purpose of a corporate code of ethics?

9. What is meant by principle-based decision making?

10. What pressures might be placed on employees that make it difficult for them to perform ethically?

11. Outline and briefly discuss a seven-step approach for ethical decision making.

12. Identify several areas in which the increased use of IT has raised ethical concerns.

Discussion Questions

1. Can you recall a situation in which you had to deal with a conflict in values? What was it? How did you resolve this issue?

2. Is every action that is legal also ethical? Can you describe an action that is legal but ethically wrong? Is every ethical action also legal? Is the law, not ethics, the only guide that business managers need to consider? Explain.

3. What is the role of the board of directors in establishing an ethical workplace?

4. Do you think it is easier to establish an ethical work environment in a nonprofit organization? Why or why not?

5. This chapter discusses four approaches to dealing with moral issues. Identify and briefly summarize each one. Do you believe one perspective is the most important? Why or why not?

6. Is it possible for an employee to be successful in the workplace without acting ethically?

7. What are the key elements of an effective corporate ethics training program?

8. Identify and briefly discuss a recent example that illustrates the negative impact of using IT unethically.

What Would You Do?

Use the seven-step approach to ethical decision making to analyze the following situations and answer the questions.

1. You and your 10 project team members are working on a new information system for your firm. You have heard a rumor that when the project is completed, the system's ongoing maintenance and support will be outsourced to another firm. The original implementation plan called for at least three team members to work full time on system support and maintenance. Should you break this news to your team? What should you do?

2. You are the customer service manager for a small software manufacturer. The newest addition to your 10-person team is Aubrey, a recent college graduate. She is a little overwhelmed by the volume of calls, but is learning fast and doing her best to keep up. Today, as you performed your monthly review of employee e-mail, you were surprised to see that Aubrey is corresponding with employment agencies. One message says, "Aubrey, I'm sorry you don't like your new job. We have lots of opportunities that I think would much better match your interests. Please call me and let's talk further." You're shocked and alarmed. You had no idea she was unhappy, and your team desperately needs her help to handle the onslaught of calls generated by the newest release of software. If you're going to lose her, you'll need to find a replacement quickly. You know that Aubrey did not intend for you to see the e-mail, but you can't ignore what you saw. Should you confront Aubrey and demand to know her intentions? Should you avoid any confrontation and simply begin seeking her replacement? Could you be misinterpreting the e-mail? What should you do?

3. While mingling with friends at a party, you mention a recent promotion that has put you in charge of evaluating bids for a large computer hardware contract. A few days later, you receive a dinner invitation at the home of an acquaintance who also attended the party. Over

cocktails, the conversation turns to the contract you're managing. Your host seems remarkably well-informed about the bidding process and likely bidders. You volunteer information about the potential value of the contract and briefly outline the criteria your firm will use to select the winner. At the end of the evening, the host surprises you by revealing that he is a consultant for several companies in the computer hardware market. Later that night your mind is racing. Did you reveal information that could provide a supplier with a competitive advantage? What are the potential business risks and ethical issues in this situation? Should you report the conversation to someone? If so, who should you talk to and what would you say?

4. You have just completed interviewing three candidates for an entry-level position in your organization. One candidate is the friend of a coworker who has implored you to "give his friend a chance." The candidate is the weakest of the three but has sufficient skills and knowledge to adequately fill the position. Would you hire this candidate?

5. A coworker calls you at 9 a.m. at work and asks for a favor. He is having trouble this morning and will be an hour late for work. He explains that he has already been late for work twice this month and that a third time will cost him four hours pay. He asks you to stop by his cubicle, turn his computer on, and place some papers on the desk so that it appears he is "in." You have worked on some small projects with this coworker and gone to lunch together. He seems nice enough and does his share of the work, but you are not sure what to tell him. What would you do?

Cases

1. Is There a Place for Ethics in IT?

On March 15, 2005, Michael Schrage published an article in *CIO* magazine entitled "Ethics, Schmethics" that stirred up a great deal of controversy in the IT community. In the article, Schrage proposed that "CIOs should stop trying to do the 'right thing' when implementing IT and focus instead on getting their implementations right." *Ethics*, Schrage argued, had become a buzzword much like *quality* in the 1980s, and that the demand for ethical behavior interferes with business efficiency.

Schrage gave a few scenarios. For example, a company is developing a customer relationship management (CRM) system, and the staff is working very hard to meet the deadline. The company plans to outsource the maintenance and support of the CRM once the system is developed. There is a good chance that two-thirds of the IT staff will be laid off. Would you disclose this information? Schrage answered, "I don't think so."

Schrage asked readers in another scenario, "How about deliberately withholding important information from your boss because you know that its disclosure would provoke his immediate counterproductive intervention in an important project?" Schrage said he would do it; business involves competing values, he argued, and trade-offs must be made to keep business operations from becoming paralyzed.

Schrage was hit with a barrage of responses accusing him of being dishonorable, short-sighted, and lazy. Other feedback provided new perspectives on his scenarios that Schrage hadn't considered. For example, Kathleen Dewey, an IT manager at Boise State University, argued that doing the right thing is good for business. Not disclosing layoffs, she argued, is a trick that only works once. Remaining employees will no longer trust the company and pursue jobs where they can feel more secure. New job applicants will think twice before joining a company with a reputation for exploiting IT staff. Other readers responded to the scenario by suggesting that the company maintain loyalty by offering incentives for those who stayed or providing job placement services for departing employees.

Addressing the second scenario, Dewey suggested that not giving the boss important information could backfire on the employee. "What if your boss finds out the truth? What if you were wrong, and the boss could have helped? Once your boss knows that you lied once, will he believe you the next time?"

Another reader, Gautam Gupta, had actually worked under an unproductive, reactive, meddling boss. He suggested confronting the boss about the problem at an appropriate time and place. In addition, as situations arose that required Gupta to convey important information that might elicit interference, he developed action plans and then made firm presentations to his boss. The boss, he assured Schrage, will adapt.

Gupta, Dewey, and others argued that CIOs must consider a company's long-term needs rather than just the current needs of a specific project. Others argued that engaging in unethical behavior, even for the best of purposes, crosses a line that eventually leads to more serious transgressions. Some readers suspected that Schrage had published the article to provoke outrage. Another reader, Maikel Marrero, agreed with Schrage, arguing that ethics has to "take a back seat to budgets and schedules" in a large organization. Marrero explained, "At the end of the day, IT is business."

Questions:

1. Discuss how a CIO might handle Schrage's scenarios using the virtue ethics approach, the fairness approach, the utilitarian approach, and the common good approach.

2. Discuss the possible short-term losses and long-term gains in implementing ethical solutions to each of Schrage's scenarios.

3. Must businesses choose between good ethics and financial benefits? Explain your answer using Schrage's scenarios or your own examples.

2. Computer Associates Is Forced to Clean up Its Act

On July 5, 2000, Erika Miller of the Nightly Business Report announced, "While most of the nation was preparing for Fourth of July fireworks, Computer Associates quietly released a bombshell of its own." The company had issued a warning that its first quarter earnings for fiscal year 2001 would fall short of expectations. The forecast had been for a $0.55 profit per share, but the company said it expected an actual profit of no more than $0.16 per share. Stock prices dropped 43 percent, from $51 to $29.50, in a single day.

Miller reported, "The company blames the shortfall on weak European sales, a slowdown in its mainframe business, and delays in several large contracts," but the real reason was quite different and would plague the company for years to come.

According to the Securities and Exchange Commission (SEC), executives at Computer Associates (CA) practiced a fraudulent accounting method between January 1998 and September 2000. CA would keep the books open after the quarter ended and report revenue from contracts that had not yet been signed. This method, called the "35-day month," allowed them to meet quarterly earnings expectations and artificially raise stock prices. The SEC maintains that first, second, third, and fourth quarter earnings for fiscal year 2000 were inflated by 25 percent, 53 percent, 46 percent, and 22 percent, respectively. When CA changed auditors and came under increasing scrutiny, they altered their accounting methods, precipitating the July 2000 crash.

The conspiracy involved an array of CA employees from the Finance and Sales departments to top executives. As news of the scandal leaked, CA was pummeled with class actions, the Standard and Poor's Rating Service placed the company on its CreditWatch list, and the SEC indicted CA executives. What seems difficult to understand is why one of the world's largest software companies would sink to criminal practices.

The answer may lie in an earlier scandal. According to a 1995 agreement, founder Charles Wang, cofounder Russel Artzt, and former CEO Sanjay Kumar were to receive 20.25 million shares if stock prices closed at or above $53.30 for 60 days within a 12-month period—by the end of the year 2000. The lure of this reward may have motivated the executives to conspire to maintain artificially high stock prices. Not only did CA institute a "35-day month," it engaged in other shady dealings that allowed it to meet estimated quarterly earnings. In early 2000, after the stock prices reached the required benchmark, the three tried to cash in their shares. Irate shareholders sued the company and forced the executives to return $560 million.

The attempt to cash in these stock options brought CA and its accounting practices under closer scrutiny. In February 2002, pressure mounted and CA elected Walter Schuetze, former SEC Chief Accountant, to the Board of Directors. He led an independent investigation that initially concluded in September 2002 that CA had not violated accepted accounting principles. By 2003, however, the Sarbanes-Oxley Act (passed in July 2002) and increased media scrutiny aroused by the Enron scandal was affecting companies across the nation. Chief financial officers (CFOs) had to answer to the CEO and to stronger boards composed of independent directors. Companies were instituting more oversight and better controls. In October 2003, Schuetze revealed that—upon further investigation—his committee *had* discovered irregularities. CA's CFO was forced to step down, and Kumar himself later resigned. The board also asked other senior executives involved in the scandal to resign.

In the meantime, the SEC and the federal government are looking to prevent accounting fraud in the future by imposing the harshest measures possible on violators. In June 2005, the SEC expanded its charges against Kumar, accusing him of paying hush money in the form of a $3.7 million contract to a CA client who had learned of CA's fraudulent accounting practices. The Sarbanes-Oxley Act provides for stiffer punishment—longer prison sentences and higher fines. CA may not see the end of the scandal for years to come, which could serve as a considerable disincentive for other companies to follow in CA's footsteps.

Questions:

1. Why do you think Walter Schuetze reversed his initial finding that Computer Associates committed no accounting irregularities?

2. Research the Web to identify other accounting irregularities employed by Computer Associates beyond the use of a "35-day month."

3. Do you think a company can commit widespread accounting fraud without the knowledge of lower-level managers in the Accounting and Finance departments?

3. McKesson HBOC Accused of Accounting Improprieties

HBO (named for founders Huff, Barrington, and Owens) & Company was formed in November 1974. The company quickly made a name for itself by delivering cost-effective patient information and hospital data collection systems. Its premiere product, MEDPRO, was designed to be the most cost-effective system in the industry. MEDPRO helped hospital administrators track patient admissions, discharges, emergency room registrations, order communications and results reporting, scheduling, and data collection.

The company went public in June 1981 under the NASDAQ stock symbol HBOC. Its fast growth in sales and profits made it a favorite of investors throughout the 1990s, and its stock rose more than a hundredfold between October 1990 and October 1998. Much of its growth came through acquisitions of other companies.

On January 12, 1999, McKesson Corporation, one of the largest distributors of prescription drugs in the United States, completed a merger with HBO & Company by exchanging shares of common stock. At the time, the two companies had a combined market value of more than $23 billion.

Just three months after the merger, McKesson HBOC, Inc. announced that its auditors had discovered accounting irregularities at HBOC during a routine annual review. The problems were uncovered when McKesson's accounting firm, Deloitte & Touche, mailed a survey to several clients and asked them to report the actual amount of goods and services they had purchased from the company. Several of the amounts returned by clients did not match what HBOC had recorded. As a result, McKesson HBOC, Inc. had to restate earnings for the last four quarters. When the restatement of earnings was announced in April 1999, shares of the company plunged from $65 to $34 in a single day. The restated results were as follows:

	REVENUE (IN MILLIONS)			NET INCOME (IN MILLIONS)		
Quarter Ending	Originally Reported	As Restated	% Overstated	Originally Reported	As Restated	% Overstated
3/98	$393.1	$376.8	4.3%	$64.9	$45.6	42.3 %
6/98	$376.7	$308.1	22.3%	$75.6	$23.5	221.7 %
9/98	$399.6	$330.5	20.9%	$83.7	$16.5	407.3 %
12/98	$469.0	$381.0	23.9%	$59.6	$8.5	601.2 %
3/99	$431.9	$402.6	7.3%			

Some believed that certain HBOC managers, seeking to ensure that the company would meet or beat analysts' expectations for sales and profits, had used several innovative approaches to report financial results. Throughout 1998, it was alleged that HBOC allowed more than a dozen hospitals to buy HBOC software or services with conditional "side letters" that enabled the hospitals to back out of the sales. It was further alleged that these side letters were not shared with the auditors and that the associated sales were reported as complete, violating accounting rules. In at least one case, a hospital canceled a purchase HBOC had booked.

Additional allegations were made that, to bolster its results, HBOC agreed to questionable sales with two other large computer companies. In September 1998, two days before the end of HBOC's quarter and just weeks before HBOC and McKesson agreed to merge, HBOC signed to buy $74 million in software from Computer Associates (CA), supposedly for resale. CA, a software manufacturer for large companies, bought $30 million in HBOC software in return, also supposedly for resale. The deals were split into separate contracts, neither of which made reference to the other. In the two years following its purchase of HBOC's software, CA neither used nor distributed any of the $30 million in HBOC software products, according to a court indictment. Similarly, HBOC had sold only a fraction of the $74 million in CA software it bought.

It was also alleged that McKesson HBOC, Inc. and Data General, a computer hardware maker based in Massachusetts, had agreed to a similar deal in March 1999. Data General disclosed to auditors that what appeared to be a simple $20 million purchase of HBOC software had a side agreement that essentially ensured Data General would never have to pay for it.

Investors and analysts had no idea these problems would be so costly. McKesson first said it had found $42 million in sales from its HBOC unit that had been improperly booked. Eventually, however, $327 million in overstated revenue and $191 million in overstated income were uncovered from 1997 to 1999.

In one of the drug industry's largest corporate shakeups, the board of McKesson HBOC, Inc. ousted some of its top executives over the accounting irregularities. In July 1999, the U.S. Attorney's Office for the Northern District of California and the SEC started investigations. In addition, 53 class actions, three derivative actions, and two individual actions were filed against the company and its current or former officers and directors. Two former top executives of HBOC were indicted in July 2000 and accused of costing investors more than $9 billion, one of the largest financial frauds in American history.

In October 2003, former HBOC president Albert Bergonzi pleaded guilty to two counts of securities fraud charges. In a plea agreement, prosecutors dropped nine other charges and Bergonzi agreed to cooperate in the government's case against another McKesson executive, Richard Hawkins. Bergonzi was the fourth former senior HBOC executive to plead guilty to charges stemming from the scandal. In January 2005, McKesson agreed to pay $960 million to settle a federal class action, but numerous other suits remain.

Questions:

1. Make a list of the parties that were hurt by the use of nonstandard accounting practices at HBOC. Identify the harm suffered by each party.

2. As you read this case, was there a clear point at which ethical wrongdoings became legal wrongdoings? If so, when?

3. What would motivate HBOC managers to use nonstandard accounting procedures to report increased revenues and earnings?

End Notes

[1] National Cancer Center Institute, www.cancer.gov/statistics.

[2] "Compliance and Integrity," Altria Web site, www.altria.com/responsibility/4_2_complianceandintegrity.asp.

[3] Weiss, Shari, "Case Study: Developing An Ethical Company, Compliance Pipeline," www.compliancepipeline.com/showArticle.jhtml?articleID=165701047, July 8, 2005.

[4] Weiss, Todd R., "Study: Global Software Piracy Losses Totaled $29B in 2003," *Computerworld*, www.computerworld.com, July 7, 2003.

[5] Lawson, Stephan, "MCI Settles Fraud Charges with SEC," *Computerworld*, www.computerworld.com, May 20, 2003.

[6] Eichenwald, Kurt, "Ebbers' Verdict A Sign: Juries Blame the CEO," *Cincinnati Enquirer*, page D1, March 16, 2005.

[7] Kadlec, Daniel, "A Wake-Up Call for Directors," *Time*, www.time.com, January 17, 2005.

[8] Shore, Sandy, "SEC Suit Charges Fraud at Qwest," *Cincinnati Enquirer*, page D2, March 16, 2005.

[9] Associated Press, "Adelphia Founder John Rigas Found Guilty," MSNBC News, www.MSNBC.com, July 8, 2004.

[10] Gross, Grant, "Update: Kumar Leaves CA," *Computerworld*, www.computerworld.com, June 4, 2004.

[11] Cowley, Stacy, "CA Shows Revenue Growth as Turbulent Fiscal Year Ends," *Computerworld*, www.computerworld.com, May 27, 2005.

[12] Full text of Sarbanes-Oxley Act, frwebgate.access.gpo.gov/cgi-bin/getdoc.cgi?dbname=107_cong_reports&docid=f:hr610.107.pdf.

[13] "Sarbanes-Oxley Act," whatis.com at searchcio.techtarget.com, March 30, 2005.

[14] Bruno, Joe Bel, "Accounting Compliance Spurs Investors' Fears," *MSN Money*, news.moneycentral.msn.com, March 17, 2004.

[15] Hoffman, Thomas, "IT Managers Brace to Meet Ongoing Sarbanes-Oxley Compliance Demands," *Computerworld*, www.computerworld.com, August 2, 2004.

[16] Holder, Tony and Mindak, Mary, "Overall Analysis of Stock Price Reactions, Disclosure Level and Significant Issues Related to SOX 404," University of Cincinnati, March 17, 2005.

[17] Pham, Duc, "Sarbanes-Oxley: Technical Enforcement of IT Controls," *Computerworld*, www.computerworld.com, July 20, 2004.

[18] Weiss, Todd R., "Accounting Problem at SunTrust Could Delay Sarbanes-Oxley Filing," *Computerworld*, www.computerworld.com, November 16, 2004.

[19] Johnson, Kenneth, "Federal Sentencing Guidelines: Enterprise Risk Management," *Ethics Today Online*, www.ethics.org/today/, Volume 3, Issue 3, December 2004.

[20] Johnson, Kenneth W., "Federal Sentencing Guidelines: Enterprise Risk Management," Ethics Resource Center 2004-09, Article ID 864, www.ethics.org/resources/article_detail.cfm?ID=864.

[21] "Two Top Execs at Fannie Mae Forced Out," *SmartPros Accounting,* accounting.smartpros. com/x46295.xml, December 22, 2004.

[22] Gordon, Marcy, "New Problems Found in Fannie Mae Accounting," *MSN Money*, news. moneycentral.msn.com, February 23, 2005.

[23] "What is an Ethics Officer," Web site of Ethics Officer Association, www.eoa.org/Whatis.asp, March 17, 2005.

[24] "What is an Ethics Officer," Web site of Ethics Officer Association, www.eoa.org/Whatis.asp, March 17, 2005.

[25] Harned, Patricia, "A Word from the President: Ethics Offices and Officers," *Ethics Today Online*, www.ethics.org/today/, Volume 3, Issue 2, October 2004.

[26] Home page of Ethics Officer Association, www.eoa.org, March 17, 2005.

[27] Intel Web site, Corporate Compliance, www.intel.com/intel/finance/docs/CBP%20-%20Ethics%20and%20Compliance.pdf, March 19, 2005.

[28] Intel Press Release, "Intel Reports on Corporate Social Responsibility Performance," www.intel. com/pressroom/archive/releases, May 27, 2004.

Sources for Case 1

Schrage, Michael, "Ethics, Shmethics," CIO, www.cio.com/archive/031505/ethics.html,
March 15, 2005.

"Ethics, Shmethics, Readers Comments," CIO, www.cio.com/comment_list.html?ID=3308.

Sources for Case 2

"Computer Associates' Earnings Warning Computes Big Losses On Wall St.," Nightly Business News Web site, www.nightlybusiness.org/transcript/2000/trnscrpt070500.htm,
July 5, 2000.

"SEC Files Securities Fraud Charges Against Computer Associates International, Inc., Former CEO Sanjay Kumar, and Two Other Former Company Executives," Press Release 2004-134, U.S. Securities and Exchange Commission Web site, www.sec.gov/news/press/2004-134.htm, September 22, 2004.

Berniker, Mark, "Computer Associates Sacks Execs," Internetnews.com, www.internetnews.com/bus-news/article.php/3089691, October 9, 2003.

Cowley, Stacy, "Prosecutors Revise Indictment of Ex-CA CEO Kumar," Network World, www. computerworld.com/governmenttopics/government/legalissues/story/0,10801,96096,00.html, June 30, 2005.

Sources for Case 3

Morrow, David J., "The Markets: Market Place; McKesson to Restate Earnings for 4 Quarters and Stock Falls 48%," www.nytimes.com, April 29, 1999.

Berenson, Alex, "Two Ex-Executives Are Indicted In Fraud Case," www.nytimes.com,
September 29, 2000.

"Suit Contends Board of McKesson Knew of Problems at HBO," www.nytimes.com,
November 15, 2000.

McKesson HBOC, Inc., Form 10-Q filed with Securities and Exchange Commission for quarter ended December 31, 2000, www.edgar-online.com.

McKesson HBOC, Inc., Form 10-K Annual Report for the year ended March 31, 1999, filed with Securities and Exchange Commission, www.edgar-online.com.

"Bergonzi Pleads Guilty to HBOC Fraud," Health Data Management, www.healthdatamanagement.com, June 30, 2005.

United States Attorney's Office, Northern District of California, www.usdoj.gov/usao/can/press/html/2003_06_04_mckesson.html, June 30, 2005.

CHAPTER 2

ETHICS FOR IT PROFESSIONALS AND IT USERS

VIGNETTE

Cleveland State University Accuses PeopleSoft of Selling Vaporware

As the computer age reached universities, their admissions, financial aid, human resources, and other offices adopted new technologies that increased efficiency. By the mid-1990s, university information systems were a mess, a collection of disconnected systems that couldn't communicate with each other. So when PeopleSoft announced that it had developed an enterprise resource planning system (ERP), the universities jumped—and none higher than Cleveland State University.

In 1997, Cleveland State hired Klaudis Consulting Group to install PeopleSoft's ERP within a year. Then-president Claire Van Ummersen admits that she approved of the decision to hire Klaudis despite the fact that it had no prior experience installing PeopleSoft applications. Cleveland State thus became the first school to implement the student administration modules, and the results were disastrous.

The IBM mainframe couldn't handle the application, so the school was forced to purchase a UNIX system. Vice president of Human Resources Joseph Nolan claimed that 35 functions the software was supposed to perform were either missing or did not work. University officials say that the system failed to process financial aid and sent out incorrect tuition bills, a mistake that hit Cleveland State with a $5 million loss. The university hired new consultants, spent an additional $7 million, and installed hundreds of fixes, but the system still didn't work as promised.

In January 2004, Cleveland State finally threw up its hands and filed a $510 million lawsuit against PeopleSoft for breach of contract, fraud, and negligent misrepresentation. The university also sued Klaudis. After Oracle acquired PeopleSoft, both parties agreed to a settlement of $4.25 million. Although the university has not recovered its losses, it has served as a cautionary tale for others.[1,2,3,4,5,6]

LEARNING OBJECTIVES

As you read this chapter, consider the following questions:

1. What key characteristics distinguish a professional from other kinds of workers, and what is the role of an IT professional?
2. What relationships must an IT professional manage, and what key ethical issues can arise in each?
3. How do codes of ethics, professional organizations, certification, and licensing affect the ethical behavior of IT professionals?
4. What are the key tenets of four different codes of ethics that provide guidance for IT professionals?
5. What are the common ethical issues that face IT users?
6. What approaches can support the ethical practices of IT users?

IT PROFESSIONALS

A **profession** is a calling that requires specialized knowledge and often long and intensive academic preparation. The United States has adopted labor laws and regulations that require a more precise definition of what is meant by a *professional* employee. The U.S. Code of Federal Regulations defines a person "employed in a professional capacity" as one who meets these four criteria:

1. One's primary duties consist of the performance of work requiring knowledge of an advanced type in a field of science or learning customarily acquired by a prolonged course of specialized intellectual instruction and study or work.
2. One's instruction, study, or work is original and creative in character in a recognized field of artistic endeavor and the result of which depends primarily on the invention, imagination, or talent of the employee.
3. One's work requires the consistent exercise of discretion and judgment in its performance.
4. One's work is predominately intellectual and varied in character, and the output or result cannot be standardized in relation to a given period of time.

In other words, professionals such as doctors, lawyers, and accountants require advanced training and experience, they must exercise discretion and judgment in the course of their work, and their work cannot be standardized. Many people would also expect professionals to contribute to society, to participate in a lifelong training program (both formal and informal), to keep abreast of developments in their field, and to help develop other professionals. In addition, many professional roles carry special rights and special responsibilities. Doctors, for example, prescribe drugs, perform surgery, and request confidential patient information.

Are IT Workers Professionals?

Many business workers have duties, backgrounds, and training that qualify them to be classified as professionals, including marketing analysts, financial consultants, and IT specialists. A partial list of IT specialists includes programmers, systems analysts, software engineers, database administrators, local area network (LAN) administrators, and chief information officers (CIOs). One could argue, however, that not every IT role requires "knowledge of an advanced type in a field of science or learning customarily acquired by a prolonged course of specialized intellectual instruction and study," to quote again from the U.S. Code's definition of a professional. From a legal perspective, IT workers are not recognized as professionals because they are not licensed. This distinction is important, for example, in malpractice lawsuits—many courts have ruled that IT workers are not liable for malpractice because they do not meet the legal definition of a professional.

Professional Relationships That Must Be Managed

IT professionals typically become involved in many different relationships, including those with employers, clients, suppliers, other professionals, IT users, and society at large. In each relationship, an ethical IT professional acts honestly and appropriately. These various relationships are discussed in the following sections.

Relationships Between IT Professionals and Employers

IT professionals and employers have a critical, multifaceted relationship that requires ongoing effort by both parties to keep it strong. An IT professional and employer discuss and agree upon fundamental aspects of this relationship before the professional accepts an employment offer. These issues include job title, general performance expectations, specific work responsibilities, drug testing, dress code, location of employment, salary, work hours, and company benefits. Many other issues are addressed in the company's policy and procedures manual or in the company code of conduct, if it exists; these issues include protection of company secrets, vacation policy, time off for a funeral or illness in the family, tuition reimbursement, and use of company resources, including computers and networks. Other aspects of the relationship develop over time as the need arises (for example, whether the employee can leave early one day if the time is made up on another day). Some aspects are addressed by law—for example, an employee cannot be required to do anything illegal, such as falsify the results of a quality assurance test. Some aspects are specific to the role of the IT professional and are established based on the nature of the work or project—for example, the programming language to be used, the type and amount of documentation to be produced, and the extent of testing to be conducted.

As the stewards of an organization's IT resources, IT professionals must set an example and enforce policies regarding the ethical use of IT. IT professionals have the skills and knowledge to abuse systems and data or to allow others to do so. **Software piracy**, the act of illegally making copies of software or enabling others to access software to which they are not entitled, is an area in which IT professionals can be tempted to violate laws and policies. Although end users get the blame when it comes to using illegal copies of commercial software, software piracy in a corporate setting is sometimes directly traceable to IT staff—either they allow it to happen or actively engage in it, often to reduce IT-related spending to meet challenging budgets. According to a study conducted by International Data Corporation (IDC), a global provider of market intelligence, advisory services, and events for the IT and telecommunications industries, 35 percent of the world's software was illegally copied in 2004. This represents a 1 percent decrease from 2003. Yet, losses due to piracy increased from $29 billion to $33 billion. Table 2-1 lists the 10 countries that had the highest software piracy rates in 2004 and the 10 countries with the lowest rates.[7]

The Business Software Alliance (BSA) is a trade group that represents the world's largest software and hardware manufacturers. Its mission is to stop the unauthorized copying of software produced by its members (see Table 2-2). More than 100 BSA lawyers and investigators prosecute thousands of cases of software piracy each year.[8] BSA investigations are usually triggered by calls to the BSA hotline (888-NO-PIRACY), reports sent to the BSA Web site, and referrals from member companies. Many of these cases are reported by disgruntled employees. When the BSA finds cases of software piracy, it assesses heavy monetary penalties. BSA is funded through dues based on member companies' software revenues and through settlements from companies that commit piracy. In 2004, for example, Red Bull North America Inc. paid the BSA $105,000 to settle claims that it had more copies of Adobe, Microsoft, and Symantec software programs on its computers than it had licenses to support.

Failure to cooperate with the BSA can be extremely expensive. The cost of criminal or civil penalties to a corporation and the people involved can easily be many times more expensive than the cost of "getting legal" by acquiring the correct number of software licenses. Penalties can be up to $100,000 per copyrighted work if a software piracy case goes to trial and the defendant loses.

TABLE 2-1 Software piracy rates of selected countries

10 countries with the highest piracy rates	2004 piracy rate	10 countries with the lowest piracy rates	2004 piracy rate
Vietnam	92%	United States	21%
Ukraine	91%	New Zealand	23%
China	90%	Austria	25%
Zimbabwe	90%	Sweden	26%
Indonesia	87%	United Kingdom	27%
Russia	87%	Denmark	27%
Nigeria	84%	Switzerland	28%
Tunisia	84%	Japan	28%
Algeria	83%	Finland	29%
Kenya	83%	Germany	29%

TABLE 2-2 Members of Business Software Alliance (as of July 2005)

Adobe	Apple	Autodesk
Avid	Bentley Systems	Borland
Cadence	Cisco Systems	CNC Software/Mastercam
Dell	Entrust	HP (Hewlett-Packard)
IBM	Intel	Internet Security Systems
Macromedia	McAfee, Inc.	Microsoft
PTC	RSA Security	SAP
SolidWorks	Sybase	Symantec
The Mathworks	UGS Corp.	VERITAS Software

Trade secrecy is another area that can cause problems between employers and IT professionals. A **trade secret** is information used in a business, generally unknown to the public, that the company has taken strong measures to keep confidential. It represents something of economic value that has required effort or cost to develop and has some degree of uniqueness or novelty. Trade secrets can include the design of new software code, hardware designs, business plans, the design of the user interface to a computer program, and manufacturing processes. Examples include the Colonel's secret recipe of 11 herbs and spices, the formula for Coke, and Intel's manufacturing process for the Pentium 4 chip. Employers fear that employees may reveal these secrets to competitors, especially when they leave the company. As a result, they require employees to sign confidentiality agreements and promise not to reveal the company's trade secrets. However, the IT industry is known for high employee turnover, and things can get complicated when an employee moves on to a competitor.

In 2005, for example, a California jury ordered Toshiba Corporation to pay $465 million in damages to Lexar, its former business partner, for theft of trade secrets. The case centered on technology used in flash memory chips, which are widely used in computers and other consumer products to retain data when their power supply is disconnected. Lexar is a manufacturer and marketer of removable flash memory cards, USB flash drives (see Figure 2-1), card readers, and controller technology solutions for the digital photography, consumer electronics, and communications markets. Lexar sued Toshiba because it used its relationship to gain access to Lexar's business plans and technology while simultaneously working with SanDisk Corporation, a major rival of Lexar, on similar flash memory technology, according to Lexar spokespeople. In addition, Lexar asked for an injunction that bars Toshiba products from sale in the United States if they include the Lexar technology.[9]

FIGURE 2-1 Lexar flash memory products

Another issue that can create friction between employers and IT professionals is whistle-blowing. **Whistle-blowing** is an effort by an employee to attract attention to a negligent, illegal, unethical, abusive, or dangerous act by a company that threatens the public interest. Whistle-blowers often have special information based on their expertise or position within the offending organization. For example, an employee of a chip manufacturing company may know that the chemical process used to make the chips is dangerous to employees and the general public. A conscientious employee would call the problem to management's attention and try to correct it by working with appropriate resources within the company. But what if the employee's attempt to correct the problem through internal channels was thwarted or ignored? The employee could then consider becoming a whistle-blower and reporting the problem to people outside the company, including state or federal agencies that have jurisdiction. Obviously, such actions could have negative consequences on the employee's job, and could even result in retaliation and firing.

In May 2005, for example, Oracle Corporation paid $8 million to settle charges that it fraudulently collected fees before providing training for clients and failed to comply with federal travel regulations in billing for travel and expenses. The charges arose from a whistle-blower lawsuit brought by a former Oracle vice president. As a result of the settlement, the whistle-blower received $1.58 million of the $8 million total settlement.[10] Whistle-blowing is discussed more fully in Chapter 8.

Relationships Between IT Professionals and Clients

A professional often provides services to clients who either work outside the professional's organization or are "internal." In relationships between IT professionals and clients, each party agrees to provide something of value to the other. Generally speaking, the IT professional provides hardware, software, or services at a certain cost and within a given time frame. For example, an IT professional might agree to implement a new accounts payable software package that meets the client's requirements. The client provides compensation, access to key contacts, and perhaps work space. This relationship is usually documented in contractual terms—who does what, when the work begins, how long it will take, how much the client pays, and so on. Although there is often a vast disparity in expertise between IT professionals and their clients, the two parties must work together to be successful.

Typically, the client makes decisions about a project on the basis of information, alternatives, and recommendations provided by IT professionals. The client trusts them to use their expertise and to think and act in the client's best interests. The IT professional must trust that the client will provide relevant information, listen to and understand what the professional says, ask questions to understand the impact of key decisions, and use the information to make wise choices between alternatives. Thus, the responsibility for decision making is shared between client and professional.

One ethical problem between IT professionals and clients involves IT consultants or auditors who recommend their own products and services or those of an affiliated vendor to remedy a problem they have detected. For example, an IT consulting firm might be hired to assess a firm's IT strategic plan. After a few weeks of analysis, the consulting firm might provide a poor rating for the existing strategy and insist that its proprietary products and services are required to develop a new strategic plan. Such findings raise questions about the vendor's objectivity and whether its recommendations can be trusted.

During a project, IT professionals might be unable to provide full and accurate reporting of the project's status if they lack the information, tools, or experience to perform an accurate assessment. The project manager may want to keep resources flowing into the project and hope that problems can be corrected before anyone notices. The project manager may also be reluctant to share status information because of contractual penalties for failure to meet the schedule or to develop certain system functions. In this situation, the client may not be informed about the problem until it has become a crisis. After the truth comes out, finger-pointing and heated discussions about cost overruns, missed schedules, and technical incompetence can lead to charges of fraud, misrepresentation, and breach of contract, as discussed in the following Legal Overview.

LEGAL OVERVIEW

Fraud, Misrepresentation, and Breach of Contract

Fraud is the crime of obtaining goods, services, or property through deception or trickery. Fraudulent misrepresentation occurs when a person consciously decides to induce another person to rely and act on the misrepresentation. To prove fraud in a court of law, prosecutors must demonstrate the following elements:

1. The wrongdoer made a false representation of material fact.
2. The wrongdoer intended to deceive the innocent party.
3. The innocent party justifiably relied on the misrepresentation.
4. The innocent party was injured.[11]

Breach of contract occurs when one party fails to meet the terms of a contract. Further, a material breach of contract occurs when a party fails to perform certain express or implied obligations that impair or destroy the essence of the contract. Because there is no clear line between a minor breach and a material breach, determination is made on a case-by-case basis.[12] "When there has been a material breach of contract, the nonbreaching party can either: (1) rescind the contract, seek restitution of any compensation paid under the contract to the breaching party, and be discharged from any further performance under the contract; or (2) treat the contract as being in effect and sue the breaching party to recover damages."[13]

When IT projects go wrong because of cost overruns, schedule slippage, lack of system functionality, and so on, aggrieved parties might charge fraud, fraudulent misrepresentation, and/or breach of contract. Trials can take years to settle, generate substantial legal fees, and create bad publicity for both parties. As a result, more than 90 percent of such disputes are settled out of court, and the proceedings and outcomes are concealed from the public. In addition, IT vendors have become more careful about protecting themselves from major legal losses by requiring that contracts place a limit on potential damages.

Most IT projects are joint efforts in which vendors and customers work together to develop a system. Assigning fault when such projects go wrong can be difficult; one side might be partially at fault while the other side is mostly at fault. Consider the following frequent causes of problems in IT projects:

- The customer changes the scope of the project or the system requirements during the effort.
- Poor communications between customer and vendor lead to performance that does not meet expectations.
- The vendor delivers a system that meets customer requirements, but a competitor comes out with a system that offers more advanced and useful features.
- The customer fails to reveal information about legacy systems or databases that make the new system extremely difficult to implement.

continued

Who is to blame in such circumstances? For example, Collin County, Texas, awarded an $8 million contract to Siemens Business Services Inc. for software applications to manage the fast-growing county's financial, human resources, and other operations. Shortly after work began, however, Siemens encountered problems meeting the contract's requirements. Eventually, Siemens said it couldn't complete the project at all. Angry county leaders sued Siemens and the software vendor's public-services unit for $10 million for fraud and breach of contract after having paid $1 million for previous work.[14]

Relationships Between IT Professionals and Suppliers

IT professionals deal with many different hardware, software, and service providers. Most IT professionals understand that building a good working relationship with suppliers encourages the flow of useful communication and the sharing of ideas. Such information can lead to innovative and cost-effective ways of using the supplier's products and services that the IT professional may never have considered.

IT professionals should develop good relationships with suppliers by dealing fairly with them and not making unreasonable demands. Threatening to replace a supplier who can't deliver needed equipment tomorrow, when the normal industry lead time is one week, is aggressive behavior that does not help a working relationship.

Suppliers also strive to maintain positive relationships with their customers to make and increase sales. Sometimes, their actions to achieve this goal might be perceived as unethical—for example, they could offer an IT professional a gift that is actually intended as a bribe. Clearly, IT professionals should not accept a bribe from a vendor, but they must be careful in considering what constitutes a bribe. For example, accepting invitations to expensive dinners or payment of entry fees for a golf tournament may seem innocent to the recipient, but may be perceived as bribery by an internal accounting auditor.

Bribery involves providing money, property, or favors to someone in business or government to obtain a business advantage. An obvious example is a software supplier that offers money to another company's employee to get its business. This type of bribe is often referred to as a "kickback" or "payoff." The person who offers a bribe commits a crime when the offer is made, and the recipient is guilty of bribery upon accepting the offer.

The U.S. Foreign Corrupt Practices Act (FCPA) makes it a crime to bribe a foreign official, a foreign political party official, or a candidate for foreign political office. The act applies to any U.S. citizen or company, or to any company with shares listed on any U.S. stock exchange. However, a bribe is not a crime if the payment was lawful under the laws of the foreign country in which it was paid. Penalties for violating the FCPA are severe—corporations face a fine of up to $2 million per violation, and individual violators may be fined up to $100,000 and imprisoned for up to five years.

The FCPA also requires corporations to meet its accounting standards by having an adequate system of internal controls, including maintaining books and records that accurately and fairly reflect their transactions. The goal of these standards is to prevent companies from using "slush funds" or other means to disguise payments to foreign officials. A firm's business practices and its accounting information systems are frequently audited both by internal and outside auditors to ensure that they meet these standards.

The FCPA permits facilitating payments that are made for "routine government actions," such as obtaining permits or licenses; processing visas; providing police protection; providing phone services, power, or water supplies; or facilitating actions of a similar nature. Thus, it is permissible under the FCPA to pay an official to perform some official function faster (for example, to speed customs clearance), but not to make a different substantive decision (for example, to award business to one's firm).

In some countries, gifts are an essential part of doing business. In fact, in some countries it would be considered rude not to bring a present to an initial business meeting. In the United States, a gift might take the form of free tickets to a sporting event from a personnel agency that wants to get on your company's list of preferred suppliers. At what point does a gift become a bribe? Who decides?

The key distinguishing factor is that no gift should be hidden. A gift may be considered a bribe if it is not declared. As a result, most companies require all gifts to be declared and that everything but token gifts must be declined. Some companies have a policy of pooling the gifts received by their employees, auctioning them, and giving the proceeds to charity.

When it comes to distinguishing between bribes and gifts, the perceptions of the donor and recipient can differ. The recipient may believe he received a gift that in no way obligates him to the donor, particularly if the gift was not cash. The donor's intentions, however, might be very different. Table 2-3 helps you distinguish between a gift and a bribe.

TABLE 2-3 Distinguishing between a bribe and a gift

Bribes	Gifts
Are made in secret, as they are neither legally nor morally acceptable	Are made openly and publicly as a gesture of friendship or goodwill
Are often made indirectly through a third party	Are made directly from donor to recipient
Encourage an obligation for the recipient to act favorably toward the donor	Come with no expectation of a future favor for the donor

Relationships Between IT Professionals and Other Professionals

Professionals feel a degree of loyalty to the other members of their profession. As a result, they are quick to help each other obtain new positions but slow to criticize each other in public. Professionals also have an interest in their profession as a whole, because how it is perceived affects how individual members are perceived and treated. (For example, politicians are not generally thought to be very trustworthy, but teachers are.) Hence, professionals owe each other an adherence to the profession's code of conduct. Experienced professionals can also serve as mentors and help develop new members of the profession.

A number of ethical problems can arise between members of the IT profession. One of the most common is **résumé inflation**, which involves lying on a résumé and claiming competence in an IT skill that is in high demand. Even though IT professionals might benefit in the short term from exaggerating qualifications, such action can hurt the profession and themselves in the long run. Customers, and society in general, might become much more skeptical of IT professionals as a result. As many as 30 percent of job applicants exaggerate their accomplishments and about 10 percent seriously misrepresent their backgrounds, according to some estimates.[15]

Another ethical issue is the inappropriate sharing of corporate information. Because of their roles, IT professionals have access to corporate databases of private and confidential information about employees, customers, suppliers, new product plans, promotions, budgets, and so on. As discussed in Chapter 1, this information is sometimes shared inappropriately. It might be sold to other organizations or shared informally during work conversations with others who have no need to know.

Relationships Between IT Professionals and IT Users

The term **IT user** distinguishes the person for whom a hardware or software product is designed from the IT professionals who develop, install, service, and support the product. IT users need the product to deliver organizational benefits or to increase their productivity.

IT professionals have a duty to understand a user's needs and capabilities and to deliver products and services that best meet those needs—subject, of course, to budget and time constraints. IT professionals also have a key responsibility to establish an environment that supports ethical behavior by users. Such an environment discourages software piracy, minimizes the inappropriate use of corporate computing resources, and avoids the inappropriate sharing of information. Later in this chapter, you will learn more about establishing an effective IT usage policy that addresses these issues.

Relationships Between IT Professionals and Society

Regulatory laws establish safety standards for products and services to protect the public. However, these laws are less than perfect, and they fail to safeguard against all negative side effects of a product or process. Often, professionals can see more clearly what effect their work will have and can take action to eliminate potential public risks. Thus, society not only expects members of a profession not to cause harm, but to provide significant benefits. One approach to meeting this expectation is to establish and maintain professional standards that protect the public.

Clearly, the actions of an IT professional can affect society. For example, a systems analyst may design a computer-based control system to monitor a chemical manufacturing process. A failure or error in the system may put workers or residents near the plant at risk. As a result, IT professionals have a relationship with others in society who may be affected by their actions. However, there is currently no single, formal organization of IT professionals that takes responsibility for establishing and maintaining standards that protect the public.

THE ETHICAL BEHAVIOR OF IT PROFESSIONALS

Chapter 1 points out that the risks associated with inappropriate ethical behavior have grown in number, complexity, likelihood, and significance. As a result, corporations are taking a number of actions to ensure good business ethics among their employees. This section focuses on actions that support the ethical behavior of IT professionals.

Professional Codes of Ethics

A **professional code of ethics** states the principles and core values that are essential to the work of a particular occupational group. Practitioners in many professions subscribe to a code of ethics that governs their behavior. For example, doctors adhere to varying versions of the 2000-year-old Hippocratic oath, which medical schools offer as an affirmation to their graduating classes. Most codes of ethics created by professional organizations have two main parts. The first outlines what the professional organization aspires to become, and the second typically lists rules and principles by which members of the organization are expected to abide. Many codes also include a commitment to continuing education for those who practice the profession. (For examples of professional codes of ethics, see Appendices B through E.)

Laws do not provide a complete guide to ethical behavior. Just because an activity is not defined as illegal does not mean it is ethical. You also cannot expect a professional code of ethics to provide the complete answer—no code can be the definitive collection of behavioral standards. However, practicing according to a professional code of ethics can produce many benefits for the individual, the profession, and society as a whole:

- *Improves ethical decision making*—Adherence to a professional code of ethics means that practitioners use a common set of core values and beliefs to serve as a guideline for ethical decision making.
- *Promotes high standards of practice and ethical behavior*—Adherence to a code of ethics reminds professionals of the responsibilities and duties that they may be tempted to compromise to meet the pressures of day-to-day business. The code also defines behaviors that are acceptable and unacceptable to guide professionals in their interactions with others. Strong codes of ethics have procedures for censuring professionals for serious violations, with penalties that can include the loss of the right to practice. Such codes are the exception, however, and few of them exist in the IT arena.
- *Enhances trust and respect from the general public*—Public trust is built on the expectation that a professional will behave ethically. People often must depend on the integrity and good judgment of a professional to tell the truth, abstain from giving self-serving advice, and offer warnings about the potential negative side effects of their actions. Thus, adherence to a code of ethics enhances trust and respect of professionals and their profession.
- *Provides an evaluation benchmark*—A code of ethics provides an evaluation benchmark that a professional can use as a means of self-assessment. Peers of the professional can also use the code for recognition or censure.

Professional Organizations

No IT professional organization has emerged as preeminent, so there is no universal code of ethics for IT professionals. However, the existence of such organizations is useful in a field that is rapidly growing and changing. IT professionals need to know about new developments in the field, which requires networking with others, seeking out new ideas, and building personal skills and expertise. Whether you are a freelance programmer or the CIO of a Fortune 500 company, membership in an organization of IT professionals enables you to associate with others of similar work experience, to develop working relationships, and to exchange ideas. Information is disseminated from these organizations through e-mail, periodicals, Web sites, meetings, and conferences. Furthermore, in recognition of the need for professional standards of competency and conduct, many of these organizations have developed a code of ethics. Four of the most prominent IT-related professional organizations are summarized in this section.

Association for Computing Machinery (ACM)

The ACM is a computing society founded in 1947 that serves more than 80,000 professionals in more than 100 countries and offers many publications for technology professionals. *Tech News*, for example, is a comprehensive news-gathering service published three times a week. ACM's *Ubiquity* publication is a forum and opinion magazine. The organization also offers a substantial digital library of bibliographic information, citations, articles, and journals. The ACM sponsors special-interest groups that focus on a variety of IT issues, including artificial intelligence, computer architecture, programming languages, computer-human interaction, and mobile communications. Each group provides publications, workshops, and conferences for information exchange.

The ACM has a code of ethics and professional conduct with supplemental explanations and guidelines. The ACM code consists of eight general moral imperatives, eight specific professional responsibilities, six organizational leadership imperatives, and two elements of compliance. The complete text of this code is provided in Appendix B.

Association of Information Technology Professionals (AITP)

The AITP has its roots in Chicago in 1951, when a group of machine accountants got together and decided that the future was bright for the TAB machines they were operating. They were members of a local group called the Machine Accountants Association (MAA), which evolved into the Data Processing Management Association in 1962 and finally the AITP in 1996.

The AITP provides quality IT-related education, information on relevant IT issues, and forums for networking with experienced peers and other IT professionals for its nearly 9000 members.[16] Its mission is to provide superior leadership and education in information technology, and one of its goals is to help members make themselves more marketable to the industry. The AITP also has a code of ethics and standards of conduct, which are presented in Appendix C. The standards of conduct are considered to be rules that no true IT professional should violate.

Computer Society of the Institute of Electrical and Electronics Engineers (IEEE-CS)

The Institute of Electrical and Electronics Engineers (IEEE) covers the broad fields of electrical, electronic, and information technologies and sciences. The IEEE-CS is one of the oldest and largest IT professional associations, with more than 100,000 members. Roughly 40 percent of its members live and work outside the United States. Founded in 1946, the IEEE-CS is the largest of the 36 societies of the IEEE. "The IEEE-CS's vision is to be the leading provider of technical information and services to the world's computing professionals. The society promotes an active exchange of information, ideas, and technological innovation among its members through its many conferences, applications-related and research-oriented journals, local and student chapters, technical committees, and standards working groups."[17]

In 1993, the IEEE-CS and the ACM formed a Joint Steering Committee for the Establishment of Software Engineering as a Profession. The initial recommendations of the committee were to define ethical standards, to define the required body of knowledge and recommended practices in software engineering, and to define appropriate curricula to acquire knowledge. The Software Engineering Code of Ethics and Professional Practice (see Appendix D) documents the ethical and professional responsibilities and obligations of software engineers.

Project Management Institute (PMI)

The Project Management Institute was established in 1969 and currently has more than 150,000 members in more than 150 countries. Its members include project managers from such diverse fields as construction, sales, finance, and production, not just information systems. It has certified more than 100,000 people as project management professionals (PMPs). Certification requires that a person meet specific education and experience requirements, agree to follow the PMP Code of Ethics, and pass the PMP exam, which is designed to assess and measure knowledge of project management. The PMI Member Code of Ethics is presented in Appendix E.

Certification

Certification indicates that a professional possesses a particular set of skills, knowledge, or abilities, in the opinion of the certifying organization. Unlike licensing, which applies only to people and is required by law, certification can also apply to products and is generally voluntary. IT-related certifications typically carry no requirement to adhere to a code of ethics, whereas such a requirement is standard with licensing.

Numerous companies and professional organizations offer certifications, and opinions are divided on their value. Many employers view them as a benchmark that indicates mastery of a defined set of basic knowledge. On the other hand, because certification is no substitute for experience and doesn't guarantee that a person will perform well on the job, some hiring managers are rather cynical on the subject. Most IT employees are motivated to learn new skills, and certification provides a structured way of doing so. For such people, completing a certification provides clear recognition and correlates with a good plan to help them advance their careers. Others view certification as just another means for product vendors to generate additional revenue with little merit attached.[18]

Vendor Certifications

Many IT vendors such as Cisco, IBM, Microsoft, Sun, and Oracle offer certification programs for their products. Workers who successfully complete a program can represent themselves as certified users of a manufacturer's product. Depending on the job market and the demand for skilled workers, some certifications might substantially improve IT workers' salaries and career prospects. Certifications that are tied to a vendor's product are relevant for narrowly defined roles or certain aspects of broader roles. Sometimes, however, vendor certifications are too focused on technical details and do not address more general concepts.

Certifications require passing a written exam, which usually contains multiple-choice questions because of legal concerns about whether other types of exams can be graded objectively. A few certifications, such as Cisco Certified Internetworking Engineer (CCIE), also require a hands-on lab exam that demonstrates skills and knowledge. It can take years to obtain the necessary experience required for some certifications. Courses and training material are available to help speed up the preparation process, but some training costs can be expensive. Depending on the certification, study materials can cost $1000 and in-class formal training courses can cost more than $10,000.

The Microsoft Certified System Engineer (MCSE) certification is one of the more demanding Microsoft certifications. Engineers, analysts, and consultants frequently obtain it for their work in designing and implementing Windows server solutions and architectures. Four networking system exams, one client operating system exam, and one design exam are required. So many people have obtained MCSE certification (more than 400,000 in the United States) that it has almost become an entry-level requirement for some companies.

Because of the rapid pace of change in the IT field, workers are commonly recertified as newer technologies become available. For example, many people who were MCSE-certified and trained on the Windows NT 4.0 operating system went through recertification when newer operating systems were developed.

Industry Association Certifications

Certifications from industry associations generally require a certain level of experience and a broader perspective than vendor certifications; however, they often lag in developing tests that cover new technologies. The trend in IT certification is to move from purely technical content to a broader mix of technical, business, and behavioral competencies, which are required in today's demanding IT roles. This trend is evident in industry association certifications that address broader roles such as e-commerce, network security, and project management.

For example, the Institute for Certification of Computing Professionals (ICCP) offers two levels of certification—Certified Associate Computing Professional and Certified Computing Professional. Since 1973, more than 50,000 IT professionals worldwide have completed ICCP certification.[19] Candidates for either certificate must take a common core exam that includes questions on organizational frameworks, systems concepts, data and information, systems development, technology, and associated disciplines.

The Associate Computing Professional (ACP) certification requires applicants to successfully complete an exam for an additional computer programming language. The ACP certification is for new members of the IT industry or recent college graduates who want professional credentials that substantiate their level of computing knowledge.

More experienced IT professionals can obtain the Certified Computing Professional (CCP) certification, which requires successful completion of exams in two of the following areas: management, procedural programming, business information systems, communications, office information systems, systems security, microcomputing and networks, systems development, software engineering, systems programming, and data resource management. In addition, CCP candidates must have four years of full-time experience in information systems or applicable college degrees with two years of full-time experience. All candidates must subscribe to a specified code of ethics, conduct, and good practice.[20]

The American Society for Quality Control (ASQC) offers certifications for software quality engineers who have eight years of professional experience and at least three years in a decision-making position. A bachelor's degree may count as four years of experience, and an advanced degree may count as an additional, fifth year of experience. In addition, engineers must have professional credentials, such as membership in a recognized professional society, an engineer's license, or statements from two professional colleagues. Engineers must successfully complete a written exam that covers software quality management, software engineering, project management, analytical methods, and quality systems.

Clearly, many IT certifications are available. Their value varies greatly depending on where people are in their career path, what other certifications they possess, and the nature of the IT job market.

Government Licensing

Government licensing is generally administered at the state level in the United States. Some professionals must be licensed to prove that they can do their work ethically and safely, including certified public accountants (CPAs), lawyers, doctors, various types of medical and day care providers, and some engineers.

Various states have enacted legislation to establish licensing requirements and protect public safety. For example, Texas passed the Engineering Registration Act after a tragic school explosion at New London, Texas, in 1937. Under the act and subsequent revisions, only duly licensed people may legally perform engineering services for the public, and public works must be designed and constructed under the direct supervision of a licensed professional engineer. People cannot call themselves engineers or professional engineers unless they are licensed, and violators are subject to legal penalties.[21] Most states have similar laws.

The Case for Licensing IT Professionals

The days of simple, stand-alone information systems are over. Modern systems are highly complex, interconnected, and critically dependent on each other. Highly integrated enterprise resource planning systems (ERPs) help multibillion-dollar companies control all their business functions, including forecasting, production planning, purchasing, inventory control, manufacturing, and distribution. Complex computers and information systems manage and control the nuclear reactors of power plants that generate electricity for cities. Medical information systems monitor the vital statistics of hospital patients on critical life support. Local, state, and federal government information systems are entrusted with generating and distributing millions of checks worth billions of dollars to the public.

As a result of the increasing importance of IT in our everyday lives, the development of reliable, effective information systems has become an area of mounting public concern. This concern has led to a debate whether the licensing of IT professionals would improve information systems. Proponents argue that licensing would strongly encourage IT professionals to follow the highest standards of the profession and practice a code of ethics, and that licensing would allow violators to be punished. Without licensing, there are no requirements for heightened care and no concept of professional malpractice.

Issues Associated with Government Licensing of IT Professionals

Australia, Great Britain, and the Canadian provinces of Ontario and British Columbia have adopted licensing for software engineers. The National Council of Engineering Examiners and Surveyors (NCEES) has developed a professional exam for electrical engineers and computer engineers.[22] However, there are few international or national licensing programs for IT professionals, for many reasons:

- *There is no universally accepted core body of knowledge.* The core body of knowledge for any profession outlines agreed-upon sets of skills and abilities that all licensed professionals must possess. At present, however, there are no universally accepted standards for licensing programmers, software engineers, and other IT professionals. Instead, various professional societies, state agencies, and federal governments have developed their own standards.

- *It is unclear who should manage the content and administration of licensing exams.* How will licensing exams be constructed, and who will be responsible for designing and administering them? Will someone who passes a license exam in one state or country be accepted in another state or country? In a field as rapidly changing as IT, professionals clearly must commit to ongoing, continuous education. If an IT professional's license expires every few years (like a driver's license), when must practitioners prove competence in new practices before they can renew their license? Such questions would normally be answered by the state agency that licenses other professionals.

- *There is no administrative body to accredit professional education programs.* Unlike the American Medical Association for medical schools or the American Bar Association for law schools, no single body accredits IT professional education programs. Furthermore, there is no well-defined, step-by-step process to train IT professionals, even for specific jobs, such as programming. There is not even broad agreement on what skills a good programmer must possess—it is highly situational, depending on the computing environment.

- *There is no administrative body to assess and ensure competence of individual professionals.* Lawyers, doctors, and other licensed professionals are held accountable to high ethical standards and can lose their license for failing to meet these standards or for demonstrating incompetence. The AITP standards of conduct state that professionals should "take appropriate action in regard to any illegal or unethical practices that come to my attention. However, I will bring charges against any person only when I have reasonable basis for believing in the truth of the allegations and without any regard to personal interest." The AITP code addresses the censure issue much more forcefully than other IT codes of ethics, although it has seldom, if ever, been used to censure practicing IT professionals.

IT PROFESSIONAL MALPRACTICE

Negligence has been defined as not doing something that a reasonable man would do, or doing something that a reasonable man would not do. **Duty of care** refers to the obligation to protect people against any unreasonable harm or risk. For example, people have a duty to keep their pets from attacking others and to operate their cars safely. Similarly, businesses must keep dangerous pollutants out of the air and water, make safe products, and maintain safe operating conditions for employees.

The courts decide whether parties owe a duty of care by applying a **reasonable person standard** to evaluate how an objective, careful, and conscientious person would have acted in the same circumstances. Likewise, defendants who have particular expertise or competence are measured against a **reasonable professional standard**. For example, in a medical malpractice suit based on improper treatment of a broken bone, the reasonable person standard would be higher if the plaintiff were an orthopedic surgeon rather than a general practitioner. In the IT arena, consider a negligence case in which an employee inadvertently destroyed millions of customer records. The reasonable person standard would be higher if the plaintiff were a licensed, Oracle-certified database administrator (DBA) with 10 years of experience, instead of an unlicensed systems analyst with no DBA experience or specific knowledge of the Oracle system.

If a court finds that a defendant actually owed a duty of care, it must then determine whether the duty was breached. A **breach of the duty of care** is the failure to act as a reasonable person would act. A breach of duty may consist of an action, such as throwing a lit cigarette into a fireworks factory and causing an explosion, or a failure to act when there is a duty to do so—for example, a police officer who does not protect a citizen from an attacker.

Professionals who breach this duty of care are liable for injuries their negligence causes. This liability is commonly referred to as **professional malpractice**. For example, a CPA who fails to use reasonable care, knowledge, skill, and judgment when auditing a client's books is liable for accounting malpractice. Professionals who breach this duty are liable to their patients or clients, and possibly to some third parties.

Courts have consistently rejected attempts to sue individual parties for computer-related malpractice. Professional negligence can only occur when people fail to perform within the standards of their profession, and software engineering is not a uniformly licensed profession in the United States. Because there are no uniform standards against which to compare a software engineer's professional behavior, he cannot be subject to malpractice lawsuits.

IT USERS

Chapter 1 outlined the general topic of how corporations are addressing the increasing risks of unethical behavior. This section focuses on improving employees' ethical use of IT, which has become an area of growing concern as more companies provide employees with PCs, access to corporate information systems and data, and the Internet.

Common Ethical Issues for IT Users

This section discusses a few common ethical issues for IT users. Other ethical issues will be discussed in future chapters.

Software Piracy

As mentioned earlier in this chapter, software piracy in a corporate setting can be directly traceable to IT professionals—they might allow it to happen or they might actively engage in it. Corporate IT usage policies and management should encourage users to report instances of piracy and to challenge its practice.

Sometimes, IT users are the ones who commit software piracy. A common violation occurs when employees copy software from their work computers for use at home. When confronted, the IT user's argument might be: "I bought a home computer partly so I could take work home and be more productive; therefore, I need the same software on my home computer as I have at work." However, this is still piracy if no one has paid for an additional license to use the software on the home computer.

Inappropriate Use of Computing Resources

Some employees use their computers to surf popular Web sites that have nothing to do with their jobs, participate in chat rooms, view pornographic sites, and play computer games. These activities eat away at worker productivity and waste time. Furthermore, viewing sexually explicit material, sharing lewd jokes, and sending hate e-mail could lead to lawsuits and allegations that a company allowed a work environment conducive to racial or sexual harassment. According to a survey conducted by Delta Consulting in 2005, half of all Fortune 500 companies dealt with at least one incident related to computer porn in the workplace in the previous 12 months. The companies handled the problem by firing the offenders in 44 percent of the cases and taking other disciplinary action 41 percent of the time.[23]

Inappropriate Sharing of Information

Every organization stores vast amounts of information that can be classified as either private or confidential. Private data describes individual employees—for example, their salary information, attendance data, health records, and performance ratings. Confidential information describes a company and its operations, including sales and promotion plans, staffing projections, manufacturing processes, product formulae, tactical and strategic plans, and research and development. An IT user who shares this information with an unauthorized party, even inadvertently, has violated someone's privacy or created the potential that company information could fall into the hands of competitors. For example, if an IT employee saw a coworker's payroll records and then discussed them with a friend, it would be a clear violation of the worker's privacy.

Supporting the Ethical Practices of IT Users

The growing use of IT has increased the potential for new ethical issues and problems, so many organizations have recognized the need to develop policies that protect against abuses. Although no policy can stop wrongdoers, it can set forth the general rights and responsibilities of all IT users, establish boundaries of acceptable and unacceptable behavior, and enable management to punish violators. Adherence to the policy can improve services to users, increase productivity, and reduce costs. Companies can take several of the following actions when creating an IT usage policy.

Defining and Limiting the Appropriate Use of IT Resources

Companies must develop, communicate, and enforce written guidelines that encourage employees to respect corporate IT resources and use them to enhance their job performance. Effective guidelines allow some level of personal use while prohibiting employees from visiting objectionable Internet sites or using company e-mail to send offensive or harassing messages.

Establishing Guidelines for Use of Company Software

Company IT managers must provide clear rules that govern the use of home computers and associated software. Some companies negotiate contracts with software manufacturers and provide PCs and software so that IT users can work at home. Other companies help employees buy hardware and software at corporate discount rates. The goal should be to ensure that employees have legal copies of all the software they need to be effective, regardless of whether they work in an office, on the road, or at home.

Structuring Information Systems to Protect Data and Information

Organizations must implement systems and procedures that limit data access to employees who need it. For example, sales managers may have total access to sales and promotion databases through a company network, but their access should be limited to products for which they are responsible. Furthermore, they should be prohibited from accessing data about research and development results, product formulae, and staffing projections if they don't need it to do their jobs.

Installing and Maintaining a Corporate Firewall

A **firewall** is a hardware or software device that serves as a barrier between a company and the outside world and limits access to the company's network based on the organization's Internet usage policy. The firewall can be configured to serve as an effective deterrent to unauthorized Web surfing by blocking access to specific, objectionable Web sites. Unfortunately, the number of such sites grows so rapidly that it is difficult to block them all. The firewall can also serve as an effective barrier to incoming e-mail from certain Web sites, companies, or users. It can even be programmed to block e-mail with certain kinds of attachments (for example, Microsoft Word documents), which reduces the risk of harmful computer viruses.

Table 2-4 presents a manager's checklist that summarizes items to consider when establishing an IT usage policy. The preferred answer in each case is *yes*.

TABLE 2-4 Manager's checklist of items to consider when establishing an IT usage policy

Questions	Yes	No
Is there a statement that explains the need for an IT usage policy?	____	____
Does the policy provide a clear set of guiding principles for ethical decision making?	____	____
Is it clear how the policy applies to the following types of workers?		
Employees	____	____
Part-time workers	____	____
Temps	____	____
Contractors	____	____
Does the policy address the following issues?		
Protection of the data privacy rights of employees, customers, suppliers, and others	____	____
Limits and control of access to proprietary company data and information	____	____
The use of unauthorized or pirated software	____	____
Employee monitoring, including e-mail, wiretapping and eavesdropping on phone conversations, computer monitoring, and surveillance by video	____	____
Respect of the intellectual rights of others, including trade secrets, copyrights, patents, and trademarks	____	____
Inappropriate use of IT resources, such as Web surfing, e-mailing, and other use of computers for purposes other than business	____	____
The need to protect the security of IT resources through adherence to good security practices, such as not sharing user IDs and passwords, use of "hard-to-guess" passwords, and frequent changing of passwords	____	____
The use of the computer to intimidate, harass, or insult others through abusive language in e-mails and by other means	____	____
Are disciplinary actions defined for IT-related abuses?	____	____
Is there a process for communicating the policy to employees?	____	____
Is there a plan to provide effective, ongoing training relative to the policy?	____	____
Has a corporate firewall been implemented?	____	____
Is the corporate firewall maintained?	____	____

Summary

1. What key characteristics distinguish a professional from other kinds of workers, and what is the role of an IT professional?

    Professionals require advanced training and experience, they must exercise discretion and judgment in the course of their work, and their work cannot be standardized. A professional is expected to contribute to society, to participate in a lifelong training program (both formal and informal), to keep abreast of developments in the field, and to help develop other professionals. From a legal standpoint, a professional has passed the state licensing requirements (if they exist) and earned the right to practice there.

2. What relationships must an IT professional manage, and what key ethical issues can arise in each?

    IT professionals typically become involved in many different relationships, each with its own set of ethical issues and potential problems. In relationships between IT professionals and employers, important issues include setting and enforcing policies regarding the ethical use of IT, the potential for whistle-blowing, and the safeguarding of trade secrets. In relationships between IT professionals and clients, the key issues revolve around defining, sharing, and fulfilling each party's responsibilities for successfully completing an IT project. A major goal for IT professionals and suppliers is to develop good working relationships in which no action can be perceived as unethical. In relationships between fellow IT professionals, the key issues are to improve the profession through such activities as mentoring inexperienced colleagues and demonstrating professional loyalty. Résumé inflation and the inappropriate sharing of corporate information are relevant problems. In relationships between IT professionals and IT users, important issues include software piracy, inappropriate use of IT resources, and inappropriate sharing of information. When it comes to the relationship between IT professionals and society at large, the main challenge is to practice the profession in ways that cause no harm to society and provide significant benefits.

3. How do codes of ethics, professional organizations, certification, and licensing affect the ethical behavior of IT professionals?

    A professional code of ethics states the principles and core values that are essential to the work of an occupational group. A code serves as a guideline for ethical decision making, promotes high standards of practice and ethical behavior, enhances trust and respect from the general public, and provides an evaluation benchmark.

    Many people believe that the licensing and certification of IT professionals would increase the reliability and effectiveness of information systems, but the question of licensing raises many issues; for example, (a) there is no universally accepted core body of knowledge on which to test people; (b) it is unclear who should manage the content and administration of licensing exams; (c) there is no administrative body to accredit professional education programs; and (d) there is no administrative body to assess and ensure competence of individual professionals.

4. What are the key tenets of four different codes of ethics that provide guidance for IT professionals?

    Several IT-related professional organizations have developed a code of ethics, including the ACM, the AITP, the IEEE-CS, and the Project Management Institute. These codes have two main parts—the first outlines what the organization aspires to become, and the second typically lists rules and principles that members are expected to live by. They also include a commitment to continuing education for those who practice the profession.

5. What are the common ethical issues that face IT users?

    Issues include software piracy, inappropriate use of corporate IT resources, and the inappropriate sharing of private and secret information.

6. What approaches can support the ethical practices of IT users?

    The development of an IT usage policy is the first step for an organization in defining appropriate and inappropriate IT user behavior. The policy should define and limit the appropriate use of IT resources and set clear guidelines for use of company software. In addition, IT professionals within the organization can structure information systems and establish corporate firewalls to support appropriate use of IT resources.

Self-Assessment Questions

1. A professional is someone who:
    a. requires advanced training and experience
    b. must exercise discretion and judgment in the course of his or her work
    c. does work that cannot be standardized
    d. all of the above

2. Many courts have ruled that IT workers are not liable for malpractice because they do not meet the legal definition of a professional. True or False?

3. According to a study conducted by IDC, what percentage of the world's software was illegally copied (pirated) in 2004?
    a. 10–20 percent
    b. 20–30 percent
    c. 30–40 percent
    d. more than 40 percent

4. A _____ is information used in a business, generally unknown to the public, that the company has taken strong measures to keep confidential. It represents something that has economic value, has required effort or cost to develop, and has some degree of uniqueness or novelty.
    a. copyright
    b. trademark
    c. trade secret
    d. patent

5. Whistle-blowing is an effort by an employee to attract attention to a negligent, illegal, unethical, abusive, or dangerous act by a company that threatens the public interest. True or False?

6. Résumé inflation is a usual and customary practice tolerated by employers. True or False?

7. Society expects professionals to act in a way that:

   a. causes no harm to society

   b. provides significant benefits

   c. establishes and maintains professional standards that protect the public

   d. all of the above

8. Laws do not provide a complete guide to ethical behavior. Just because an activity is not defined as illegal does not mean that it is ethical. True or False?

9. _____ is a process that one undertakes voluntarily to prove competency in a set of skills.

   a. Licensing

   b. Certification

   c. Registering

   d. all of the above

10. There are many international and national licensing programs for IT professionals. True or False?

11. A policy on the use of information technology can:

   a. set forth the general rights and responsibilities common to all users of information technology

   b. establish the boundaries of acceptable and unacceptable behavior

   c. enable management to take action against those who violate the policy

   d. all of the above

12. A device that serves as a barrier between a company and the outside world and limits access to the company's computer network based on the organization's Internet usage policy is a(n):

   a. Internet service provider

   b. firewall

   c. encryption device

   d. malware

Review Questions

1. What criteria would you use to define a person employed in a professional capacity? How would you define the term *IT professional*?

2. What are the six relationships in which an IT professional becomes involved? Identify at least one key issue for each of these relationships.

3. What is software piracy? What role does the Business Software Alliance take in combating software piracy?

4. How do you define a trade secret? What actions might an organization take to protect its trade secrets?

5. Identify three ethical issues that can arise in the relationship between an IT professional and a client.

6. What is a professional code of ethics? How is it different from the corporate code of conduct discussed in Chapter 1?

7. List three benefits associated with adherence to a code of professional ethics.

8. Identify four prominent IT-related professional organizations. What are some benefits of membership in each of these organizations?

9. What is the difference between vendor and industry association certification?

10. Identify three benefits of government licensing for IT professionals.

11. Identify and discuss four issues associated with the government licensing of IT professionals.

12. What is negligence? What must a plaintiff show to prove negligence?

13. What are some common ethical issues encountered by IT users? What negative impact does unethical behavior have in each of these areas?

14. What are four actions that can strengthen the ethical practices of IT users?

Discussion Questions

1. How do you prove that fraud has been committed? How do you prove breach of contract? What is the difference between the two?

2. Discuss the following topic: Laws do not provide a complete guide to ethical behavior. An activity can be legal but not ethical.

3. What is professional malpractice? Should a software engineer ever be sued for professional malpractice? Why or why not?

4. Review the ACM code of ethics in Appendix B. The code covers many of the issues an IT professional is likely to face, but not all. Identify two key issues not addressed by the ACM code of ethics.

5. What can IT professionals do to ensure that the projects they lead meet the client's expectations and do not lead to charges of fraud, fraudulent misrepresentation, and breach of contract?

6. Should all IT professionals either be licensed or certified? Why or why not?

7. What commonalities do you find among the IT professional codes of ethics discussed in this chapter? What differences are there? Are any issues that are important to you not addressed by these codes of ethics?

What Would You Do?

1. As the vice president of marketing, you hired a software contractor to build a Web site for your home building supply firm. Unfortunately, the project did not go well. From your perspective, the contractor overcommitted and underdelivered. The initial Web site was completed three months late at a cost of $2 million over the estimated $5 million. Before your site went online, a competing manufacturer launched a Web site with unique features. You demanded that these features be added to your site. This took another four months and cost an additional $1 million. To top things off, once your Web site went online, it was slow

and buggy and got hacked by a former employee of the software contractor. The contractor has estimated that it will cost another $2 million to fix these problems. You are considering suing the software contractor; however, you are concerned that this will cause further delays in the delivery of a workable Web site. What would you do?

2. You are in charge of awarding all PC service contracts for your employer. In recent e-mails with the company's PC service contractor, you casually exchanged ideas about home landscaping, your favorite pastime. You also said you would like to have a few Bradford pear trees in your yard. Upon returning from a vacation, you discover three mature trees in your yard, along with a thank-you note in your mailbox from the PC service contractor. You really want the trees, but you didn't mean for the contractor to buy them for you. You suspect that the contractor interpreted your e-mail comment as a hint that you wanted him to buy the trees. You also worry that the contractor still has the e-mail. If the contractor sent your boss a copy, it might look as if you were trying to solicit a bribe. Can the trees be considered a bribe? What would you do?

3. In Italy, *raccomandazione* is the custom of seeking and receiving special treatment from people in power or from people who are close to power. The ability to solicit favors from someone in a higher place, be it through the chief of police or the chief's chauffeur, has been part of the Italian art of getting things done for more than 2000 years. In April 2001, Italy's highest court of appeal ruled that influence peddling is not a crime. The judges did rule, however, that it is a crime to overstate one's power to exert influence.

   Your firm is opening a new sales office in Rome and will be using a local employment agency to identify and screen candidates, who will then undergo employment testing and interviews by members of your organization. What guidelines would you provide to the agency regarding the practice of raccomandazione to ensure that the agency operates ethically and effectively?

4. Jacob is the vice president of sales and an important ally of your IT department. He's gone to bat for you before the CEO on important IT projects, such as the customer relationship management system, and has valuably assisted in advocating for the use of the latest software packages within the sales organization. Jacob has played a major role in your success so far. However, you've just learned that Jacob and his support staff are using an unlicensed Lotus software suite on their desktops, while the rest of the company is standardized using Microsoft Office. You've talked to him and the rest of the company's leadership team about the need for standardized software and the risks the company runs if it uses unlicensed software, but no action has been taken. What would you do?

5. You are the new CIO at a small manufacturing company with a total of 500 employees at one plant, two warehouses, and a headquarters building. Your manager is the chief financial officer (CFO), who wants you to make it a high priority to establish a set of policies and guidelines on the use of IT resources—the firm currently has none. How would you proceed?

Cases

1. When Certification Is Justified

On June 13, 2005, Don Tennant, editor-in-chief of *Computerworld*, published an editorial in favor of IT certification and was promptly hit with a barrage of angry responses from IT professionals. They argued that testable IT knowledge does not necessarily translate into quality IT work. A professional needs good communication and problem-solving skills as well as perseverance to get the job done well. Respondents explained that hard-working IT professionals focus on skills and knowledge that are related to their current projects and don't have time for certifications that will quickly become obsolete. They suspected vendors of offering certification as a marketing ploy and a source of revenue. They accused managers without technical backgrounds of using certification as "a crutch, a poor but politically defensible substitute for knowing what and how well one's subordinates are doing."

Any manager would certainly do well to review these insightful points, yet they beg the question: what useful purposes *can* certification serve within an organization?

Robert Tekiela, vice president of technology at Sapient Corporation, asserts that many employers use certification as a means of training employees and increasing skill levels within the company. Some companies are even using certification as a perk to attract and keep good employees. American Century Investments is taking this a step further by offering a job-rotation program through which workers can acquire experience as well.

Employers are also making good use of certification as a hiring gate both for entry-level positions and for jobs that require specific core knowledge. For example, a company with a Windows Server 2003 network might run an ad for a systems integration engineer and require a Microsoft Systems Engineer (MCSE) certification. A company that uses Siebel customer relationship management software may require a new hire to have a certification in the latest version of Siebel.

In addition, specific IT fields such as project management and security have a greater need for certification. As the speed and complexity of production increase within the global marketplace, people from all industries are showing an increasing interest in project management certification. With mottos like "Do It, Do It Right, Do It Right Now," the Project Management Institute has already certified more than 100,000 people. As the IT industry recovers from declines in IT spending that followed the 2000 recession, industry employers are beginning to encourage and sometimes require project management certification.

Calls for training in the field of security management go beyond certification. The demand for security professionals is expected to double in the next three years in the face of growing threats. Spam, computer viruses, spyware, and identity theft have businesses and government organizations worried. They want to make sure that their security managers can protect their data, systems, and resources.

The best recognized security certification is the CISSP, awarded by the International Information Systems Security Certification Consortium (ISC2). Yet the CISSP examination, like so many other IT certification examinations, is multiple choice. Employers and IT professionals alike have begun to recognize the limitations of these types of examinations. They want to ensure that examinees not only have core knowledge, but know how to use that knowledge—and a multiple-choice exam, even a six-hour, 250-question exam like CISSP, can't provide this assurance.

As a result, security professionals in the UK have formed the Information Security Professionals Working Group with the purpose of raising security training to the level of other professional training. They plan to accredit academic and professional development courses and to set up a mentoring program. ISC2 also plans to run master courses and mentoring programs.

In the meantime, other organizations are catching on. Sun Computers requires the completion of programming or design assignments for some of its certifications. So, while there is no universal need for certification or a uniform examination procedure that answers all needs within the IT profession, certifying bodies are beginning to adapt their programs to better fulfill the evolving needs for certification in IT.

Questions:

1. How are Sun Computers and other vendors discussed in the chapter changing their certification programs to test for skills as well as core knowledge?

2. What are the central arguments against certification, and how can certifying bodies change their programs to overcome their shortcomings?

3. What are the benefits of certification? How can certification programs change in the future to better serve the needs of the IT community?

2. Antibribery Laws Force U.S. Companies to Raise the Bar on Business Ethics

In the mid-1970s, investigations by the U.S. Securities and Exchange Commission (SEC) revealed that more than 400 companies in the United States had made illegal or questionable payments to foreign sources. To clean up the United States' image overseas, Congress enacted the Foreign Corrupt Practices Act (FCPA), which allows the U.S. Department of Justice (DOJ) and the SEC to prosecute businesses and company personnel who bribe governments, politicians, or political parties abroad. Companies can be fined up to $2 million and be barred from doing business with the U.S. government, receiving an export license, and engaging in the securities business. People can be imprisoned for up to five years and fined twice the amount they hoped to receive as a result of the bribery.

So, executives of a large IT company today wouldn't dare bribe foreign officials, say, to obtain a large government contract. Or would they?

In the wake of the Enron scandal, a Saudi Arabian telecommunications company called National Group for Communications and Computers filed a lawsuit in a New York District Court against Lucent Technologies, claiming that the telecommunications giant, along with the Swiss company ACEC, had bribed a former Saudi Arabian minister. The telecommunications minister, Ali Al-Johani, allegedly persuaded a government-controlled company to purchase Lucent and ACEC equipment. In return, company officials purportedly gave cash gifts, paid medical and hotel bills, and made available private jets to Al-Johani between 1995 and 2002. The suit claims that these favors are worth approximately $15 million.

An amended complaint later named former Lucent CEO Richard McGinn and former chief protocol officer, Robert W. Frye, as having approved two checks totaling more than $2 million to a Seattle cancer center where Al-Johani was being treated. The complaint also fingered the CEO of Lucent's spin-off company Avaya, Donald Peterson, claiming that he signed the checks to the cancer center.

In response to these accusations against its former highest-ranking officials, Lucent launched an internal audit in 23 of its overseas operations and reported potential FCPA violations to the DOJ and the SEC. In April 2004, Lucent made headlines again when it dismissed four top Chinese officials, including President Jason Chi and Chief Operating Officer Michael Kwan. Kwan spoke out to the press, denying wrongdoing and accusing Lucent of damaging his reputation. The Chinese government subsequently failed to prosecute the executives.

In China, where certain types of bribery are pervasive, this outcome is not surprising, and the question arises: is the FCPA damaging the competitiveness of U.S. companies abroad by preventing them from securing awards that foreign companies can acquire without fear of repercussion? Congress certainly thought so in 1988, when it requested that the executive branch take measures to ensure that the United States' major trading partners adopt antibribery laws similar to the FCPA.

The FCPA further provides for affirmative defenses, the assertion that a payment considered unlawful in the United States is in fact legal in the country where it occurred. Although the Department of Justice warns that lawfulness of a payment may be difficult to prove, some acts that the FCPA would consider unlawful are perfectly legal in China. For example, if an executive pays a Chinese government official $1000 to facilitate approval procedures, the executive has acted legally according to Chinese law. To constitute criminal bribery, the value of the bribe would have to surpass a threshold of $1208. Commercial bribes under this threshold value are permitted as long as their purpose is not linked to the sale of goods or services.

Questions:

1. Lucent purportedly gave cash payments, paid medical and hotel bills, and made available private jets to Al-Johani. Under what circumstances would these actions be considered gifts? Under what circumstances would these actions be considered bribes?

2. The SEC is considering taking civil action against Lucent's former CEO Richard McGinn and Lucent's former head of Saudi Arabia operations, John Heindel. What would they have to prove to make an affirmative defense of their actions?

3. In 2004, IBM dismissed several senior executives in Korea after they were indicted by the Seoul District Prosecutor's Office, which charged that the executives used a $2.5 million slush fund to obtain contracts worth $55 million. Compare this case to Lucent's situation in China.

3. IT Usage Policy

Read the following policy on the use of IT technology for the University of Cincinnati, and use the manager's checklist in Table 2-4 to answer the following questions:

Questions:

1. Are all of the key issues covered by this policy? If not, which ones need to be addressed?

2. Is the statement of enforcement clear and strong? If not, how would you reword this section of the policy?

3. How would you ensure that this policy is communicated and understood by the broad group of IT users at the university—students, professors, research people, administrative support staff, contractors, and part-time workers?

4. Examine the IT usage policy in effect at your school. Write a paragraph identifying its strengths and weaknesses.

Ethics for IT Professionals and IT Users

General Policy on the Use of Information Technology for the University of Cincinnati

As an institution of higher learning, the University both uses information technology and supplies it to the members of the University community. This policy sets forth the general rights and responsibilities common to all uses of information technology, from the simple stand-alone PC to the complex systems that create virtual classrooms, workplaces, and recreational facilities in the University.

This policy applies to all members of the University community, including guests who have been given accounts on the University's information technology systems for specific purposes. It also applies whether access is from the physical campus or from remote locations. In addition, there may be specific policies issued for individual systems, departments, colleges, and the like. While these policies must be consistent with this general policy, they provide more detailed guidance about what is allowed and what is prohibited on each system. All members of the University community are responsible for familiarizing themselves with any applicable policy prior to use.

Guiding Principles

The primary guiding principle is that the rules are the same for information technology as for other aspects of University life. The rights and responsibilities governing the behavior of members of the University community are the same on both the virtual and physical campuses, and the same disciplinary procedures will be followed when the rules are violated. There is nothing special about the virtual campus that makes it distinctly different.

The University has a strong commitment to the principles of free speech, open access to knowledge, and respect for a diversity of opinions. The rights as well as the restrictions governing these principles on the physical campus apply fully to the virtual campus.

Specific Areas

1. Applicable Laws and Regulations

    All members of the University community must obey:

    *   All relevant federal, state, and local laws. These include laws of general application such as libel, copyright, trademark, privacy, obscenity, and child pornography laws, as well as laws that are specific to computers and communication systems, such as the Computer Fraud and Abuse Act and the Electronic Communications Privacy Act.

    *   All relevant University rules and regulations. These include the Rules of the University, the Student Code of Conduct, the various collective bargaining agreements between the University and its employees, and all other University policies, including the policy against sexual and racial harassment.

    *   All contracts and licenses applicable to the resources made available to users of information technology.

    *   This policy as well as other policies issued for specific systems.

2. Resource Limits

    Information technology resources are often limited; what is used by one person is no longer available to others. Many systems have specific limits on several kinds of resources, such as storage space or connect time. All users must comply with these limits and not attempt to circumvent them. Moreover, users are expected not to be wasteful of resources, whether or not there are specific limits placed on them. Unreasonable use of resources may be curtailed.

3. Privacy

    Members of the University community shall not attempt to access the private files of others. The ability to access a file does not, by itself, constitute authorization to do so.

    The University does not routinely monitor or inspect individual accounts, files, or communications. There are situations, however, in which the University has a legitimate need to do so: (1) system managers may access user accounts, files, or communications when there is reason to believe that the user is interfering with the performance of a system; (2) authorized investigators may access accounts, files, or communications to obtain relevant information when there is a reasonable suspicion that the user has violated either laws or University policies; (3) coworkers and supervisors may need to access accounts, files, or communications used for University business when an employee becomes unavailable; and (4) when required by law. All monitoring and inspection shall be subject to authorization, notification, and other requirements specified in the IT Management Policy.

    Though the University will attempt to prevent unauthorized access to private files, it cannot make any guarantees. Because the University is a public entity, information in an electronic form may be subject to disclosure under the Ohio Public Records Act just as paper records are. Information also can be revealed by malfunctions of computer systems, by malicious actions of hackers, and by deliberate publication by individuals with legitimate access to the information. Users are urged to use caution in the storage of any sensitive information.

4. Access

    Some portions of the virtual campus, such as public Web pages, are open to everyone. Other portions are restricted in access to specific groups of people. No one is permitted to enter restricted areas without authorization or to allow others to access areas for which they are not authorized. The ability to access a restricted area does not, by itself, constitute authorization to do so.

    Individual accounts are for the use of the individual only; no one may share individual accounts with anyone else, including members of the account holder's family. Joint access to resources when needed should be provided from separate accounts.

5. Security

    All members of the University community must assist in maintaining the security of information technology resources. This includes physical security, protecting information, and preventing and detecting security breaches. Passwords are the keys to the virtual campus and all users are responsible for the security of their passwords. Users must report all attempts to breach the security of computer systems or networks to an appropriate official.

6. Plagiarism and Copyright

    Intellectual honesty is of vital importance in an academic community. You must not represent the work of others as your own. You must respect the intellectual rights of others and not violate their copyright or trademark rights. It is especially important that you obey the restrictions on using software or library resources for which the University has obtained restricted licenses to make them available to members of the University community.

7. Enforcement

    Anyone who becomes aware of a possible violation of this policy or the more specific regulations of the systems that comprise the virtual campus should notify the relevant department head or system administrator. The administrator will investigate the incident and determine whether further action is warranted. The administrator may resolve minor issues by obtaining the agreement that the inappropriate action will not be repeated. In those cases that warrant disciplinary action, the system administrator will refer the matter to the appropriate authorities. These include Public Safety for violations of criminal law, the Office of Student Affairs for violations by students, the appropriate Provost for violations by faculty, and the Office of Human Resources for violations by staff members.

    System administrators can act to block access and disable accounts when necessary to protect the system or prevent prohibited activities, but such actions cannot be used as punishments. Users must be notified promptly of the action and the restrictions must be removed unless the case is referred for disciplinary action.

End Notes

[1] "Settlement Reached Over PeopleSoft Implementation," *On Campus*, www.csuohio.edu/oncampus/2005/0307e.html, March 07, 2005.

[2] Olsen, Florence, "Delays, Bugs, and Cost Overruns Plague PeopleSoft's Services," *The Chronicle of Higher Education*, http://chronicle.com/free/v46/i05/05a03101.htm, September 24, 1999.

[3] Olsen, Florence, "Cleveland State U. Sues Consulting Company That Managed Its PeopleSoft Installation," *The Chronicle of Higher Education*, chronicle.com/free/2002/05/2002051302t.htm, May 13, 2002.

[4] Songini, Marc L., "University Pins $510M Lawsuit on PeopleSoft," *Computerworld*, www.computerworld.com/softwaretopics/erp/story/0,10801,91720,00.html, March 29, 2004.

[5] Stedman, Craig, "ERP Problems Plague College," *Computerworld*, www.computerworld.com/news/1999/story/0,11280,37669,00.html, November 22, 1999.

[6] Wailgum, Thomas, "Big Mess on Campus," *CIO*, www.cio.com/archive/050105/college.html, May 1, 2005.

[7] Business Software Alliance, "2nd Annual BSA and IDC Global Software Piracy Study," www.bsa.org/globalstudy/, July 8, 2005.

[8] "Business Software Alliance," Wikipedia, en.wikipedia.org/wiki/Business_Software_Alliance, July 8, 2005.

[9] Williams, Martyn, "Jury Awards Lexar Another $84 Million in Toshiba Case," *Computerworld*, www.computerworld.com, March 25, 2005.

[10] Evers, Jorvis, "Oracle Pays $8 Million to Settle Suit Over Training Charges," *Computerworld*, www.computerworld.com, May 16, 2005.

[11] Cheeseman, Henry R., *Contemporary Business Law*, Prentice-Hall, 2000, page 249.

[12] Cheeseman, Henry R., *Contemporary Business Law*, Prentice-Hall, 2000, page 292.

[13] Cheeseman, Henry R., *Contemporary Business Law*, Prentice-Hall, 2000, page 292.

[14] Chabrow, Eric, "See You In Court," *InformationWeek*, www.informationweek.com, July 25, 2005.

[15] Adler, Edward C., *The Complete Reference Checking Book*, AMACOM, a division of the American Management Association, 2003.

[16] AITP Web site, www.aitp.org, July 12, 2005.

[17] About the Computer Society, IEEE-CS Web site, www.computer.org/portal/site/ieeecs/, July 12, 2005.

[18] Tekiela, Robert, "How To Get Value from Technology Certifications," *Computerworld*, www.computerworld.com, December 22, 2004.

[19] Institute for Certification of Computing Professionals, www.iccp.org, July 18, 2005.

[20] "Certified Computing Professional," Institute for Certification of Computing Professionals, www.iccp.org, July 18, 2005.

[21] "Licensing Information—Who Should Be Licensed," Texas Board of Professional Engineers, www.tbpe.state.tx.us, July 25, 2005.

[22] "PE Electrical and Computer Exam," www.ncees.org/exams/professional/pe_electrical_exams.php, July 16, 2005.

[23] Martens, China, "Survey: Computer Porn Remains Issue at U.S. Corporations," *Computerworld*, www.computerworld.com, June 21, 2005.

Sources for Case 1

"About the Profession," Project Management Institute Web site, www.pmi.org/info/PP_AboutProfessionOverview.asp?nav=0501.

"Project Managers," *ComputerWeekly.com*, www.computerweekly.com/Jobs/2005/08/02/11071/Project+Managers.htm, August 09, 2005.

Goodwin, Bill, "Blueprint for Professionalism in IT Security," *ComputerWeekly.com*, www.computerweekly.com/Articles/2005/01/19/207802/BlueprintforprofessionalisminITsecurity.htm, January 19, 2005.

Hoffman, Thomas, "Demand for IT Certifications on the Rise," *Computerworld*, www.computerworld.com/careertopics/careers/story/0,10801,99903,00.html, February 21, 2005.

Nicolle, Lindsay, "Open the Door to a Secure Career," *ComputerWeekly.com*, www.computerweekly.com/Articles/2005/02/07/208200/Openthedoortoasecurecareer.htm, February 7, 2005.

Tekiela, Robert, "How to Get Value from Technology Certifications," *Computerworld*, www.computerworld.com/careertopics/careers/story/0,10801,98449,00.html, December 22, 2004.

Tennant, Don, "Certifiably Concerned," *Computerworld*, www.computerworld.com/careertopics/careers/training/story/0,10801,102394,00.html, June 13, 2005.

Tennant, Don, "Certifiably Mad?" *Computerworld*, www.computerworld.com/careertopics/careers/story/0,10801,102564,00.html, June 20, 2005.

Weiss, Todd, "Profile: American Century Investments," *Computerworld*, www.computerworld.com/careertopics/careers/story/0,10801,102712,00.html, June 27, 2005.

Sources for Case 2

"Firing of Lucent Executives Highlights Calls for Transparency in China," *China High Tech PR*, www.chinahightechpr.com/fullArticle.cfm?code=318, April 2004.

"Foreign Corrupt Practices Act Antibribery Provisions," U.S. Department of Justice Web site, www.usdoj.gov/criminal/fraud/fcpa/dojdocb.htm.

Associated Press, "Saudi Suit Claims Lucent OK'd Bribes," *The Boston Globe*, www.boston.com/business/technology/articles/2004/03/19/saudi_suit_claims_lucent_okd_bribes?mode=PF, March 19, 2004.

IDG News Service, "Lucent Fires Top Chinese Executives for Bribery," *ITWorld.com*, www.itworld.com/Man/2698/040407lucentchina/, April 7, 2004.

Leyden, John, "Saudi Firm Accuses Lucent of Bribery," *The Register*, www.theregister.co.uk/2003/08/13/saudi_firm_accuses_lucent/, August 13, 2003.

Reuters, "4 Lucent Execs to Leave After Saudi Probe," *USA Today*, www.usatoday.com/money/industries/telecom/2004-04-06-lucent_x.htm, April 6, 2004.

Reuters, "Lucent Names Shen to Head China After Ousters," *The Union Tribune*, www.signonsandiego.com/news/business/20040913-0251-telecoms-lucent-shen.html, September 13, 2004.

Tjoa, Laetitia, Jianyu, Ouyang, and Pykstra, Like, "Complying with PRC Antibribery Laws," The China Business Review, www.chinabusinessreview.com/public/0503/wong.html, March-April 2005.

CHAPTER 3

COMPUTER AND INTERNET CRIME

VIGNETTE

Treatment of Sasser Worm Author Sends Wrong Message

Unleashed in April 2004, the Sasser worm hit IT systems around the world hard and fast. Unlike most computer viruses before it, the Sasser worm didn't spread through e-mail, but moved undetected across the Internet from computer to computer. It exploited a weakness in Microsoft Windows XP and Windows 2000 operating systems. By the first weekend in May, American Express, the Associated Press, the British Coast Guard, universities, and hospitals reported that the Sasser worm had swamped their systems. Computer troubles led Delta Airlines to cancel 40 flights and delay many others.

Microsoft quickly posted a $250,000 reward, and by mid-May, authorities apprehended Sven Jaschen, a German teenager. Jaschen confessed and was convicted after a three-day trial. Jaschen could have received up to five years in prison, but because he was tried as a minor, the court suspended his 21-month sentence, leaving him with only 30 hours of community service.

Authorities said that once Jaschen realized the havoc the Sasser worm was causing, he tried to author a new version that reversed the damage. His real intent, they said, had simply been to gain fame as a programmer and perhaps to increase business for his mother, who owned a computer shop in his hometown.

Although Jaschen's sentence seemed like a crime to many in the IT industry, the real injustice occurred just a few months after Jaschen's indictment, when Securepoint, a German IT security company, hired Jaschen as a programmer. It appeared that the teen responsible for 70 percent of all computer virus infections during the first six months of 2004 got exactly what he wanted all along.[1, 2, 3, 4]

LEARNING OBJECTIVES

As you read this chapter, consider the following questions:

1. What key trade-offs and ethical issues are associated with the safeguarding of data and information systems?
2. Why has there been a dramatic increase in the number of computer-related security incidents in recent years?
3. What are the most common types of computer security attacks?
4. What are some characteristics of common computer criminals, including their objectives, available resources, willingness to accept risk, and frequency of attack?
5. What are the key elements of a multilayer process for managing security vulnerabilities, based on the concept of reasonable assurance?
6. What actions must be taken in response to a security incident?

IT SECURITY INCIDENTS: A WORSENING PROBLEM

The security of information technology used in business is of utmost importance. Confidential business data and private customer and employee information must be safeguarded, and systems must be protected against malicious acts of theft or disruption.

Although the necessity of security is obvious, it often must be balanced against other business needs and issues. Business managers, IT professionals, and IT users all face a number of ethical decisions regarding IT security:

- If their firm is a victim of a computer crime, should they pursue prosecution of the criminals at all costs, should they maintain a low profile to avoid the negative publicity, must they inform their affected customers, or should they take some other action?
- How much effort and money should be spent to safeguard against computer crime (how safe is safe enough)?
- If their firm produces software with defects that allow hackers to attack customer data and computers, what actions should they take?
- What tactics should management ask employees to use to gather competitive intelligence without doing anything illegal?
- What should be done if recommended computer security safeguards make life more difficult for customers and employees, resulting in lost sales and increased costs?

Unfortunately, the number of IT-related security incidents is increasing—not only in the United States, but around the world. To deal with these incidents, the Computer Emergency Response Team Coordination Center (CERT/CC) was established in 1988 at the Software Engineering Institute (SEI), a federally funded research and development center at Carnegie Mellon University in Pittsburgh, Pennsylvania. It is charged with coordinating communication among experts during computer security emergencies and helping to prevent future incidents. CERT/CC employees study Internet security vulnerabilities, handle computer security incidents, publish security alerts, research long-term changes in networked systems, develop information and training to help organizations improve security at their sites, and conduct an ongoing public awareness campaign.

The number of security problems reported to CERT/CC skyrocketed between 1997 and 2003, from 2134 to 137,529. There are many reasons for this increase, as outlined in the following sections. (Given the widespread use of automated attack tools, attacks against Internet-connected systems have become so commonplace that counts of the reported incidents provide little information that can assess their scope and impact. Therefore, as of 2004, CERT/CC no longer publishes the number of incidents reported.)

Increasing Complexity Increases Vulnerability

The computing environment has become enormously complex. Networks, computers, operating systems, applications, Web sites, switches, routers, and gateways are interconnected and driven by hundreds of millions of lines of code. This environment continues to increase in complexity day by day. The number of possible entry points to a network expands continually as more devices are added, increasing the possibility of security breaches.

Higher Computer User Expectations

Today, time means money, and the faster that computer users can solve a problem, the sooner they can be productive. As a result, computer help desks are under intense pressure to provide fast responses to users' questions. Under duress, help desk personnel sometimes forget to verify users' identities or to check whether they are authorized to perform a requested action. In addition, even though they have been warned against doing so, some computer users share their login ID and password with other coworkers who have forgotten their own passwords. This can enable workers to gain access to information systems and data for which they are not authorized.

Expanding and Changing Systems Introduce New Risks

Business has moved from an era of stand-alone computers, in which critical data was stored on an isolated mainframe computer in a locked room, to a network era in which personal computers connect to networks with millions of other computers, all capable of sharing information. Businesses have moved quickly into e-commerce, mobile computing, collaborative work groups, global business, and interorganizational information systems. Information technology has become ubiquitous and is a necessary tool for organizations to achieve their goals. However, it is increasingly difficult to keep up with the pace of technological change, successfully perform an ongoing assessment of new security risks, and implement approaches for dealing with them.

Increased Reliance on Commercial Software with Known Vulnerabilities

In computing, an **exploit** is an attack on an information system that takes advantage of a particular system vulnerability. Often, this attack is due to poor system design or implementation. Once the vulnerability is discovered, software developers quickly create and issue a "fix" or patch to eliminate the problem. Users of the system or application are responsible for obtaining and installing the patch, which they can usually download from the Web. (These fixes are in addition to other maintenance and project work that software developers perform.) Any delay in installing a patch exposes the user to a security breach. Figure 3-1 illustrates the growth in the number of vulnerabilities reported to CERT/CC. The rate of discovering software vulnerabilities exceeds 10 per day, creating a serious work overload for developers who are responsible for security fixes. Clearly, it can be difficult to keep up with all the required patches.

A **zero-day attack** takes place before the security community or a software developer knows about a vulnerability or has been able to repair it. Although the potential for damage from zero-day exploits is great, few such attacks have been documented as of this writing.[5] Unfortunately, malicious hackers are getting better and faster at exploiting flaws. The SQL Slammer worm appeared in January 2004, eight months after the vulnerability it targeted was first disclosed.[6] The Blaster worm was released to the Internet in the summer of 2004, barely a month after Microsoft released a patch for the software flaw it exploited. In August 2005, the Zotob computer worm and its variants began targeting corporate networks that ran the Windows 2000 operating system, less than a week after Microsoft released a critical patch addressing the vulnerability.[7] In an attempt to avoid further attacks and the ultimate zero-day attack, computer security firms and software manufacturers are paying hackers to identify vulnerabilities before they can be exploited.[8]

U.S. companies increasingly rely on commercial software with known vulnerabilities. Even when vulnerabilities are exposed, many corporate IT organizations prefer to use already installed software "as is" rather than implement security fixes that will make the

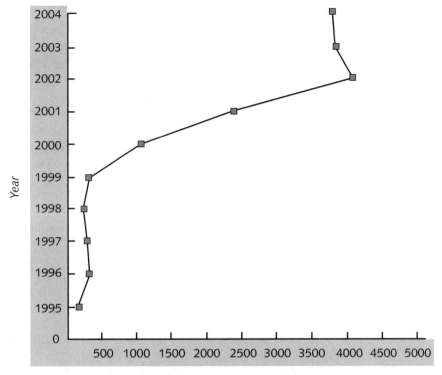

Number of vulnerabilities reported

FIGURE 3-1 Number of vulnerabilities reported to CERT/CC

software harder to use or eliminate "nice-to-have" features. The following vulnerabilities were all announced in the spring of 2005:

- Security researchers at the Austrian security consulting firm SEC Consult Unternehmensberatung GmbH discovered a bug in the Microsoft Internet Explorer browser that an attacker could use to run unauthorized software on a user's machine.[9]
- RealNetworks Inc. issued patches to four vulnerabilities in its RealPlayer media software, including RealPlayer on the Windows, Macintosh, and Linux operating systems, and Version 3 of the Rhapsody player for RealNetworks' online music service.[10]
- Sun Microsystems Inc. issued alerts about vulnerabilities in its Java Web Start and Java Runtime Environment that could allow applications to grant themselves permission to write local files or execute other applications. This problem could let an attacker gain backdoor access to victims' computers without any visible symptoms.[11]

In such an environment of increasing complexity, higher user expectations, expanding and changing systems, and increased reliance on software with known vulnerabilities, it is no wonder that the number of security incidents is increasing dramatically.

Types of Attacks

Security incidents can take many forms, but one of the most frequent is an attack on a networked computer from an outside source. There are numerous types of attacks, and new varieties are being invented all the time. Most attacks involve a virus, worm, Trojan horse, or denial of service.

Viruses

Computer virus has become an umbrella term for many types of malicious code. Technically, a **virus** is a piece of programming code, usually disguised as something else, that causes some unexpected and usually undesirable event. Often, a virus is attached to a file so that when the infected file is opened, the virus executes. Other viruses sit in a computer's memory and infect files as the computer opens, modifies, or creates them. Most viruses deliver a "payload" or malicious act. For example, the virus may be programmed to display a certain message on the computer's display screen, delete or modify a certain document, or reformat the hard drive.

A true virus does not spread itself from computer to computer. To propagate to other machines, it must be passed on to other users through infected e-mail document attachments, programs on diskettes, or shared files. In other words, it takes action by the computer user to spread a virus.

Macro viruses have become the most common and easily created viruses. Attackers use an application macro language (such as Visual Basic or VBScript) to create programs that infect documents and templates. After an infected document is opened, the virus is executed and infects the user's application templates. Macros can insert unwanted words, numbers, or phrases into documents or alter command functions. After a macro virus infects a user's application, it can embed itself in all future documents created with the application.

Worms

Unlike a computer virus, which requires users to spread infected files to other users, **worms** are harmful programs that reside in the active memory of the computer and duplicate themselves. They differ from viruses because they can propagate without human intervention, sending copies of themselves to other computers by e-mail or Internet Relay Chat (IRC). For example, the W32.Sober-K@mm variation of the Sober worm originated in Germany in February 2005. Within hours, it had spread to most of Europe and the United States via an attachment to e-mail messages from infected computers. The Visual Basic source code of the worm was cleverly written to create random subject lines and body text in either English or German, depending on the e-mail addresses it found on the infected computer. This made it difficult to warn users what to look for. It also generated fake messages to entice the recipient to open the attached .zip file. When an unfortunate recipient did open the attachment, the worm created several executable files and modified the registry key so that the files executed on start-up. The worm made infected computers perform slowly and clogged e-mail servers and networks with fake messages, trying to infect even more computers.[12]

The negative impact of a virus or worm attack on an organization's computers can be considerable—lost data and programs, lost productivity because workers cannot use their computers, additional lost productivity as workers attempt to recover data and programs, and lots of effort for IT workers to clean up the mess and restore everything as close to normal as possible. The cost to repair the damage done by each of the Code Red, SirCam, Melissa, and ILOVEYOU worms was estimated to exceed $1 billion, as shown in Table 3-1.

TABLE 3-1 Cost impact of worms

Name	Year released	Worldwide economic impact
ILOVEYOU	2000	$8.75 billion
Code Red	2001	$2.62 billion
SirCam	2001	$1.15 billion
Melissa	1999	$1.10 billion

Trojan Horses

A **Trojan horse** is a program that a hacker secretly installs on a computer. The program's harmful payload can allow the hacker to steal passwords or Social Security numbers, or spy on users by recording keystrokes and transmitting them to a server operated by a third party. The data may then be sold to criminals who use the information to obtain credit cards or pilfer bank accounts. (Spyware, a particularly insidious and widespread form of such software, is discussed in detail in Chapter 4.)

Users are often tricked into installing Trojan horses. For example, the Opanki worm disguised itself as a file coming from Apple Computer's popular online iTunes music service. It was distributed via an instant message that read "This picture never gets old." An unsuspecting user who clicked a link in the message would install the virus. Another worm, the Trojan Abwiz.C, infected a computer when a user either opened an e-mail attachment or visited a malicious Web site.[13]

Another type of Trojan horse is a **logic bomb**, which executes under specific conditions. For example, logic bombs can be triggered by a change in a particular file, by typing a specific series of keystrokes, or by a specific time or date.

Denial-of-Service (DoS) Attacks

A **denial-of-service attack** is one in which a malicious hacker takes over computers on the Internet and causes them to flood a target site with demands for data and other small tasks (see Figure 3-2). A denial-of-service attack does not involve a break-in at the target computer; instead, it just keeps the target machine so busy responding to a stream of automated requests that legitimate users cannot get in—the Internet equivalent of dialing a telephone number repeatedly so that all other callers hear a busy signal. The target machine "holds the line open" while waiting for a reply that never comes, and eventually the requests exhaust all resources of the target.

The software to initiate a denial-of-service attack is simple to use and readily available at hacker sites. A tiny program is downloaded from the attacker's computer to dozens, hundreds, or even thousands of computers all over the world. Based on a command

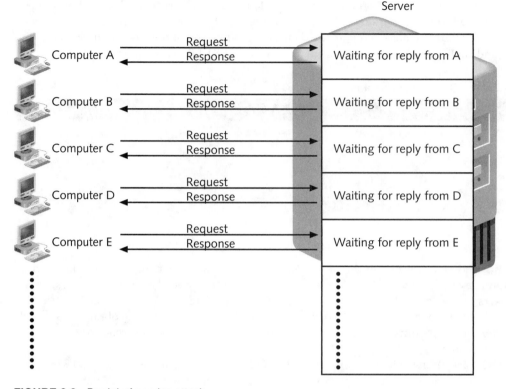

FIGURE 3-2 Denial-of-service attack

by the attacker or a preset time, these computers (called **zombies**) go into action, each sending a simple request for access to the target site again and again—dozens of times a second.

The zombies are often programmed to put false return addresses on the packets they send out (known as **spoofing**) so that the sources of the attack are obscured and cannot be identified and turned off. However, this approach actually provides an opportunity to prevent such attacks. Internet service providers (ISPs) can prevent incoming packets with false IP addresses from being passed on by a process called **ingress filtering**. Corporations with Internet connections can ensure that spoofed packets don't leave their corporate network using a process called **egress filtering**. However, such checking of addresses takes a tremendous amount of processing by Internet routers. As the number of packets increases, more and more processing capacity is required to check the IP address on each packet. Companies would have to deploy faster and more powerful routers and switches to maintain the same level of performance, which would be expensive. As a result, few ISPs or corporations perform this checking, although such capabilities may be built into the next generation of network equipment.

The zombies involved in a denial-of-service attack are seriously compromised and are left with more enduring problems than their target. As a result, zombie machines need to be inspected to ensure that the attacker software is completely removed from the system. In addition, system software must be reinstalled from a reliable backup to reestablish the system's integrity, and an upgrade or patch must be implemented to eliminate the vulnerability that originally allowed the attacker to enter the system.

In 2004, the MyDoom e-mail worm raced through the Internet, overwhelmed e-mail systems with infected messages, and installed a Trojan horse program on the computers it infected. The zombies it created were used to launch a denial-of-service attack against the SCO Web site (originally founded as the Santa Cruz Operation), with some 25,000 to 50,000 computers taking part. This network of infected machines became a valuable resource for the criminal community to launch future attacks, distribute future MyDoom variants, and generate spam. As a result, many believe that "the MyDoom e-mail worm signaled the end of the amateur virus writers and the clear emergence of sophisticated, new virus authors with criminal ties and a hunger for illicit profit from spam and online extortion."[14]

Denial-of-service attacks involving extortion of e-commerce companies have escalated. Authorize.Net, which provides credit card payment services for about 100,000 e-commerce firms, was the target of a massive attack on its payment gateway service that disrupted customers for several days. Prior to the attack, the company received an extortion note demanding payment of a large amount of money. The company frequently receives demands for payments of up to several million dollars. Its policy is not to pay and to get law enforcement agencies involved immediately.[15]

Perpetrators

Computer criminals often have the same motives as other types of criminals—thrill seekers wanting a challenge, common criminals looking for financial gain, industrial spies trying to gain a competitive advantage, and terrorists seeking to cause destruction to further their cause. Each type of perpetrator has different objectives and access to varying resources, and each is willing to accept different levels of risk to accomplish the objective. Knowing the profile of each set of likely attackers, as shown in Table 3-2, is the first step toward establishing effective countermeasures.

TABLE 3-2 Classifying perpetrators of computer crime

Type of perpetrator	Objectives	Resources available to perpetrator	Level of risk acceptable to perpetrator	Frequency of attack
Hacker	Test limits of system and gain publicity	Limited	Minimal	High
Cracker	Cause problems, steal data, and corrupt systems	Limited	Moderate	Medium
Insider	Make money and disrupt company's information systems	Knowledge of systems and passwords	Moderate	Low

TABLE 3-2 Classifying perpetrators of computer crime (continued)

Type of perpetrator	Objectives	Resources available to perpetrator	Level of risk acceptable to perpetrator	Frequency of attack
Industrial spy	Capture trade secrets and gain competitive advantage	Well funded and well trained	Minimal	Low
Cyber-criminal	Make money	Well funded and well trained	Moderate	Low
Cyber-terrorist	Destroy key infrastructure components	Not necessarily well funded or well trained	Very high	Low

Hackers and Crackers

Hackers test the limitations of systems out of intellectual curiosity—to see whether they can gain access and how far they can go. They have at least a basic understanding of information systems and security features, and much of their motivation comes from a desire to learn even more. Today's hacker commonly is male, in his mid-20s or younger, has lots of spare time, has minimal financial resources, and is a social outsider. Some hackers are smart and talented, but many are technically inept and are referred to as **lamers** or **script kiddies** by more skilled hackers. Surprisingly, hackers have a wealth of available resources to hone their skills—online chat groups, Web sites, downloadable hacker tools, and even hacker conventions (such as Defcon, an annual gathering in Las Vegas).

For example, in 2005, a hacker broke into ApplyYourself, an admissions management system used by many colleges and universities. The hacker then posted the procedure in a *Business Week* online forum that more than 100 people used to gain access to the admission-decision page before their school intended it to be published. School officials identified the people who broke into the system and said that their actions would have a strong impact on the acceptance decision.[16]

Another example involves Microsoft, which is often the target of hackers because its software is so widely used. To update Microsoft software, users must go through the Windows Genuine Advantage program to validate that they're running a legitimate copy of the Windows operating system. But hackers discovered that, before downloading software through the Windows Update service, they could bypass the validation check by entering a simple JavaScript command in the Web browser's address bar and pressing the Enter key, which took them directly to the download.[17]

Cracking is a form of hacking that is clearly criminal activity. **Crackers** break into other people's networks and systems, deface Web pages, crash computers, spread harmful programs or hateful messages, and write scripts and automated programs that let other people do the same things. For example, in 2005, a cracker stole data for 200,000 credit and debit accounts from CardSystems Solutions, an Atlanta-based company that handles credit card transactions for merchants and banks.[18] Credit card theft is so common that public online exchanges have formed in which stolen numbers are bought, sold, and traded.

Malicious Insiders

The top security concern for companies is the malicious insider—an ever-present adversary. An estimated 85 percent of all fraud is perpetrated by employees, who account for more than $660 billion per year in losses, according to the Association of Certified Fraud Examiners (ACFE).[19] Companies are exposed to a wide range of fraud risks, including diversion of company funds, theft of assets, fraud connected with bidding processes, invoice and payment fraud, computer fraud, and credit card fraud. Not surprisingly, fraud that occurs within an organization is usually due to weaknesses in its internal control procedures. As a result, many frauds are discovered by chance and by outsiders—via tips, through resolving payment issues with contractors or suppliers, or during a change of management—rather than through control procedures. Often, frauds involve some form of **collusion** or cooperation between an employee and an outsider. For example, an employee in Accounts Payable may engage in collusion with a company supplier. Each time the supplier submits an invoice, the Accounts Payable employee adds $1000 to the amount approved for payment. The inflated payment is received by the supplier and the two split the extra money.

Insiders are not necessarily employees; they can also be consultants and contractors. However, "the typical employee who commits fraud has many years with the company, is an authorized user, is in a nontechnical position, has no record of being a problem employee, uses legitimate computer commands to commit the fraud, and does so mostly during business hours."[20] Their risk tolerance depends on whether they are motivated by financial gain, revenge on their employers, or publicity.

Malicious insiders are extremely difficult to detect or stop because they're often authorized to access the very systems they abuse. Although insiders are less likely to attack systems than outside hackers or crackers are, the company's systems are far more vulnerable to them. Most computer security measures are designed to stop external attackers, but they are nearly powerless against insiders. Insiders have knowledge of individual systems, which often includes the procedures to gain access to login IDs and passwords. Insiders know how the systems work and where the weak points are. Their knowledge of organizational structure and security procedures helps them avoid investigation of their actions.

Industrial Spies

Industrial spies use illegal means to obtain trade secrets from competitors of their firm. Trade secrets are protected by the Economic Espionage Act of 1996, which makes it a federal crime for people to use a trade secret for their own benefit or another's benefit. Trade secrets are most often stolen by insiders, such as disgruntled employees and ex-employees.

Competitive intelligence uses legal techniques to gather information that is available to the public. Participants gather and analyze information from financial reports, trade journals, public filings, and printed interviews with company officials. Industrial espionage involves using illegal means to obtain information that is not available to the public. Participants might place a wiretap on the phones of key company officials, bug a conference room, or break into a research and development facility to steal confidential test results. An unethical firm may spend a few thousand dollars to hire an industrial spy to steal trade secrets that can be worth one hundred times that amount. The industrial spy avoids taking risks that would expose his employer—the employer's reputation (an intangible but valuable item) would be damaged considerably if the espionage was discovered.

Industrial espionage can involve the theft of new product designs, production data, marketing information, or new software source code. For example, Shekhar Verma was employed by Geometric Software Solutions Ltd. (GSSL), an Indian company that provides outsourcing services, including software development. GSSL was awarded a contract to debug the source code of SolidWorks 2001 Plus, a popular computer-aided design software package. Verma eventually was fired from GSSL; he allegedly stole the source code and offered it to several of SolidWorks' U.S. competitors for $200,000. (The value of the source code has been estimated to exceed $50 million.) A competitor contacted the FBI, a sting was set up, and Verma was arrested. However, Indian law at the time did not recognize misappropriation of trade secrets, so technically, Verma did not steal from his employer, as the source code belonged to SolidWorks. Prosecutors were forced to charge Verma with simple theft; the outcome of the case has yet to be settled.[21]

Cybercriminals

Information technology provides a new and highly profitable venue for **cybercriminals**. They hack into corporate computers and steal, often by transferring money from one account to another to another—leaving a hopelessly complicated trail for law enforcement officers to follow. Cybercriminals also engage in all forms of computer fraud—stealing and reselling credit card numbers, personal identities, and cell phone IDs. They can spend large sums of money to buy the technical expertise and access they need from unethical insiders.

The use of stolen credit card information is a favorite ploy of computer criminals. Estimates of online credit card fraud rates vary greatly, from 0.06 percent to 4 percent of all online transactions. Fraud rates are highest for merchants who sell downloadable software or expensive items such as electronics and jewelry (because of their high resale value). Credit card companies are so concerned about making consumers feel safe while shopping online that many are marketing new and exclusive zero-liability programs, although the Fair Credit Billing Act limits consumer liability to only $50 of unauthorized charges. When a charge is made fraudulently in a retail store, the bank that issued the credit card must pay the fraudulent charges. For credit card transactions over the Internet, the Web merchant absorbs the cost.

A high rate of disputed transactions, known as **chargebacks**, can greatly reduce a Web merchant's profit margin. However, the permanent loss of revenue caused by lost customer trust has far more impact than the costs of fraudulent purchases and bolstering security. Most companies are afraid to admit publicly that they've been hit by online fraud or hackers because they don't want to hurt their reputations.

To reduce the potential for online credit card fraud, most e-commerce Web sites use some form of encryption technology to protect information as it comes in from the consumer. Some also verify the address submitted online against the one the issuing bank has on file, although the merchant may inadvertently throw out legitimate orders as a result—for example, a consumer might place a legitimate order but request shipment to a different address because it is a gift. Another security technique is to ask for a card verification value (CVV), the three-digit number above the signature panel on the back of a credit card. This technique makes it impossible to make purchases with a credit card number stolen online. An additional security option is transaction-risk scoring software, which keeps track of a customer's historical shopping patterns and notes deviations from the norm. For example, say that you have never been to a casino and your credit card turns up at Caesar's Palace at 2 a.m. Your transaction-risk score would go up dramatically, so much so that the transaction might be declined.

Some card issuers are implementing **smart cards**, which contain a memory chip that is updated with encrypted data every time the card is used. This encrypted data might include the user's account identification and the amount of credit remaining. To use a smart card for online transactions, consumers must purchase a card reader that attaches to their personal computers and enter a personal identification number to gain access to the account. Although smart cards are used widely in Europe, they are not as popular in the United States because of the changeover costs for merchants.

From time to time, various local, state, or federal governments establish various acts intended to improve the conduct of business. Unfortunately, some of these acts have unintended consequences that create new opportunities for criminals to commit fraud. The Legal Overview explains one such act.

LEGAL OVERVIEW

The Check Clearing for the 21st Century Act

Until recently in the banking industry, the check-clearing and settlement process was based on methods created in the 1950s—magnetic ink encoding was used to route actual paper checks via truck and plane to the banks that issued them, where data was manually entered to credit and debit the proper accounts. This process required one to four days to clear checks.[22]

The Check Clearing for the 21st Century Act, or Check 21, requires that banks accept paper documents with check images and data related to transactions in lieu of original paper checks (image replacement documents, or IRDs). This new process not only eliminates costs associated with the physical transport of actual checks (estimated to be as much as 2 cents per check), it speeds the clearing of checks, thus improving the availability of funds. As an added benefit, customers can view and print their cleared checks over the Internet from a link to their bank's Web page. Although not required, most large banks are expected to convert to this new process to take advantage of the efficiencies of electronic IRD processing.[23]

Bank executives also expect Check 21 to reduce the $20 billion that banks lose to check fraud every year. With Check 21, when anyone presents a check, the bank takes a digital image of the check and sends it to the issuing bank within hours. The issuing bank looks at key pieces of information on the check—such as the check number, the payee, handwriting, and the dollar amount—to quickly determine if the check is valid. Under the paper check-based process, the issuing bank wouldn't receive the paper check

continued

for as long as four days, giving criminals plenty of time to escape and cover their tracks. However, some experts argue that Check 21 creates new opportunities for criminals to conduct check fraud:

- Criminals can steal information from a customer's online bank account and view a customer's check images, thus obtaining access to information that can enable them to evade detection.
- The speed with which checks are processed puts additional time pressure on the bank's fraud examiners to identify a fraudulent check.
- The check image is made on a gray scale, which does not reveal details as well as a physical check. Details in the gray image that could tip off banks to a fraud may be lost.
- Most banks plan to destroy the paper check after the image is created, thereby destroying any evidence not captured by the image, such as fingerprints or detailed security features.[24]

"All these new vulnerabilities lead Frank Liddy, partner in the North American banking practice at the consultancy Unisys, to conclude that bankers have yet to fully realize the extent to which they are susceptible to increased fraud with Check 21. 'This is bigger than any bank or banker can realize,' he warns."[25]

Cyberterrorists

Cyberterrorists intimidate or coerce a government or organization to advance their political or social objectives by launching computer-based attacks against other computers, networks, and the information stored on them. Such attacks could include sending a virus or worm or launching a denial-of-service attack. Because of the Internet, attacks can easily originate from foreign countries, making detection and retaliation much more difficult.

Three years before the September 11, 2001, terrorist attacks, the U.S. government considered the threat of cyberterrorism serious enough that it established the National Infrastructure Protection Center, which served as the focal point for threat assessment, warning, investigation, and response for threats or attacks against American infrastructures. (This function was later transferred to the Homeland Security Department's Information Analysis and Infrastructure Protectorate.) These infrastructures include telecommunications, energy, banking and finance, water, government operations, and emergency services. Specific targets might include telephone-switching systems, an electric power grid that serves major portions of a geographic region, or an air traffic control center that ensures airplanes can take off or land safely. Successful cyberattacks on such targets could cause widespread and massive disruptions to society. Some computer security experts believe that cyberterrorism attacks could be used to further complicate matters following a major act of terrorism by reducing the ability of fire and emergency teams to respond.

Cyberterrorists seek to cause harm rather than gather information, and they use techniques that destroy or disrupt services. They are extremely dangerous, consider themselves to be at war, have a very high acceptance of risk, and seek maximum impact. Despite their extremism, however, many experts believe that terrorist groups pose only a limited threat to information systems. Currently, terrorists prefer bomb attacks to computer attacks, and most operate on a limited budget. Unfortunately, the threat from cyberterrorist attacks can be expected to increase as younger, computer-savvy terrorists are recruited.

Here are a few documented cases of cyberterrorism:

- During the Kosovo conflict in 1999, NATO computers were blasted with e-mail containing harmful attachments and hit with denial-of-service attacks by cyberterrorists who were protesting NATO bombings in Kosovo.[26]
- In 2000, the FBI in Houston searched the home of a person who allegedly created a computer worm that sought out computers on the Internet. The worm erased the hard drives of randomly selected computers. Other computers whose hard drives were not erased actively scanned the Internet for more computers to infect and then forced the computers to use their modems to dial 911. Because each infected computer could scan approximately 2550 computers at a time, the worm had the potential to create a denial-of-service attack against the emergency 911 system. Although the cyberterrorist's motives were not clear, the potential for serious impact on the emergency response systems was obvious.[27]
- In 2002, an Australian man used an Internet connection to direct computer-controlled equipment to release 1 million gallons of raw sewage along Queensland's Sunshine Coast after he was turned down for a government job. Police discovered that he had worked for the company that designed the sewage treatment plant's control software.[28]

As of this writing, there is no evidence that al Qaeda operatives have tried to execute a cyberattack. However, they have used computers to plan attacks and store information about potential targets:

- U.S. troops recovered al Qaeda laptops in Afghanistan with structural and engineering software, electronic models of a dam, and information on computerized water systems, nuclear power plants, and U.S. and European stadiums.[29]
- Al Qaeda operatives arrested in Pakistan had extensive information on the construction of the New York Stock Exchange and the World Bank, floor plans of meeting rooms, and the configuration of parking garages. The amount of detail was shocking; it included traffic patterns, possible escape routes, details about security guards, and discussions about what kinds of explosives could do the most damage to each of the buildings mentioned.[30]

REDUCING VULNERABILITIES

The security of any system or network is a combination of technology, policy, and people, and it requires a wide range of activities to be effective. A strong security program begins by assessing threats to the organization's computers and network, identifying actions that address the most serious vulnerabilities, and educating users about the risks involved and the actions they must take to prevent a security incident. The IT security group must lead the effort to implement security policies and procedures, along with hardware and software tools to help prevent security breaches. However, no security system is perfect, so systems and procedures must be monitored to detect a possible intrusion. If an intrusion occurs, there must be a clear reaction plan that addresses notification, evidence protection, activity log maintenance, containment, eradication, and recovery.

Risk Assessment

A **risk assessment** is an organization's review of potential threats to its computers and network and the probability of those threats occurring. Its goal is to identify investments in time and resources that can best protect the organization from its most likely and serious threats. No amount of resources can guarantee a perfect security system, so organizations frequently have to balance the risk of a security breach with the cost of preventing one. The concept of **reasonable assurance** recognizes that managers must use their judgment to ensure that the cost of control does not exceed the system's benefits or the risks involved. Table 3-3 illustrates a risk assessment for a hypothetical organization.

A completed risk assessment identifies the most dangerous threats to a company and helps focus security efforts on the areas of highest payoff. For each risk area, the estimated probability of an attack occurring is multiplied by the estimated cost of a successful attack. The product is the expected cost impact for that risk area. Organizations can then assess the current level of protection against the probability of an event occurring—poor, good, or excellent. The risk areas with the highest estimated cost and the poorest level of protection are where security measures need to be improved.

TABLE 3-3 Risk assessment for a hypothetical company

Risk	Estimated probability of such an event occurring	Estimated cost of a successful attack	Probability × cost = expected cost impact	Assessment of current level of protection	Relative priority to be fixed
Denial-of-service attack	80%	$500,000	$400,000	Poor	1
E-mail attachment with harmful worm	70%	$200,000	$140,000	Poor	2
Harmful virus	90%	$50,000	$45,000	Good	3
Invoice and payment fraud	10%	$200,000	$20,000	Excellent	4

Establishing a Security Policy

A **security policy** defines an organization's security requirements and the controls and sanctions needed to meet those requirements. A good policy delineates responsibilities and expected behavior by members of the organization. A security policy outlines *what* needs to be done, but not *how* to do it. It should refer to procedure guides instead of outlining the procedures.[31]

Whenever possible, automated system policies should mirror an organization's written policies. These policies can often be put into practice using the configuration options in a software program. For example, if a written policy states that passwords must be changed every 30 days, then all systems should be configured to enforce this policy automatically. There are always trade-offs when applying system security restrictions, such as the conflict between ease of use and increased security; however, when a decision is made to favor ease of use, security incidents sometimes increase.

The use of e-mail attachments is a critical issue. Sophisticated attackers can try to penetrate a network via e-mail and its attachments, regardless of the existence of a firewall and other security measures. As a result, many companies have chosen to block all incoming mail that has executable file attachments. This greatly reduces their vulnerability. Some companies allow employees to receive and open e-mail with attachments, but only if the e-mail is expected and from someone known by the recipient. However, such a policy can be risky, because many worms use the address book of their victims to generate e-mails to a target audience.

Another growing area of concern is the use of wireless devices to access corporate e-mail, store confidential data, and run critical applications such as inventory management and sales force automation. Mobile devices such as smart phones are ripe territory for viruses, worms, and other potential security threats. As a result, wary companies require special security measures for handheld and mobile users. In some cases, employees must use a virtual private network (VPN) to gain access to their corporate network. A **VPN** works by using the Internet to relay communications, but maintains privacy through security procedures and tunneling protocols that encrypt data at the sending end and decrypt it at the receiving end. An additional level of security involves encryption of the originating and receiving network addresses. Because of the ease of loss or theft, it is also a good idea to encrypt all sensitive corporate data stored on handhelds. Unfortunately, it is hard to apply a single, simple approach to securing all handheld devices because so many manufacturers and models exist.[32]

Educating Employees, Contractors, and Part-Time Workers

Employees, contractors, and part-time workers must be educated about the importance of security so they will be motivated to understand and follow the security policy. Often, this can be accomplished by discussing recent security incidents that affected the organization. Users must understand that they are a key part of the security system and that they have certain responsibilities. For example, users must help protect an organization's information systems and data by doing the following:

- Guarding their passwords to protect against unauthorized access to their accounts
- Not allowing others to use their passwords
- Applying strict access controls (file and directory permissions) to protect data from disclosure or destruction
- Reporting all unusual activity to the organization's IT security group

Unfortunately, insufficient measures are sometimes taken to protect critical passwords—often with serious consequences. For example, the Kurtztown Area School District in Pennsylvania approved a program that gave every high school student an Apple iBook laptop. Students and parents were required to sign a code of conduct and acceptable use policy, which contained warnings of legal action if the rules were broken. A filtering program was installed on each computer to limit Internet access, along with software that let school administrators see what the students were viewing. In what proved to be a serious lapse in security procedures, the administrative password that allowed students to reconfigure computers and obtain unrestricted Internet access was taped to the back of each computer. Some students used the password to download prohibited programs, such as the popular iChat instant messaging tool, and to turn off the monitoring function. Some of the more computer-savvy students reconfigured the monitoring software so they could view the administrators' own screens. Eventually, 13 students were charged with computer trespass and subjected to a wide range of sanctions, including juvenile detention, probation, and community service, for violating the district's computer usage policy. The students were reported to police only after detentions, suspensions, and other punishments failed to get the students to conform.[33]

Prevention

No organization can ever be completely secure from attack. The key is to implement a layered security solution to make computer break-ins harder than the attacker is willing to work, so that if an attacker breaks through one layer of security, there is another layer to overcome. These layers of protective measures are explained in more detail in the following sections.

Installing a Corporate Firewall

Installation of a corporate firewall is the most common security precaution taken by businesses. As discussed in Chapter 2, a **firewall** stands guard between your organization's internal network and the Internet and limits network access based on the organization's access policy (see Figure 3-3). Any Internet traffic that is not explicitly permitted into the internal network is denied entry. Similarly, internal network users can be blocked from gaining access to certain Web sites based on content such as sex, violence, and so on. The firewall can also block instant messaging, access to newsgroups, and other Internet activities.

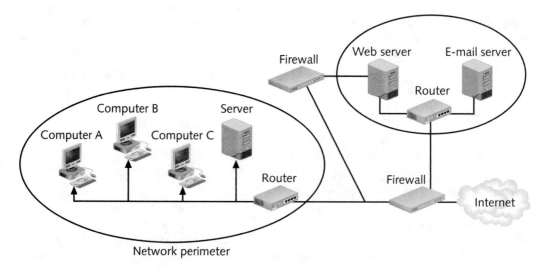

FIGURE 3-3 Firewall protection

Installing a firewall can lead to another serious security issue—complacency. For example, a firewall does nothing to protect a Web site from a denial-of-service attack. A firewall also cannot prevent a worm from entering the network as an e-mail attachment. Most firewalls are configured to allow e-mail and benign-looking attachments to reach their intended recipient.

Table 3-4 lists some of the top-rated firewall software used to protect home personal computers. Typically, the software sells for $30 to $60 for a single user license.

TABLE 3-4 Popular firewall software for personal computers

Software	Vendor
Norton Personal Firewall	Symantec
Tiny Personal Firewall	Tiny Software
BlackICE Defender	Network Ice Corporation
ZoneAlarm Pro	Zone Labs
Personal Firewall	McAfee

Installing Antivirus Software on Personal Computers

Antivirus software should be installed on each user's PC to regularly scan a computer's memory and disk drives for viruses. Antivirus software scans for a specific sequence of bytes, known as the **virus signature**. If it finds a virus, the antivirus software informs the user and may clean, delete, or quarantine any files, directories, or disks affected by the malicious code. Good antivirus software checks vital system files when the system is booted up, monitors the system continuously for viruslike activity, scans disks, scans memory when a program is run, checks programs when they are downloaded, and scans e-mail attachments before they are opened. Two of the most widely used antivirus software products are Norton Antivirus (see Figure 3-4) and Dr. Solomon's Antivirus from McAfee.

CERT has long served as a clearinghouse for news on new viruses, worms, and other computer security topics. CERT says that almost all the major attacks the team analyzes use previously known virus or worm programs. Thus, it is crucial that antivirus software be continually updated with the latest virus detection information, called **definitions**. In most corporations, the network administrator is responsible for monitoring network security Web sites at least once a week and downloading updated antivirus software as needed.

Implementing Safeguards Against Attacks by Malicious Insiders

Corporate security managers believe some of their worst security breaches come from corporate users who access information they are not authorized to see. Another potential problem is leaving user accounts active after employees leave the company. To reduce the threat of attack by malicious insiders, IT staff must promptly delete the computer accounts, login IDs, and passwords of departing employees.

Organizations also need to carefully define employee roles to properly separate key responsibilities, so that a single person is not responsible for accomplishing a task that has high security implications. For example, it would not make sense to allow an employee

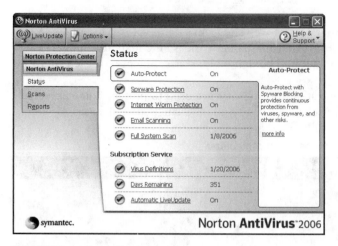

FIGURE 3-4 Norton Antivirus

both to initiate a purchase order and approve an invoice for its payment. Such an employee would be able to place a large dollar volume of orders with a "friendly vendor," approve the invoices for payment, and then disappear from the company to split the money with the vendor. In addition to separating duties, many organizations frequently rotate people in sensitive positions to prevent potential insider crimes.

Another important safeguard is to create roles and user accounts so that employees have the authority to perform their responsibilities and no more. For example, members of the Finance Department will have different authorizations from members of the Human Resources Department. An accountant should not be able to review the pay and attendance records of an employee, and a member of Human Resources should not know how much was spent to modernize a piece of equipment. Even within a given department, not all members should be given the same capabilities. Within the Finance Department, for example, some users may be able to approve invoices for payment, but not all users will. An effective administrator will identify the similarities among users and create profiles associated with these groups.

Addressing the Most Critical Internet Security Threats

The overwhelming majority of successful computer attacks are made possible by taking advantage of well-known vulnerabilities. Computer attackers know that many organizations are slow to fix problems, which makes scanning the Internet for vulnerable systems an effective attack strategy. "The easy and destructive spread of worms, such as Blaster, Slammer, and Code Red, can be traced directly to exploitation of unpatched vulnerabilities."[34]

Both the SANS (System Administration, Networking, and Security) Institute and CERT/CC regularly update a summary of the most frequent, high-impact vulnerabilities reported to them. You can read these summaries at *www.sans.org/top20/* and *www.us-cert.gov/current*, respectively. The actions required to address these issues usually involve installing a known patch to software. Those responsible for computer security must make it a priority to prevent attacks that exploit common vulnerabilities.

Verifying Backup Processes for Critical Software and Databases

It is imperative to back up critical applications and data regularly. However, many organizations have implemented inadequate backup processes and found they could not fully restore the original data. All backups should be created with enough frequency to enable a full and quick restoration of data if an attack destroys the original. This process should be tested to confirm that it works.

Conducting Periodic IT Security Audits

A security audit evaluates whether an organization has a well-considered security policy in place and if it is being followed. For example, if a policy says that all users must change their passwords every 30 days, the audit must check how the policy is implemented and whether it is truly happening. The audit will also review who has access to particular systems and data and what level of authority each user has. It is not unusual for an audit to reveal that too many people have access to critical data and that many people have capabilities beyond those needed to perform their jobs. One result of a good audit is a list of items to correct to ensure that the security policy is being met.

A thorough security audit should also test system safeguards to ensure that they are operating as intended. Such tests might include trying the default system passwords that are active when software is first received from the vendor. The goal of such a test is to ensure that all "known" passwords have been changed.

U.S. government agencies must also maintain security for their information systems and data to prevent data tampering, disruptions in critical operations, fraud, and the inappropriate disclosure of sensitive information. Title III of the E-Government Act, entitled the Federal Information Security Management Act (FISMA), "requires each federal agency to develop, document, and implement an agency-wide program to provide information security for the information and information systems that support the operations and assets of the agency, including those provided or managed by another agency, contractor, or other source."[35]

The annual Federal Computer Security Report Card is based on evaluations defined in FISMA and compiled by the House Government Reform Committee from information provided by each agency's inspector general. The overall security of federal government computer systems earned only a D+ average on the 2004 security report card. A year earlier, the average grade was a D.[36] Results for selected agencies are shown in Table 3-5.

TABLE 3-5 Summary of federal computer security report card for 2004 and 2003

Federal agency	2004	2003
Department of Transportation	A-	D+
Nuclear Regulatory Agency	B+	A
Environmental Protection Agency	B	C
Department of Justice	B-	F
Department of Interior	C+	F
Department of State	D+	F

TABLE 3-5 Summary of federal computer security report card for 2004 and 2003 (continued)

Federal agency	2004	2003
Department of Treasury	D+	D
Department of Defense	D	D
NASA	D-	D-
Department of Commerce	F	C-
Department of Energy	F	F
Department of Homeland Security	F	F

In a subsequent July 2005 audit, the Government Accountability Office found that "pervasive weaknesses exist in almost all areas of information security controls at 24 major agencies, threatening the integrity, confidentiality, and availability of information and information systems."[37] Clearly, it takes action in addition to assessment to plug holes in the security defenses of information systems. The following sections outline the necessary steps.

Detection

Even when preventive measures are implemented, no organization is completely secure from a determined attack. Thus, organizations should implement detection systems to catch intruders in the act. Organizations often employ an intrusion detection system, intrusion prevention system, or a "honeypot" to minimize the impact of intruders.

Intrusion Detection Systems

An **intrusion detection system** monitors system and network resources and activities, then notifies the proper authority when it identifies possible intrusions from outside the organization or misuse from within the organization. There are two fundamentally different approaches to intrusion detection—knowledge-based approaches and behavior-based approaches. Knowledge-based intrusion detection systems contain information about specific attacks and system vulnerabilities and watch for attempts to exploit these vulnerabilities. Examples include repeated failed login attempts, attempts to download a program to a server, or other symptoms of possible mischief. When such an attempt is detected, an alarm is triggered. A behavior-based intrusion detection system models normal behavior of a system and its users from reference information collected by various means. The intrusion detection system compares current activity to this model and generates an alarm if it finds a deviation. Examples include unusual traffic at odd hours or a user in the Human Resources Department who accesses an accounting program that she never used before.

Intrusion Prevention Systems

Intrusion prevention systems (IPSs) evolved from network intrusion detection systems; they work to prevent an attack by blocking viruses, malformed packets, and other threats from getting into the company network. The IPS sits directly behind the firewall and examines all the traffic passed by it. A firewall and a network IPS are complementary. "Firewalls block everything except what you explicitly allow through; an IPS lets everything through except what it is told to block." [38]

Honeypots

The idea of a network-based **honeypot** is to provide would-be hackers with fake information about a network by means of a decoy server to confuse them, trace them, or keep a record for prosecution. The honeypot is well-isolated from the rest of the network and can extensively log the activities of intruders. The concept is still relatively new, but a few companies have installed honeypots.

Typically, reconnaissance probes occur prior to a real attack; the probes enable an attacker to obtain necessary information about the network resources he wants to attack. A honeypot identifies this reconnaissance activity, and when the network responds back to the potential attacker, it provides fictitious data that mimics the type of information the attacker would get from legitimate network resources. In the future, if the honeypot observes network traffic based on the tagged information, it recognizes that the attacker has returned to mount a break-in attempt. Action is initiated to block the session and capture information about the attack, such as dates, time, and all keystrokes entered.

A pilot project at the Cyber Incident Detection Data Analysis Center (CIDDAC) at the University of Pennsylvania provides intrusion detection services to critical industries as a defense against cyberattacks. The CIDDAC service is based on specially built Remote Cyber Attack Detection Sensor (RCADS) appliances installed outside the corporate network of pilot participants. The RCADS are essentially honeypots that look like part of the network an intruder is trying to enter. When the RCADS are attacked, CIDDAC workers monitor the event and collect real-time data that can be forwarded to law enforcement officials. The officials then compile an attack signature—"a unique arrangement of information that can be used to identify an attacker's attempt to exploit a known operating system or application vulnerability"[39]—so cyberthreats can be more quickly located and neutralized.[40]

Response

An organization should be prepared for the worst—a successful attack that defeats all defenses and damages data and information systems. A response plan should be developed well in advance of any incident and be approved by both the Legal Department and senior management. An already developed response plan helps keep the incident under technical and emotional control.

In a security incident, the primary goal must be to regain control and limit damage, not to attempt to monitor or catch an intruder. Some system administrators have taken the discovery of an intruder as a personal challenge and lost valuable time they should have used to restore data and information systems to normal.

Incident Notification

A key element of any response plan is to define who to notify and who *not* to notify. Within the company, who needs to be notified and what information does each person need to have? Under what conditions should the company contact major customers and suppliers? How does the company inform them of a disruption in business without unnecessarily alarming them? When should local authorities, CERT/CC, or even the FBI be contacted?

Most security experts recommend against releasing specific information about a security compromise in public forums, such as news reports, conferences, professional meetings, and online discussion groups. All parties working on the problem need to be kept informed and up to date. At the same time, the company needs to avoid using systems that are connected to the compromised system. The intruder may be monitoring these systems and e-mail to learn what is known about the security breach.

Protecting Evidence and Activity Logs

An organization should document all details of a security incident as it works to resolve the incident. Documentation captures valuable evidence for a future prosecution and provides data to help during the incident eradication and follow-up phases. It is especially important to document all system events, the specific actions taken, and all external conversations in a log book. Because this may become court evidence, an organization should establish a set of document-handling procedures using the Legal Department as a resource.

Incident Containment

Often, it is necessary to act quickly to contain an attack and to keep a bad situation from becoming even worse. The response plan should clearly define the process for deciding if an attack is dangerous enough to shut down critical systems or disconnect them from the network. How such decisions are made, how quickly they are made, and who makes them are all elements of an effective response plan. In a true security incident, as many defined procedures and policies should be used as possible.

Incident Eradication

Before the IT security group begins the eradication effort, it must collect and log all possible criminal evidence from the system, and then verify that all necessary backups are current, complete, and free of any virus. Creating a disk image of each compromised system on write-only media for later study, and as evidence, can be very useful. After virus eradication, the group must create a new backup. Throughout this process, a log should be kept of all actions taken. This will prove helpful during the follow-up phase and ensure that the problem does not recur.

Incident Follow-up

Of course, an essential part of follow-up is to determine how the organization's security was compromised so that it can be prevented from happening again. Often, the fix is as simple as getting a software patch from a product vendor. However, it is important to look deeper than the immediate fix and discover why the incident occurred. For example, if a simple software fix could have prevented the incident, then why wasn't the fix installed *before* the incident occurred?

A review should be conducted after the incident to determine exactly what happened and to evaluate how the organization responded. One approach is to write a formal incident report that includes a detailed chronology of events and the impact of the incident. This report identifies any mistakes so they are not repeated in the future. The experience from this incident should be used to update and revise the security incident response plan.

Creating a detailed chronology of all events will also document the incident in great detail for later prosecution. To this end, it is critical to develop an estimate of the monetary damage. Costs include loss of revenue, loss of productivity, and salaries of people working to address the incident, along with the cost of replacing data, software, and equipment.

Another important issue is the amount of effort that should be put into capturing the perpetrator. If a Web site was simply defaced, it would be easy to fix or restore the site's HTML (Hypertext Markup Language, the code that describes to your browser how a Web page should look). However, what if the intruders inflicted more serious damage, such as erasing proprietary program source code or the contents of key corporate databases? What if they stole company trade secrets? Expert crackers can conceal their identities, and tracking them down can take tremendous corporate resources and a long time.

The potential for negative publicity must also be considered. Discussing security attacks through public trials and the associated publicity not only has enormous potential costs in public relations, but real monetary costs. For example, a brokerage firm might lose many customers who learn of an attack and then think their money or records aren't secure. Even if a company decides that the negative publicity risk is worth it and goes after the perpetrator, documents of proprietary information that are provided to the court could cause even greater security threats in the future.

Legal precedent has been set for courts to hold organizations accountable for their own IT security weaknesses. This is particularly true of Internet service providers. For example, a Maine court forced Verizon Communications Inc. to issue rebates to many of its customers for outages they experienced during the outbreak of the Slammer worm. The court reasoned that Verizon had not exercised due care to protect against the worm. To protect themselves, organizations must show that they do exercise due care. It helps if an organization has a well-written, reasonable security policy and computer logs that show it follows the policy during an incident.[41] Table 3-6 provides a manager's checklist for assessing an organization's ability to prevent and deal with an Internet security incident. *Yes* is the preferred answer for each question.

TABLE 3-6 Manager's checklist for evaluating an organization's readiness for an Internet security incident

Questions	Yes	No
Has a risk assessment been performed to identify investments in time and resources that can protect the organization from its most likely and most serious threats?	____	____
Have senior management and employees involved in implementing security measures been educated about the concept of reasonable assurance?	____	____
Has a security policy been formulated and broadly shared throughout the organization?	____	____
Have automated systems policies been implemented that mirror the written policies?	____	____
Does the security policy address:		
E-mail with executable file attachments?	____	____
Wireless networks and devices?	____	____
Use of smart phones deployed as part of corporate rollouts or bought by end users?	____	____
Is there an effective security education program for employees, contractors, and part-time employees?	____	____
Has a layered security solution been implemented to prevent break-ins?	____	____
Has a firewall been installed?	____	____
Is antivirus software installed on all personal computers?	____	____
Is the antivirus software frequently updated?	____	____
Have precautions been taken to limit the impact of malicious insiders?	____	____
Are the accounts, passwords, and login IDs of former employees promptly deleted?	____	____
Is there a well-defined separation of employee responsibilities?	____	____
Are individual roles defined so that users have authority to perform their responsibilities and no more?	____	____
Is it a requirement to review the most critical Internet security threats and implement safeguards against them?	____	____

TABLE 3-6 Manager's checklist for evaluating an organization's readiness for an Internet security incident (continued)

Questions	Yes	No
Has it been verified that backup processes for critical software and databases work correctly?	____	____
Are periodic IT security audits conducted?	____	____
Has an intrusion detection system been implemented to catch intruders in the act—both in the network and on critical computers on the network?	____	____
Has an intrusion prevention system been implemented to thwart intruders in the act?	____	____
Has the company considered installing a honeypot to confuse and detect intruders?	____	____
Has a comprehensive incident response plan been developed?	____	____
Has the plan been reviewed and approved by legal and senior management?	____	____
Does the plan address all of the following areas:		
Incident notification?	____	____
Protection of evidence and activity logs?	____	____
Incident containment?	____	____
Incident eradication?	____	____
Incident follow-up?	____	____

Summary

1. What key trade-offs and ethical issues are associated with the safeguarding of data and information systems?

   Business managers, IT professionals, and IT users face a number of ethical decisions regarding IT security, such as determining which information systems and data most need protection, how much effort and money to spend implementing safeguards against computer crime, whether to pursue prosecution of computer criminals at all costs or maintain a low profile to avoid negative publicity, what actions to take if their firm produces defective software that allows hackers to attack customers' data and computers, what tactics management should ask employees to take in gathering competitive intelligence, and what should be done if recommended computer security safeguards make life more difficult for customers and employees, resulting in lost sales and increased costs.

2. Why has there been a dramatic increase in the number of computer-related security incidents in recent years?

   The increasing complexity of the computing environment, higher user expectations, expanding and changing systems, and increased reliance on software with known vulnerabilities led to a 65-fold increase in the number of reported IT security incidents from 1997 to 2003.

3. What are the most common types of computer security attacks?

   Most incidents involve a virus, worm, Trojan horse, or denial-of-service attack.

4. What are some characteristics of common computer criminals, including their objectives, available resources, willingness to accept risk, and frequency of attack?

   Perpetrators include hackers who want to test the limits of a system, crackers who want to cause system problems, insiders who are seeking financial gain or revenge, industrial spies trying to gain a competitive advantage, cybercriminals looking for financial gain, and cyberterrorists seeking to cause destruction to bring attention to their cause. Each type of perpetrator has access to varying resources and is willing to accept different levels of risk to accomplish the objective. Knowing the profile of each set of likely attackers is the first step toward establishing effective countermeasures.

5. What are the key elements of a multilayer process for managing security vulnerabilities, based on the concept of reasonable assurance?

   A strong security program begins with an assessment of threats to the organization's computers and network. This assessment identifies actions that must be taken to address the most serious vulnerabilities. Educating users to the risks involved and the actions they must take to prevent a security incident is a key part of any successful security program.

   The IT group must lead the effort to implement security policies and procedures, along with hardware and software tools, to help prevent security breaches. No security system is perfect, so systems and procedures must also be monitored to evaluate their effectiveness.

6. What actions must be taken in response to a security incident?

   If an intrusion occurs, there must be a clear response plan that addresses notification, protection of evidence and security logs, containment, eradication, and follow-up. Knowledge gained from a security incident should be used to prevent or lessen the negative effects of a future incident.

Self-Assessment Questions

1. Computer crime is rapidly increasing in the United States but not in the rest of the world. True or False?

2. An attack on an information system that takes advantage of a vulnerability is called a(n) _____ .

3. Which of the following is harmful code that can propagate without human intervention by sending copies of itself to other computers via e-mail or Internet Relay Chat?

   a. virus

   b. logic bomb

   c. worm

   d. all of the above

4. A person who tests the limitations of computer systems out of intellectual curiosity, to see whether he can gain access and how far he can go, is called a(n) _____ .

5. Competitive intelligence relies on any means available to gain information about competing firms. True or False?

6. Concern over potential cyberterrorism began with the attacks of 9/11. True or False?

7. A(n) _____ is an organization's review of potential threats to its computers and network and the probability of those threats occurring.

8. The security requirements of an organization and the controls and sanctions used to meet those requirements are known as a _____ .

   a. risk assessment

   b. security policy

   c. firewall

   d. none of the above

9. A(n) _____ works by using the Internet to relay communications, but maintains privacy through security procedures and tunneling protocols that encrypt data at the sending end and decrypt it at the receiving end.

10. Implementation of a strong firewall provides adequate security for any network. True or False?

11. The overwhelming majority of successful computer attacks are made possible by taking advantage of well-known vulnerabilities. True or False?

12. A(n) _____ provides would-be hackers with fake information about a network by means of a decoy server.

    a. firewall

    b. intrusion prevention system

    c. honeypot

    d. none of the above

Review Questions

1. What is a computer exploit? How do software manufacturers address exploits when they are found?

2. Identify the key characteristics of six types of computer criminals. Which type is considered the top security risk for most companies?

3. What is a zero-day attack? What are some software manufacturers doing to avoid such an event?

4. What is CERT/CC, and what role does it play?

5. What is the difference between a cyberterrorist and an industrial spy?

6. How is a virus different from a worm? How is a worm different from a Trojan horse?

7. What is spoofing? Who does it and why?

8. Describe a denial-of-service attack. How serious are such attacks?

9. Outline the key elements of a cost-effective program to reduce Internet security incidents.

10. Why must antivirus software be constantly updated?

11. Why is it important to verify the backup process for critical software and databases?

12. What is the difference between a computer firewall and intrusion prevention software?

13. What activities are covered by a thorough security incident recovery plan?

Discussion Questions

1. A successful denial-of-service attack requires downloading software that turns unprotected computers into zombies under control of the malicious hacker. Should the owners of the zombie computers be fined as a means of encouraging people to better safeguard their computers? Why or why not?

2. Research the Web to find any recent examples of cyberterrorism. Would you say that cyberterrorism is a serious threat? Why or why not?

3. How can installation of a firewall give an organization a false sense of security?

4. Some IT security people believe that their organizations should always employ whatever resources are necessary to capture and prosecute computer criminals. Do you agree? Why or why not?

5. You have been assigned to be a computer security trainer for all of your firm's 2000 employees, contractors, and part-time workers. What are the key elements of your training program?

6. What do you think motivates a person to attempt to hack into computers, databases, and networks? What motivates a "white hat" hacker to become a "black hat" cracker?

What Would You Do?

1. You are the CEO of a three-year-old software manufacturer that has several products and annual revenues in excess of $500 million. You've just received a recommendation from the manager of software development to hire three notorious crackers to probe your software products in an attempt to identify any vulnerabilities. The reasoning is that if anyone can find a vulnerability in your software, they can. This will give your firm a head start on developing patches to fix the problems before anyone can exploit them. You're not sure, and feel uneasy about hiring people with criminal records and connections to unsavory members of the hacker/cracker community. What would you do?

2. You have just been hired as an IT security consultant to "fix the security problem" at Acme United Global Manufacturing. The company has been hacked mercilessly over the last six months, with three of the attacks making headlines for the negative impact they have had on the firm and its customers. You have been given 90 days and a budget of $1 million. Where would you begin, and what steps would you take to fix the problem?

3. You are the CFO of a midsized manufacturing firm. You have heard nothing but positive comments about the new CIO you hired three months ago. As you observe her outline what needs to be done to improve the firm's computer security, you are impressed with her energy, enthusiasm, and presentation skills. However, your jaw drops when she states that the total cost of the computer security improvements will be $300,000. This seems like a lot of money for security, given that your firm has had no major incident. Several other items in the budget will either have to be dropped or trimmed back to accommodate this project. In addition, the $300,000 is above your spending authorization and will require approval by the CEO. This will force you to defend the expenditure, and you are not sure how to do this. You wonder if this much spending on security is really required. How can you sort out what really needs to be done without appearing to be micromanaging or discouraging the new CIO?

4. Your friend just told you that he is developing a worm to attack the administrative systems at your college. The worm is "harmless" and will simply cause a message—"Let's party!"— to be displayed on all workstations on Friday afternoon at 3 p.m. By 4 p.m., the virus will erase itself and destroy all evidence of its presence. What would you say or do?

5. You are the vice president of application development for a small but rapidly growing software company that produces patient billing applications for doctors' offices. During work on the next release of your firm's one and only software product, a small programming glitch has been uncovered in the current release that could pose a security risk to users. The probability of the problem being discovered is low, but if exposed, the potential impact on your firm's 100 or so customers could be substantial: hackers could access private patient data and change billing records. The problem will be corrected in the next release, but you are concerned about what should be done for the users of the current release.

   The problem has come at the worst possible time. The firm is seeking approval for a $10 million loan to raise enough cash to continue operations until revenue from the sales of its just-released product offsets expenses. In addition, the effort to communicate with users, to develop and distribute the patch, and to deal with any fallout will place a major drain on your small development staff, delaying the next software release at least one month. You have a meeting with the CEO this afternoon; what course of action will you recommend?

Cases

1. UK Government Comes under Trojan Horse Attack

On June 16, 2005, the National Infrastructure Security Co-ordination Centre (NISCC) in the United Kingdom (UK) issued a briefing that immediately attracted international media attention. Approximately 300 government agencies and companies deemed critical to the national infrastructure had recently come under attack by more than 75 different Trojan horses. The majority of these cyberattacks were directed against the central government. Coming just a few weeks prior to the London bombings, could the attacks have been part of a concerted action by cyberterrorists?

Although most Trojan horses are directed against the general public, these attacks were more sophisticated, targeting specific employees who had access to sensitive data. The e-mails were socially engineered to spoof the sender's e-mail address so that the recipient believed a coworker, a government department, or a news agency had sent the e-mail. The subject line often included news headlines of particular interest to the recipient. The employee clicked on an attachment or a link, and the Trojan horse launched a remote monitoring program, gathering data from the employee's computer and sending it back to the attacker. This information included passwords as well as financially or commercially sensitive information. The compromised computers were then used to launch new attacks.

The NISCC followed the trail of the attacks back to the Far East through IP addresses and e-mail header information, but security experts do not know whether these machines were intermediaries or the original source of the attack. The NISCC did not release any additional information about the perpetrators, and the press was left to speculate who might be behind the assaults. The pattern did not resemble those of mafia cyberattacks from Eastern Europe. The cyber mob focuses on identity theft and cyberextortion. The perpetrators primarily used variations of Trojan horses already in circulation, making them very difficult to trace. Sophos, an IT security company hired by the NISCC, felt that these attacks may not have been specifically targeted at the national infrastructure.

Other experts have long predicted a move toward targeted social engineering that exploits variations of current viruses to access information of key organizations and industries worldwide. These Trojans, they believe, are a sign of things to come. Certainly, the purpose of the NISCC report was to alert key industry leaders to the problem so that they could brace themselves for future onslaughts.

The NISCC warned that antivirus software and firewalls do not offer complete protection against these Trojans. Because these viruses have been modified, they can penetrate antivirus software and communicate with the perpetrators through common ports. To fend off these assaults in the future, organizations will have to revise their monitoring and detecting techniques, secure computers that handle sensitive information, and train their employees to view e-mails in plain text and not open attachments unless the request is consistent with previous communication.

Even if these incidents were not the first of the predicted targeted attacks, NISCC's warning and advice and the accompanying media attention should help the UK and other countries prepare for what is to come.

Questions:

1. Why were the Trojan attacks so successful at infiltrating the security measures that most UK companies and government agencies had in place?

2. Why did the NISCC believe that these attacks were different from general computer viruses and other cyberassaults that have been launched in the past?

3. What kind of information did the perpetrators gather from their targets? What did the information indicate about the nature of their assault?

2. Whistle-Blower Divides IT Security Community

As a member of the X-Force, Mike Lynn analyzed online security threats for Internet Security Systems (ISS), a company whose clients include businesses and government agencies across the world. In early 2005, Lynn began investigating a flaw in the Internet operating system (IOS) used by Cisco routers. Through reverse engineering, he discovered that it was possible to create a network worm that could propagate itself as it attacks and takes control over routers across the Internet. Lynn's discovery was momentous, and he decided that he had to speak out and let IT security professionals and the public know about the danger.

"What politicians are talking about when they talk about the Digital Pearl Harbor is a network worm," Lynn said during a presentation. "That's what we could see in the future, if this isn't fixed."

Lynn had informed ISS and Cisco of his intentions to talk at the Black Hat conference, a popular meeting of computer hackers, and all three parties entered discussions with the conference managers to decide what information Lynn would be allowed to convey. On July 25, two days before the presentation, Cisco and ISS pulled the plug. Cisco employees tore out 10 pages from the conference booklet and ISS asked that Lynn speak on a different topic, Voice Over Internet Protocol (VoIP) security.

In a dramatic move, Lynn resigned from ISS on the morning of the conference and decided to give the presentation as originally planned. Within a few hours of his presentation, Cisco had filed suit against Lynn, claiming that he had stolen information and violated Cisco's intellectual property rights.

"I feel I had to do what's right for the country and the national infrastructure," Lynn explained.

And Lynn's words might have held more credibility had his presentation *not* been titled "The Holy Grail: Cisco IOS Shellcode and Remote Execution" and had Lynn *not* chosen a Black Hat annual conference as the venue for his crucial revelation.

Rather than speak to a gathering of Cisco users who would respond to the revelation by installing Cisco's patch and putting pressure on Cisco to find additional solutions, Lynn chose an audience that may well have included hackers who view the search for the flaw as a holy crusade. "Black hats" are hackers or crackers who break into systems with malicious intent. By contrast, "white hats" are hackers who reveal vulnerabilities to protect systems. Black Hat is a company that provides IT security consulting, briefings, and training. By coincidence, the CEO of Black Hat, Jeff Moss, also founded Defcon, an annual meeting of underground hackers who gather together to drink, socialize, and talk shop. During the Defcon conference, which followed the Black Hat conference, hackers worked late into the night trying to find the flaw.

"What Lynn ended up doing was describing how to build a missile without giving all the details. He gave enough details so people could understand how a missile could be built, and they could take their research from there," said one Defcon hacker.

Once well-defined, the line between white hat and black hat has become blurry. Security professionals, law-enforcement officials, and other white hats have infiltrated the ranks of the black-garbed renegades at Defcon annual conferences. IT companies hire hackers as IT security experts. Microsoft has declared that it plans to host annual hacker conferences that it will call Blue Hat conferences. The respectable IT giants have invited the black hats into the industry, and they have accepted the invitation, in large part.

Yet Cisco's handling of Lynn had the black hats up in arms. "The whole attempt at security through obscurity is amazing, especially when a big company like Cisco tries to keep a researcher quiet," exclaimed Marc Maiffret, chief hacker for eEye Digital Security. Maiffret felt that Cisco would have to mend some bridges with the IT security community.

White hats, in the meantime, bombarded the IT media with opinion pieces reminding people that a similar Black Hat disclosure about Microsoft precipitated the creation of the Blaster worm, which tore across the Internet and cost billions of dollars in damage.

Questions:

1. How did Mike Lynn explain his purpose in exposing the vulnerability in the Cisco Internet operating system? What other purposes might Lynn have had in making his presentation?

2. What types of perpetrators might be interested in exploiting the vulnerability in the Cisco Internet operating system?

3. What factors do the government and industry have to consider in dealing with the hacker and cracker community?

3. Cybercrime: Even Microsoft Is Vulnerable

On October 27, 2000, Microsoft acknowledged that its security had been breached and that outsiders using a Trojan horse virus had been able to view source code for computer programs under development. A Microsoft spokesperson, Rick Miller, called the break-in "a deplorable act of industrial espionage." The incident was discovered by the company on October 25 and was reported to the FBI the next day. The attack was believed at first to have initiated from St. Petersburg, Russia.

Customers as well as competitors were amazed that Microsoft was vulnerable. Most assumed that the world-renowned software leader had bulletproof security and was untouchable. The incident sent a loud wake-up call to organizations everywhere to step up security—if it could happen to Microsoft, it could happen to anyone.

Initially, Microsoft reported that the hacking had gone on over a six-week period. Microsoft later amended its initial reports and said the incident lasted only a week. Microsoft's corporate security officer said that the confusion stemmed from initial uncertainty over whether routine virus incidents in September were related.

There was also uncertainty over just what the hackers had seen or done. Initially, Microsoft said that the hackers were able to view, but not modify, source code of several major products. Later, Microsoft amended this statement and said that the incident was not as serious as it originally feared—the intruder did not gain access to the source code of any strategic major products.

However, the Washington Post, citing a source close to the investigation, reported that the targeted material was related to Microsoft's .NET strategy, its far-reaching plan to build the Internet into all its software.

One thing that did seem certain is that passwords to access Microsoft systems were stolen. Security analysts believe that the passwords were sent to an e-mail account in China (not St. Petersburg, as initially reported).

The attackers apparently used a worm known as QAZ to break into Microsoft's network. Once inside, the worm broadcast its location to the cracker, who then took administrative control of the system without the user's knowledge. This enabled the attacker to do the same things an authorized user was permitted to do. The worm was also programmed to deliver passwords from one computer to another.

Experts said that an attack with a worm such as QAZ shouldn't have been possible if Microsoft had properly configured its firewall and antivirus software and kept them updated. Antivirus software vendors were familiar with QAZ and had updated their packages to detect QAZ months before the attack. A description of the worm was even posted on their Web sites, including steps that users could take to protect themselves from QAZ.

One scenario of how the attack may have happened is that a Microsoft employee logged on to the company network from a home computer and inadvertently revealed system and network passwords to a hacker watching online. (The odds of a worm being downloaded on a home computer are much greater than on an office-based computer, because home security is frequently less stringent and harder to monitor.) The intruders were then able to send e-mails to Microsoft computers laced with the QAZ worm program. The QAZ program then stole additional passwords while providing the intruders with unauthorized access to the computer system.

This scenario illustrates why it is critical for companies to ensure that their employees don't unknowingly provide hackers with access into the corporate network by logging on to corporate computers from unsecured home computers. If organizations allow access to corporate networks from employees at home, there must be appropriate measures and procedures to protect their home computers. As a result, many organizations require that firewalls be installed on their employees' home computers to reduce this vulnerability. Such firewalls are especially critical when a home user is always connected to the Internet via a Digital Subscriber Line (DSL), cable network, or some other permanent connection.

Questions:

1. Some people think that Microsoft downplayed the seriousness of the break-in. What would be the motivation for such a strategy?

2. Why would Microsoft have allowed the attackers access to their network and systems for several days after first detecting the break-in?

3. Imagine that you are in charge of recommending new and improved security measures for Microsoft. What recommendations would you make? How would you justify the costs of implementing your ideas?

End Notes

[1] "Sasser Creator Avoids Jail Term," BBC News, news.bbc.co.uk/2/hi/technology/4659329.stm, July 8, 2005.

[2] "Teen Responsible for Sasser and Netsky Worm Attacks Hired by German Security Firm," Sophos Web site, www.sophos.com/virusinfo/articles/jaschanjob.html, September 20, 2004.

[3] Blau, John, "Lawyers Disagree over Punishment in Sasser Trial," *Computerworld*, www.computerworld.com/securitytopics/security/story/0,10801,103005,00.html, July 7, 2005.

[4] Roberts, Paul, "Sasser Infections Hit Hard," *PC World*, www.pcworld.com/news/article/0,aid,115979,00.asp, May 4, 2005.

[5] Joshi, Abray, "How to Protect Your Company from 'Zero-Day' Attacks," *Computerworld*, www.computerworld.com, March 1, 2004.

[6] Ubois, Jeff and Betts, Mitch, "Dual Curses: "Viruses and Spam," *Computerworld*, www.computerworld.com, February 4, 2004.

[7] Sandoval, Greg, "Windows Worm Hits Local Firms," *Cincinnati Enquirer*, page D1, August 18, 2005.

[8] Grow, Brian, and Hamm, Steve, "From Black Market to Free Market," *Business Week*, pages 28–31, August 22/29, 2005.

[9] McMillian, Robert, "IE Bug Can Crash Browser," *Computerworld*, www.computerworld.com, July 1, 2005.

[10] McMillian, Robert, "Patches Issued for RealPlayer Flaw," *Computerworld*, www.computerworld.com, June 29, 2005.

[11] Cowley, Stacy, "Sun Patches Critical Java Flaws," *Computerworld*, www.computerworld.com, June 15, 2005.

[12] Niccolai, James, "New Sober Worm Moving Fast, Security Firm Warns," *Computerworld*, www.computerworld.com, February 21, 2005.

[13] Gonsalves, Antone, "Virus Writers Adopting a Stealth Strategy," *InformationWeek*, www.informationweek.com, July 22, 2005.

[14] McMillian, Robert, "McAfee: More Attackers Seeking Money, Zombies," *Computerworld*, July 12, 2005.

[15] Vijayan, Jaikumar, "Update: Credit Card Firm Hit by DDoS Attack," *Computerworld*, www.computerworld.com, September 22, 2004.

[16] Rosencrance, Linda, "Hacker Helps Applicants Breach Security at Top Schools," *Computerworld*, www.computerworld.com, March 4, 2005.

[17] The Associated Press, "Hackers Hit Microsoft Windows Genuine Advantage," *InformationWeek*, www.informationweek.com, July 31, 2005.

[18] Bergstein, Brian, "Theft from Card Processor Indicates Cracker's Increasing Power," *InformationWeek*, June 22, 2005.

[19] Holmes, Alan, "Invitation to Steal," *CIO*, www.cio.com, February 1, 2005.

[20] Holmes, Alan, "Invitation to Steal," *CIO*, www.cio.com, February 1, 2005.

21 "Information Without Borders," *CSO Magazine*, www.csoonline.com/read/010104/machine. html, January 2004.

22 Richardson, Carey, "Check 21: Check Clearing for the 21st Century," www.denovobanks.com, January 22, 2004.

23 Mearian, Lucas, "Checkpoint for Check 21," *Computerworld*, www.computerworld.com, October 25, 2004.

24 Holmes, Alan, "Invitation to Steal," *CIO*, www.cio.com, February 1, 2005.

25 Holmes, Alan, "Invitation to Steal," *CIO*, www.cio.com, February 1, 2005.

26 Hulme, George V., and Wallace, Bob, "Top of the Week Hacktivism," *InformationWeek*, pages 22–24, November 13, 2000.

27 "E911 Virus," FBI National Infrastructure Protection Center Web site, www.npic.gov, May 30, 2001.

28 Weinmann, Gerald, "Cyberterrorism: How Real is the Threat?" *CIO*, Special Report 119, www.usip.org/pubs/specialreports/sr119.html, December 2004.

29 Weinmann, Gerald, "Cyberterrorism: How Real is the Threat?" *CIO*, Special Report 119, www.usip.org/pubs/specialreports/sr119.html, December 2004.

30 "Laptops Lead to al Qaeda Plot," Rediff.com, www.rediff.com/news/2004/aug/03alq.htm, August 3, 2004.

31 Gartenberg, Marc, "How to Develop an Enterprise Security Policy," *Computerworld*, www.computerworld.com, January 13, 2005.

32 Vijayan, Jaikumar, "Handheld Risks Prompt Push for Usage Policy," *Computerworld*, www.computerworld.com, February 21, 2005.

33 Rubinkam, Michael, "Kurtztown 13 Just Having Fun, or Felonious Hackers?" *Cincinnati Enquirer*, page A2, August 10, 2005.

34 "The SANS Top 20 Internet Vulnerabilities," www.sans.org/top20, August 22, 2005.

35 "Background, FISMA Implementation Project," csrc.nist.gov/sec-cert/ca-background.html, August 13, 2005.

36 Vijayan, Jaikumar, "Federal Agencies Get a D+ in Computer Security," *Computerworld* , www.computerworld.com, February 21, 2005.

37 "Information Security," United States Government Accountability Office, www.gao.gov/new. items/d05552.pdf, July 2005.

38 Robb, Drew, "Intrusion Prevention Systems: Erecting Barriers," *Computerworld*, www.computerworld.com, March 21, 2005.

39 "Attack Signatures," Symantec Web page, securityresponse.symantec.com/avcenter/attack_ sigs/.

40 Weiss, Todd R., "U.S. Gets New Cyberterrorism Security Center," *Computerworld,* www. computerworld.com, April 21, 2005.

41 Verton, Dan, "Sidebar: Fighting Back, Legally," *Computerworld*, www.computerworld.com, January 17, 2004.

Sources for Case 1

NISCC Briefing 08/2005, National Infrastructure Security Co-ordination Centre, www.niscc.gov. uk/niscc/docs/ttea.pdf, June 16, 2005.

Keizer, Gregg, "U.K. Under Cyber Attack, Security Center Says," Network Computing's Desktop Pipeline, nwc.desktoppipeline.com/news/164900444, June 16, 2005.

Roberts, Paul, "UK Government Warns of Massive Trojan Attack," e-Week.com, www.eweek.com/ article2/0,1895,1828863,00.asp, June 16, 2005.

Sources for Case 2

Lemos, Robert, "Cisco, ISS File Suit Against Rogue Researcher," *The Register*, www.theregister. co.uk/2005/07/28/cisco_iss_sue_vuln_whistleblower/page2.html, July 28, 2005.

Lemos, Robert, "Settlement Reached in Cisco Flaw Dispute," *The Register*, www.theregister.co. uk/2005/07/29/cisco_settles_rogue_researcher_dispute/, July 29, 2005.

Sullivan, Andy, "Hackers at Defcon Race to Expose Cisco Internet Flaw," *Computerworld*, www. computerworld.com/securitytopics/security/hacking/story/0,10801,103607,00.html, August 1, 2005.

Winkler, Ira, "Black Hat Researcher Lynn No Hero to Global Security," SearchSecurity.com, searchsecurity.techtarget.com/columnItem/0,294698,sid14_gci1112773,00.html, August 4, 2005.

Zetter, Kim, "Whistle-Blower Faces FBI Probe," *Wired News*, www.wired.com/news/privacy/ 0,1848,68356,00.html?tw=rss.TOP, July 29, 2005.

Sources for Case 3

Vaughn-Nichols, Steven J., "Microsoft Can't Spin This Worm," www.msnbc.com, October 27, 2000.

Glascock, Stuart, and Wagner, Mitch, "Microsoft, FBI, Security Experts Probe Hacking," www. TechWeb.com, October 27, 2000.

Brunker, Mike, "Hackers Boldly Break Into Microsoft," www.msnbc.com, October 28, 2000.

Vijayan, Jaikumar, and Sliwa, Carol, "Microsoft Break-In Points to Security Holes," *Computerworld*, www.computerworld.com, November 6, 2000.

Verton, Dan, "Worm Thought to Have Been Used Against Microsoft Had Links to Chinese System," *Computerworld*, www.computerworld.com, January 17, 2001.

PRIVACY

VIGNETTE

Have You Been Swiped?

You walk into a bar or a convenience store to purchase alcohol or cigarettes. The cashier asks for your driver's license. Without asking your permission, maybe even without you noticing, the cashier swipes your license with an electronic scanner. By the time the card is back in your hand, the store owners and management know your name, address, phone number, age, height, weight, and maybe even your Social Security number. Depending on what state you live in, all that information and more may be encoded on the back of your license. Tie this data to information that can be retrieved from commercial data-mining companies, and a complete stranger can know your salary, your profession, and whether you're married.

Bars and restaurants openly acknowledge that they use information scanned from licenses to do more than check for fake IDs. They assemble marketing databases by tracking your purchases. A sports bar can check the male-to-female ratios of its clientele. A restaurant that offers live entertainment can figure out what kind of crowd a band attracts to its establishment.

Although this may seem innocuous enough, privacy rights groups are concerned that bars and restaurants might sell this information to list brokers or government agencies. An employee could use the information to stalk a customer or commit identity theft.

In 2005, a group of performance artists toured the country to increase awareness about swiping. They passed out labels that people could stick to the back of their licenses, which read: "I stop shopping when you start swiping." They also created an interactive map where you can find out what data each state encodes on the back of driver's licenses. See *http://turbulence.org/Works/swipe/state_analysis.html.*[1,2,3]

LEARNING OBJECTIVES

As you read this chapter, consider the following questions:

1. What is the right of privacy, and what is the basis for protecting personal privacy under the law?
2. What are some of the laws that authorize electronic surveillance by the government, and what are the associated ethical issues?
3. What are the two fundamental forms of data encryption, and how does each work?
4. What is identity theft, and what techniques do identity thieves use?
5. What are the various strategies for consumer profiling and the associated ethical issues?
6. What must organizations do to treat consumer data responsibly?
7. Why and how are employers increasingly using workplace monitoring?
8. What is spamming, and what ethical issues are associated with its use?
9. What are the capabilities of advanced surveillance technologies, and what ethical issues do they raise?

PRIVACY PROTECTION AND THE LAW

The use of information technology in business requires balancing the needs of those who use the information against the rights and desires of the people whose information may be used.

On one hand, information about people is gathered, stored, analyzed, and reported because organizations can use it to make better decisions. Some of these decisions, including whether to hire a job candidate, approve a loan, or offer a scholarship, can profoundly affect people's lives. In addition, the global marketplace and increased competitiveness have increased the importance of knowing consumers' purchasing habits and financial condition. Companies use this information to target marketing efforts to consumers who are most likely to buy their products and services. Organizations also need basic information about customers to serve them better. It is hard to imagine an organization having a relationship with its customers without having data about them. Thus, organizations want systems that collect and store key data from every interaction they have with a customer.

On the other hand, many object to the data collection policies of government and business on the basis that they strip people of the power to control their own personal information. They feel that the existing hodgepodge of privacy laws and practices fails to provide adequate protection. Instead, it causes confusion that fuels a sense of distrust and skepticism, as illustrated by the opening vignette on card swiping. According to U.S. Census data, privacy is a key concern of Internet users and a top reason why nonusers still avoid the Internet.[4,5]

A combination of approaches—new laws, technical solutions, and privacy policies—is required to balance the scales. Reasonable limits must be set on government and business access to personal information, new information and communication technologies must be designed to protect rather than diminish privacy, and appropriate corporate policies must be developed to set baseline standards for people's privacy. Education and communication are essential as well.

This chapter helps you understand the right to privacy and presents an overview of information technology developments that affect this right. The chapter also addresses a number of ethical issues related to gathering data about people.

First, it is important to gain a historical perspective on the right to privacy. When the U.S. Constitution took effect in 1789, the drafters were concerned that a powerful government would intrude on the privacy of individual citizens. As a result, they added the Bill of Rights. So, although the Constitution does not contain the word *privacy*, the U.S. Supreme Court has ruled that the concept of privacy is protected by a number of amendments in the Bill of Rights. For example, the Supreme Court has stated that American citizens are protected by the Fourth Amendment when there is a "reasonable expectation of privacy."

The Fourth Amendment is as follows:

- The right of the people to be secure in their persons, houses, papers, and effects, against unreasonable searches and seizures, shall not be violated, and no Warrants shall issue, but upon probable cause, supported by Oath or affirmation, and particularly describing the place to be searched, and the persons or things to be seized.

Very importantly, though, without a reasonable expectation of privacy, there is no privacy right to protect.

Today, in addition to protection from government intrusion, people need privacy protection from private industry. Few laws actually provide such protection; most people assume that they have greater privacy rights than the law provides. Some people believe only those with something to hide should be concerned about the loss of privacy; however, everyone should be concerned. As the U.S. Privacy Protection Study Commission found in 1977 (when the computer age was in its infancy), "The real danger is the gradual erosion of individual liberties through the automation, integration, and interconnection of many small, separate record-keeping systems, each of which alone may seem innocuous, even benevolent, and wholly justifiable."[6]

The Right of Privacy

A broad definition of the right of privacy is "the right to be left alone—the most comprehensive of rights, and the right most valued by a free people" (Justice Louis Brandeis, dissenting in *Olmstead v. U.S., 1928*). Another definition of privacy that is particularly useful in discussing the impact of IT is "the right of individuals to control the collection and use of information about themselves."[7]

As a legal concept, the right to privacy has the following four aspects:

- Protection from unreasonable intrusion upon one's isolation, such as the gathering of details about a person's Web surfing habits
- Protection from appropriation of one's name or likeness, such as identity theft, which involves stealing credit cards or Social Security numbers
- Protection from unreasonable publicity given to one's private life, such as the revealing of details about a medical condition
- Protection from publicity that unreasonably places one in a false light before the public, such as false information published about a person on a Web site[8]

The use of information technology can lead to violations of all four aspects of the right to privacy.

Recent History of Privacy Protection

This section outlines a number of legislative acts passed over the past 40 years that affect a person's privacy. Note that most of these actions address invasion of privacy by the government. Legislation that protects people from data privacy abuses by corporations is almost nonexistent. In addition, although a number of independent laws and acts have been implemented over time, no single, overarching national data privacy policy has been developed. As a result, existing legislation is sometimes inconsistent and even conflicting at times.

Communications Act of 1934

The Communications Act of 1934 restricted the government's ability to secretly intercept communications. However, under a 1968 federal statute, law enforcement officers can use wiretapping—the interception of telephone or telegraph communications for purpose of espionage or surveillance—if they first obtain a court order.

FOIA

The Freedom of Information Act (FOIA), passed in 1966 and amended in 1974, provides the public with the means to gain access to certain government records. The public can use well-defined FOIA procedures to find out the spending patterns of an agency, the agency's policies and the reasoning behind them, and the agency's mission and goals. The act is often used by whistle-blowers to obtain records that they otherwise would be unable to get. Citizens have also used the FOIA to find out what information the government has about them. Exemptions to the FOIA bar disclosure of information that could compromise national security or interfere with an active law enforcement investigation. Another exemption prevents disclosure of records if it would invade someone's privacy. In this case, a balancing test is applied to evaluate whether the privacy interests at stake are outweighed by competing public interests. Some legislators are calling for an additional exemption to protect shared information between the government and private industry regarding attacks against computer and information systems. The sharing of such information, they say, is important for the federal government to protect the nation's critical IT infrastructure from attack. However, private companies are reluctant to share such data. They fear that the FOIA doesn't offer enough assurances that their proprietary data will remain secret.

Fair Credit Reporting Act

The Fair Credit Reporting Act of 1970 regulates the operations of credit-reporting bureaus, including how they collect, store, and use credit information. It is designed to promote accuracy, fairness, and privacy of information in the files of credit-reporting companies (such as Experian, Equifax, and Trans Union) and to check verification systems that gather and sell information about people. The act outlines who may access your credit information, how you can find out what is in your file, how to dispute inaccurate data, how long data is retained, and so on. The manual procedures and information systems of the credit-reporting bureaus must implement and support all these regulations.

Privacy Act

The Privacy Act of 1974 declares that no agency of the U.S. government can conceal the existence of any personal data record-keeping system, and that any agency that maintains such a system must publicly describe both the kinds of information in it and the manner in which the information will be used. The law also outlines 12 requirements that each record-keeping agency must meet, including issues that address openness, individual access, individual participation, collection limitation, use limitation, disclosure limitation, information management, and accountability. The purpose of the act is to provide certain safeguards for people against invasion of personal privacy by federal agencies. The Central Intelligence Agency (CIA) and law enforcement agencies are excluded from this act; nor does it cover the actions of private industry. The following Legal Overview explains how the Privacy Act has raised issues with the Department of Homeland Security over the screening of airline passengers.

LEGAL OVERVIEW

The Privacy Act

The Secure Flight airline safety program was developed under the auspices of the Department of Homeland Security to combat terrorism by comparing the names and information of 1.4 million daily U.S. airline passengers with data on known or suspected terrorists. An earlier version of the Secure Flight system, called Computer-Assisted Passenger Pre-screening System, was created in 2003 and then abandoned in 2004 because of concerns over violations of the Privacy Act. Unlike the previous system, Secure Flight only looks for terrorists, not other criminals. In addition, Secure Flight includes a redress mechanism for people who believe they have been unfairly or incorrectly selected for additional screening.[9]

In their initial investigation of the Secure Flight system, the Government Accountability Office (GAO) found that passenger information was used without proper disclosure, as required by the Privacy Act. Furthermore, the GAO said that Secure Flight passed only one of 10 tests required for certification.[10]

Within weeks of the GAO findings, the Department of Homeland Security proposed changes to its own funding bill that would allow Secure Flight to use commercial databases to perform background checks and profiling that could help determine if an airline passenger is a terrorist, despite not being on a terror watch list. It also proposed dropping the requirement for an independent congressional investigation to determine the impact of such verification on aviation security. Instead, it would be left up to the Homeland Security chief to make that determination. However, both the House and Senate passed the funding bill with identical provisions that prohibited Secure Flight from using commercial databases or computer software to profile passengers. The Department of Homeland Security plans to use Secure Flight to take over all commercial passenger screening in 2006.[11]

Another agency that is concerned with privacy is the Organization for Economic Cooperation and Development (OECD), an international organization consisting of 30 member countries, including Australia, Canada, France, Germany, Italy, Japan, Mexico, New Zealand, the United Kingdom, and the United States. Its goal is to set policies and make agreements in areas where multilateral consensus is necessary for individual countries to make progress in a global economy. Dialogue, consensus, and peer pressure are essential to make these policies and agreements "stick."[12] The 1980 privacy guidelines set by the OECD—also known as the Fair Information Practices—are often held up as a model of ethical treatment of consumer data for organizations to adopt. These guidelines are composed of the eight principles summarized in Table 4-1.

TABLE 4-1   Summary of the 1980 OECD privacy guidelines

Principle	Guideline
Collection limitation	Limit the collection of personal data. All such data must be obtained lawfully and fairly with the subject's consent and knowledge.
Data quality	Personal data should be accurate, complete, current, and relevant to the purpose for which it is used.
Purpose specification	The purpose for which personal data is collected should be specified and should not be changed.
Use limitation	Personal data should not be used beyond the specified purpose without a person's consent or by authority of law.
Security safeguards	Personal data should be protected against unauthorized access, modification, or disclosure.
Openness principle	Data policies should exist and a "data controller" should be identified.
Individual participation	People should have the right to review their data, to challenge its correctness, and to have incorrect data changed.
Accountability	A "data controller" should be responsible for ensuring that the above principles are met.

Source: *OECD Guidelines on the Protection of Privacy and Transborder Flows of Personal Data,*
pages 14–18, ©2002.

COPA

Congress passed the Children's Online Protection Act (COPA) in October 1998. According to the COPA law, a Web site that caters to children must offer comprehensive privacy policies, notify their parents or guardians about its data collection practices, and receive parental consent before collecting any personal information from children under 13 years of age. The law has had a major impact, requiring many companies to spend hundreds of thousands of dollars to make their sites compliant, while others eliminated preteens as a target audience. In 2004, the Federal Trade Commission (FTC) accused Bonzi Software Inc. and UMG Recordings Inc. of collecting personal information from children online without their parents' consent, and settled with them for penalties of $75,000 and $400,000, respectively. Bonzi Software, which distributes a free software download called Bonzi-Buddy, was the first company charged for privacy violations over a download. UMG Recordings, which operates music-related Web sites, was charged with collecting birth dates from children through its online registration process.[13]

European Community Directive 95/46/EC

The European Community Directive 95/46/EC of 1998 requires any company that does business within the borders of 15 Western European nations to implement a set of privacy directives on fair and appropriate use of information. Basically, this directive requires member countries to ensure that data transferred to non-European Union (EU) countries is protected. It also bars the export of data to countries that do not have data privacy protection standards comparable to the European Union's. Initially, EU countries were concerned that the largely voluntary system of data privacy in the United States did not meet the EU directive's stringent standards. Eventually, the U.S. Department of Commerce worked out an agreement with the European Union; American companies that could certify adherence to certain safe harbor principles were allowed to import and export personal data. The following list summarizes these European data privacy principles.[14]

- Notice—Tell all customers what is done with their information.
- Choice—Give customers a way to opt out of marketing.
- Onward transfer—Ensure that suppliers comply with the privacy policy.
- Access—Give customers access to their information.
- Security—Protect customer information from unauthorized access.
- Data integrity—Ensure that information is accurate and relevant.
- Enforcement—Independently enforce the privacy policy.

BBB Online and TRUSTe

The European philosophy of addressing privacy concerns with strict government regulation, including enforcement by a set of commissioners, differs greatly from the U.S. philosophy of having no federal privacy policy. The United States instead relies on self-regulation, which is overseen by the Department of Commerce, the Better Business Bureau Online (BBB Online), and TRUSTe.

BBB Online and TRUSTe are independent, nonprofit initiatives that favor an industry-regulated approach to data privacy. They are concerned that strict government regulation could have a negative impact on the Internet's use and growth, and that such regulation would be costly to implement and difficult to change or repeal. A Web site operator can apply for either the BBB Online or TRUSTe data privacy seal to demonstrate that a site adheres to a high level of data privacy. Having one of the seals can increase consumer confidence in the site operator's desire and ability to manage data responsibly. The seals also help users make more informed decisions about whether to release personal information (such as phone numbers, addresses, and credit card numbers) to a Web site.

An organization must join the Better Business Bureau and pay an annual fee ranging from $200 to $7000, depending on annual sales, before applying for the BBB Online privacy seal. Because the seal program identifies online businesses that honor their own stated privacy protection policies, the Web site must also have adopted and posted an online privacy notice. For a Web site to receive the TRUSTe seal, its operators must demonstrate that it adheres to established privacy principles. They must also agree to comply with TRUSTe's oversight and consumer resolution process, and pay an annual fee. The privacy principles require the Web site to openly communicate what personal information it gathers, how it will be used, with whom it will be shared, and whether the user has an option to control its dissemination.

From time to time, the FTC and various members of Congress have proposed privacy-related bills. A key issue in these bills is the policy of opting out or opting in for information gathering. An **opt-out** policy assumes that consumers approve of companies collecting and storing their personal information, and it requires consumers to actively opt out by specifically telling companies, one by one, not to collect data about them. An **opt-in** policy requires data collectors to get specific permission from consumers before collecting any of their data. Data collectors favor opt-out policies, and consumer groups favor opt-in policies.

Gramm-Leach-Bliley Act

One example of a law that controls opt-out information gathering is the 1998 Gramm-Leach-Bliley Act, which required all financial-services institutions to communicate their data privacy policies and honor customer data-gathering preferences by July 1, 2001. Its goal was to require all financial institutions to take appropriate actions to protect and secure customers' nonpublic data from unauthorized access or use. To comply, institutions resorted to mass mailings to contact their customers with privacy-disclosure forms. As a result, many people received a dozen or more similar-looking forms—one from each financial institution with which they do business. However, most people did not take the time to read the long forms, which were printed in small type and were full of legalese. Rather than making it easy for customers to opt out, the documents required that consumers send one of their own envelopes to a specific address and state in writing that they wanted to opt out—all this rather than sending a simple prepaid postcard that allowed customers to check off their choice. As a result, most customers threw out the forms without grasping their full implications. These customers thus opted in and gave financial institutions the right to sell information, such as annual earnings, net worth, employers, specific investments, loan amounts, and Social Security numbers, to other financial institutions.

The United States has neither enacted the legislation to establish minimal privacy standards nor established an advisory agency that could recommend acceptable privacy practices to businesses. Instead, there are laws that address potential abuses by the government, with little or no restrictions for private industry. You can track the status of privacy legislation at the Electronic Privacy Information Center's Web site (*www.epic.org*).

KEY PRIVACY AND ANONYMITY ISSUES

The rest of this chapter discusses a number of current and important privacy issues, including government electronic surveillance, data encryption, identity theft, customer profiling, the need to treat customer data responsibly, workplace monitoring, spamming, and advanced surveillance techniques.

Governmental Electronic Surveillance

This section discusses laws that address government electronic surveillance, including the Federal Wiretap Act, the Electronic Communications Privacy Act, the Foreign Intelligence Surveillance Act, Executive Order 12333, the Communications Assistance for Law Enforcement Act, and the USA Patriot Act.

Federal Wiretap Act

The **Federal Wiretap Act** (U.S. Code Title 18 Part I, Chapter 119, Wire and Electronic Communications Interception and Interception of Oral Communications), sometimes referred to as Title III, was adopted in 1968 and expanded in 1986. It outlines processes to obtain court authorization for surveillance of all kinds of electronic communications, including e-mail, fax, Internet, and voice, in criminal investigations. Before a wiretap can commence, a judge must issue a court order based on probable cause. The act specifies that drug trafficking, hijackings, terrorist bombings, and other violent activities are crimes for which wiretaps can be ordered. Judges almost never deny government requests for Title III wiretap orders. The act also provides the government with "roving tap" authority to obtain a court order that does not name a specific telephone line or e-mail account, but allows the government to tap any phone line, cell phone, or Internet account that a suspect uses. There is also a provision requiring Internet service providers (ISPs) to take necessary steps to preserve electronic records and other evidence, pending the issuance of a court order or other legal process. The order is a letter from a law enforcement agency requiring an ISP to preserve logs in preparation for a subpoena or search warrant. ISPs commonly receive such orders.

Figure 4-1 shows the number of Title III wiretaps approved from 1991 to 2004. During this period, less than three dozen wiretap requests were refused.

ECPA

The **Electronic Communications Privacy Act of 1986 (ECPA)**, U.S. Code Title 18, Part II, Chapter 206, set standards for access to stored e-mail and other electronic communications and records. The ECPA amended Title III of the Omnibus Crime Control and Safe Streets Act of 1968, extending Title III's prohibitions against the unauthorized interception, disclosure, or use of a person's oral or electronic communications. In *Smith v. Maryland* (1979), the Supreme Court held that there is no constitutionally protected privacy interest in the numbers you dial to initiate a telephone call. Accordingly, the provisions in ECPA establish minimum standards for court-approved law enforcement access to electronic or other impulses that identify the numbers dialed for outgoing calls (**pen register**) and the originating number for incoming calls (**trap and trace**). To obtain such an order, the government only needs to certify that "the information likely to be obtained is relevant to an ongoing criminal investigation." A prosecutor does not have to justify the request, and judges are required to approve every request.

A recording of every telephone number dialed and the source of every call received can provide an excellent profile of a person's associations, habits, contacts, interests, and activities. Federal law enforcement agencies conduct roughly 10 times as many pen register and trap-and-trace surveillances as they do wiretaps. So much of this information is collected that Justice Department agencies have developed several generations of software to enhance its analysis.

This type of surveillance is highly controversial, especially as it relates to the collection of computer data sent over the Internet, including e-mail addresses, e-mail header information, Internet provider addresses, dial-up numbers, and e-mail logs.

The ECPA failed to address emerging technologies such as wireless modems, radio-based electronic mail, and cellular data networks; thus, these communications can still be legally intercepted.

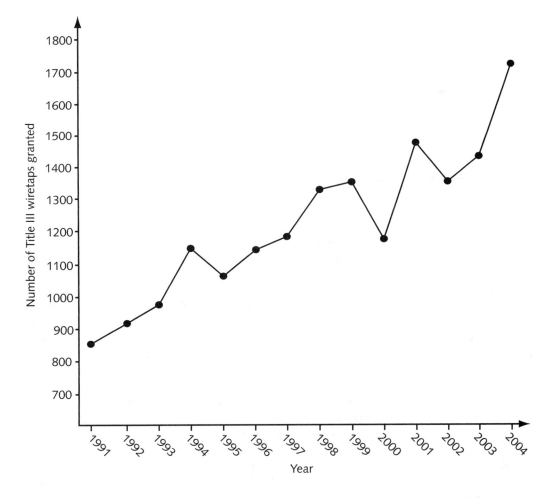

Source: Administrative Office of the U.S. Courts, www.uscourts.gov/wiretap.html for 1997-2004 wiretap reports.

FIGURE 4-1 Number of Title III wiretaps granted

FISA

The **Foreign Intelligence Surveillance Act of 1978 (FISA)** allows wiretapping of aliens and citizens in the United States, based on a finding of probable cause that the target is a member of a foreign terrorist group or an agent of a foreign power. For U.S. citizens and permanent resident aliens, there must also be probable cause that the person is engaged in criminal activities. Suspicion of illegal activity is not required in the case of aliens. For them, membership in a terrorist group is enough, even if their activities on behalf of the group are legal. Both Title III and FISA allow the government to carry out wiretaps without a court order in emergencies that involve national security, risk of death, or risk of serious injury.

Executive Order 12333

Executive Order 12333 is the legal authority for electronic surveillance outside the United States. Issued by President Reagan in 1982, this order permits intelligence agencies to intercept communications outside the United States without a court order. If a U.S. citizen or permanent resident alien is targeted for surveillance abroad, the executive order requires the approval of the Attorney General, who must find probable cause that the U.S. citizen targeted by the surveillance is an agent of a foreign power, as defined in FISA. Decisions to target non-U.S. citizens are left to the intelligence community.

CALEA

The **Communications Assistance for Law Enforcement Act (CALEA)** was passed by Congress in 1994. CALEA was a hotly debated law because it required the telecommunications industry to build tools into its products that federal investigators can use—after getting court approval—to eavesdrop on conversations. At the time, law enforcement and the FBI in particular argued that the law was necessary to preserve their capacity to engage in electronic communication surveillance, and assured Congress that they were not seeking any additional power.

The provision covering radio-based data communication grew from a realization that the Electronic Communications Privacy Act of 1986 failed to protect emerging technologies such as wireless modems, radio-based electronic mail, and cellular data networks. The 1986 statute outlawed the unauthorized interception of wire-based digital traffic on commercial networks, but the law's drafters did not foresee the growing interest in wireless data networks. Section 203 of the new legislation fixed that oversight by effectively covering all publicly available "electronic communication."

The Federal Communications Commission (FCC) responded to appeals from the Justice Department and other law enforcement officials and required providers of Internet phone services and broadband services to ensure that their equipment can allow police wiretaps. This equipment includes voice over Internet (VoIP) technology, which shifts calls away from the traditional phone network of wires and switches to technology based on converting sounds into data and transmitting them over the Internet. The decision has created a controversy among many who fear that opening VoIP to access by law enforcement agencies will create additional points of attack and security holes that hackers can exploit.[15]

USA Patriot Act

The **USA Patriot Act of 2001** (Uniting and Strengthening America by Providing Appropriate Tools Required to Intercept and Obstruct Terrorism) was passed just after the terrorist attacks of September 11, 2001. It gave sweeping new powers both to domestic law enforcement and international intelligence agencies. Although the act was more than 340 pages and quite complex (it changed more than 15 existing statutes), it was passed into law just five weeks after being introduced. Legislators rushed to get the act approved in the House and Senate, arguing that law enforcement authorities needed more power to help track down terrorists and prevent future attacks. Critics argue that the law removes many checks and balances that previously gave courts the opportunity to ensure that law enforcement agencies did not abuse their powers. They also argue that many of its provisions have nothing to do with fighting terrorism.

The Patriot Act contains several "sunset" provisions that terminated on December 31, 2005; these sections increase authority for law enforcement searches and electronic surveillance. (A sunset provision terminates or repeals a law or portions of it after a specific date unless further legislative action is taken to extend the law.) Because the law gave government much more surveillance capability, a sunset date was crucial to determine how well the tools work and how responsibly the government has applied the laws. The provisions that were subject to sunset were primarily those that modified Title III and FISA. As of this writing, President Bush was pressing for permanent renewal of the Patriot Act, but Congress passed only a temporary extension to allow more time to consider protections of civil liberties. Table 4-2 summarizes the key provisions of the Patriot Act that are subject to sunset.

TABLE 4-2   Key provisions of the USA Patriot Act subject to sunset

Section	Issue addressed	Summary
201	Wiretapping in terrorism cases	Added several crimes for which federal courts may authorize wiretapping of people's communications
202	Wiretapping in computer fraud and abuse felony cases	Added computer fraud and abuse to the list of crimes the FBI may obtain a court order to investigate under Title III
203 b	Sharing wiretap information	Allows the FBI to disclose evidence obtained under Title III to other federal officials, including "law enforcement, intelligence, protective, immigration, national defense, [and] national security" officials
203 d	Sharing foreign intelligence information	Provides for disclosure of threat information obtained during criminal investigations to "appropriate" federal, state, local, or foreign government officials for the purpose of responding to the threat
204	FISA pen register/trap-and-trace exceptions	Exempts foreign intelligence surveillance from statutory prohibitions against the use of pen register or trap-and-trace devices, which capture "addressing" information about the sender and recipient of a communication. It also exempts the U.S. government from general prohibitions against intercepting electronic communications and allows stored voice-mail communication to be obtained by the government through a search warrant rather than more stringent wiretap orders.
206	FISA roving wiretaps	Expands FISA to permit "roving wiretap" authority, which allows the FBI to intercept any communications to or by an intelligence target without specifying the telephone line, computer, or other facility to be monitored
207	Duration of FISA surveillance of non-U.S. agents of a foreign power	Extends the duration of FISA wiretap orders relating to an agent of a foreign power from 90 days to 120 days, and allows an extension in 1-year intervals instead of 90-day increments

TABLE 4-2 Key provisions of the USA Patriot Act subject to sunset (continued)

Section	Issue addressed	Summary
209	Seizure of voice-mail messages pursuant to warrants	Enables the government to obtain voice-mail messages under Title III using just a search warrant rather than a wiretap order, which is more difficult to obtain. Messages stored on an answering machine tape, however, remain outside the scope of this section.
212	Emergency disclosure of electronic surveillance	Permits providers of communication services (such as telephone companies and Internet service providers) to disclose consumer records to the FBI if they believe immediate danger of serious physical injury is involved. Communication providers cannot be sued for such disclosure.
214	FISA pen register/trap-and-trace authority	Allows the government to obtain a pen register/trap-and-trace device "for any investigation to gather foreign intelligence information." It prohibits the use of FISA pen register/trap-and-trace surveillance against a U.S. citizen when the investigation is conducted "solely on the basis of activities protected by the First Amendment."
215	FISA access to tangible items	Permits the FBI to compel production of any record or item without showing probable cause. People served with a search warrant issued under FISA rules may not disclose, under penalty of law, the existence of the warrant or the fact that records were provided to the government. It prohibits investigation of a U.S. citizen when it is conducted solely on the basis of activities protected by the First Amendment.
217	Interception of computer trespasser communications	Creates a new exception to Title III that permits the government to intercept the "communications of a computer trespasser" if the owner or operator of a "protected computer" authorizes it. It defines a protected computer as any computer "used in interstate or foreign commerce or communication" (because of the Internet, this effectively includes almost every computer).
220	Nationwide service of search warrants for electronic evidence	Expands the geographic scope where the FBI can obtain search warrants or court orders for electronic communications and customer records
223	Civil liability and discipline for privacy violations	Provides that people can sue the government for unauthorized disclosure of information obtained through surveillance
225	Provider immunity for FISA wiretap assistance	Provides immunity from lawsuits for people who disclose information to the government pursuant to a FISA wiretap order, physical search order, or an emergency wiretap or search

Data Encryption

Cryptography is the science of encoding messages so that only the sender and the intended receiver can understand them. Cryptography was once a mysterious science, employed mainly during wartime, to make it impossible for enemies to understand each others' communications. For example, during World War II, the United States sent radio messages between troops in the native language of the Navajo Indians, a "code" that the Axis powers were unable to crack. Today, cryptography is a key tool for ensuring the confidentiality, integrity, and authenticity of electronic messages and online business transactions.

Encryption is the process of converting an electronic message into a form that can be understood only by the intended recipients. In cryptography, an encryption **key** is a variable value that is applied using an algorithm to encrypt or decrypt text. The length of the key used to encode and decode messages determines the strength of the encryption algorithm. Encryption methods rely on the limitations of computing power for their security—if breaking a code requires too much computing power, even the most determined code crackers will not be successful.

A **public key encryption system** uses two keys to encode and decode messages. One key of the pair, the message receiver's public key, is readily available to the public; anyone can use it to send a person encrypted messages. The second key, the message receiver's private key, is kept secret and is known only by the receiver (see Figure 4-2). Its owner uses the private key to decrypt messages—in other words, to convert an encoded message back into the original message.

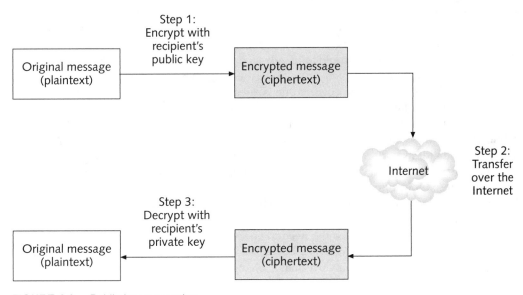

FIGURE 4-2   Public key encryption

For example, if Company A wants to send an encrypted message to Company B, it must first obtain Company B's public key from what is known as a certifying authority, a kind of Department of Motor Vehicles for encryption users. These certifying authorities maintain a list of companies and their public keys. Then, Company A encrypts the message to Company B using Company B's public key. Once the message is encrypted, only Company B can read the message by decrypting it using its private key. The public key-private key combinations are unique, so only one private key can open a message encrypted with its corresponding public key. Obviously, a company must keep its private key well protected to ensure the security of encrypted messages. An analogy to public key encryption is the use of two keys: one that locks your front door, and one that can unlock it. You could keep a "public key" under your doormat to lock your front door, but you would keep possession of the "private key" that could unlock it. Knowing a person's public key does not enable you to decrypt an encoded message to that person.

RSA is a public key encryption algorithm that has been available since 1978. Named for its inventors, Ronald Rivest, Adi Shamir, and Leonard Adleman, RSA is the basis for much of the security that protects Web consumers and merchants. Pretty Good Privacy (PGP) is software that uses 128-bit encryption, the de facto standard for Internet e-mail encryption. The key itself is 128 bits long, meaning that it can represent a total of 2^{128} possible keys.

A **private key encryption system** uses a single key to both encode and decode messages. Obviously, both the sender and receiver must know the key to communicate. It is critical that no one else learns the key, or else all messages between the two can be decoded by others. Indeed, the issue of distributing the private key without others learning its value is a problem in such systems. An analogy to private key encryption is the use of a single key that both locks and unlocks your front door. Whoever possesses the key can gain access to your home. The Digital Encryption Standard (DES), the standard for commercial private key encryption for nearly 30 years, employs a 56-bit key; anyone who wants to crack the code might have to try more than 7.2×10^{16} different combinations.[16] In 1998, computers were able to crack messages sent using the DES code in less than three days. Experts today estimate that it would take only a couple of hours to break the key using state-of-the-art computers and so-called brute force methods, which essentially try every possible key combination until the correct one is guessed. As a result, the National Institute of Standards and Technology (NIST) has proposed that DES lose its certification for use in software products sold to the government. NIST is proposing that federal agencies use the Triple Data Encryption algorithm, also known as Triple DES, or the even stronger and faster Advanced Encryption Standard (AES) algorithm instead.[17] The AES algorithm would require code crackers to try as many as 1.1×10^{77} combinations, which would take current computers more than 149 trillion years to decode a message encrypted using AES.[18] Future advances in computer technology can be expected to render even this code "crackable," but at least not for a decade or two.

Although most people can see the need to encrypt data sent over a public network, many IT managers thought until recently that storing unencrypted data was a better option than encrypting it. The process can involve major changes in how data is stored, accessed, and backed up. Wholesale data encryption can change how applications interact with one another. The management and administration of encryption can become a major headache—what do you do when a user forgets the encryption password? But recent developments are forcing IT managers to reconsider their position, thanks to concern over unauthorized access and use of data by employees, the rapid growth of identity theft, increasing regulatory requirements for data confidentiality, the potential for data theft by insiders, and embarrassing lapses in the handling of backup tapes.[19] Most people now agree that encryption eventually must be built into networks, file servers, and tape backup systems.

For example, in 2005, Seagate Technology announced a hard drive for laptops and other mobile devices that automatically encrypts all data as it enters and exits the drive. Nothing on the drive is accessible unless you know the password; so, even if you lose your laptop, you won't lose your data.[20]

In recognition of the importance of data encryption, the U.S. Arms Export Control Act classified encryption algorithms, hardware, and software as export-controlled material that cannot be released to foreign concerns without government approval. These limitations are designed to prevent the technology from falling into the hands of the nation's enemies. If found guilty of violating the Arms Export Control Act, a person can be jailed for 10 years and fined $1 million. Once extremely restrictive, these limitations have been loosened to increase exports. For example, U.S. companies can export encryption products to customers in the European Union, Australia, the Czech Republic, Hungary, Japan, New Zealand, Norway, Poland, and Switzerland. In spite of these restrictions, intelligence agencies say that terrorist groups such as Osama bin Laden's al Qaeda network have used encryption to protect their phone messages and e-mail.

Identity Theft

Identity theft occurs when someone steals key pieces of personal information to gain access to a person's financial accounts. This information includes name, address, date of birth, Social Security number, passport number, driver's license number, and mother's maiden name. Using this information, an identity thief may apply for new credit or financial accounts, rent an apartment, set up utility or phone service, and register for college courses—all in someone else's name.

Although estimates of the number of incidents vary greatly, identity theft is widely recognized as the fastest growing form of fraud in the United States. Consumer Sentinel, the complaint database developed and maintained by the FTC, received 246,000 identity theft complaints in 2004.[21] However, other sources estimated that the number of victims was much higher—as many as 10 million per year.[22] Taking on another identity can be easy and extremely lucrative, enabling the thief to use a victim's credit cards, siphon off money from bank accounts, and even obtain Social Security benefits. The victim is left with a credit history in shambles. In addition, the average victim spends at least 600 hours over several years recovering from identity theft, according to a 2003 survey by the San Diego-based Identity Theft Resource Center.[23]

An alarming number of incidents involve the hacking of large databases to gain personal identity information. Here is a partial list of major incidents from 2005:

- In February 2005, Check Point, keeper of more than 19 million public records, revealed that hackers stole data on more than 147,000 consumers.[24]
- In March 2005, Reed Elsevier, the parent company of LexisNexis, announced that hackers had compromised its massive databases, stealing information on more than 300,000 people.[25]
- In March 2005, Retail Ventures Inc. reported the theft of credit card data and other personal information of 1.4 million customers from its DSW Store Warehouse stores.[26]
- In March 2005, Bank of America disclosed that it lost computer tapes containing credit card account records of 1.2 million federal employees.[27]

- Visa USA and American Express announced that they were terminating their contract with CardSystems Solutions after a hacker accessed as many as 40 million credit card numbers from the company's computers in June 2005.[28]
- In June 2005, computer tapes that contained personal information from 3.9 million CitiFinancial Branch Network customers was lost by UPS while in transit to a credit bureau.[29]

Not only is the number of incidents alarming, so is the lack of initiative by some companies in informing people whose data was stolen. Executives from LexisNexis and ChoicePoint revealed in a Senate Judiciary Committee hearing that they did not report some data breaches to potential victims until a California law requiring notification went into effect in 2003.[30]

Three approaches that are frequently used by identity thieves include hacking databases, phishing, and spyware. **Phishing** is an attempt to steal personal identity data by tricking users into entering the information on a counterfeit Web site; this data includes credit card numbers, account usernames, passwords, and Social Security numbers. Spoofed e-mails, such as the one shown in Figure 4-3, lead consumers to counterfeit Web sites designed to trick them into divulging personal data. Savvy users become suspicious and refuse to enter data into the fake Web sites, but often, just accessing the Web site can trigger an automatic and transparent download of malicious software to a computer. eBay Inc. and its online payment unit PayPal are the two Web sites that phishers spoof most frequently, and Citibank is the most frequent target among banks.[31] Spear-phishing is a variation in which employees are sent phony e-mails that look like they came from high-level executives within their organization. Employees are again directed to a fake site disguised to look like a corporate Web site and then asked to enter personal information.[32] In a 12-month period beginning in May 2004, the number of people who received phishing e-mail grew 28 percent; about 1.2 million U.S. consumers suffered phishing-related losses of more than $900 million in the same period.[33]

Spyware is a term for keystroke-logging software that is downloaded to users' computers without adequate notice, consent, or control for the user.[34] It is often promoted as a spouse monitor, child monitor, or surveillance tool. Spyware creates a record of the keystrokes entered on the computer, enabling the capture of account usernames, passwords, credit card numbers, and other sensitive information. The spy can even view the Web sites visited and transcripts of chat logs. Spyware operates even if the infected computer is not connected to the Internet, continuing to record each keystroke until the next time the user connects to the Internet. Then, the data captured by the spyware is e-mailed directly to the spy or is posted to a Web site where the spy can view it. Spyware frequently employs sophisticated methods to avoid detection by popular software packages that are specifically designed to combat it. Consumers' fear of spyware has become so widespread that many people now delete e-mail from unknown sources without even opening the messages. This trend is seriously damaging the effectiveness of e-mail as a means for legitimate companies to communicate with customers.[35]

Congress passed the Identity Theft and Assumption Deterrence Act of 1998 (full text at *www.ftc.gov/os/statutes/itada/itadact.htm*) to fight identity fraud, making it a federal felony punishable by a prison sentence of three to 25 years. The act also appoints the FTC to help victims restore their credit and erase the impact of the imposter. Although

FIGURE 4-3 E-mail used by phishers

people have been convicted under this act, researchers estimate that fewer than one in 700 identity crimes leads to a conviction.[36]

Consumer Profiling

Companies openly collect personal information about Internet users when they register at Web sites, complete surveys, fill out forms, or enter contests online. Many companies also obtain information about Web surfers without their manual input through the use of **cookies**, a text file that a Web site puts on your hard drive so that it can remember your information later. Companies also use tracking software to allow their Web sites to analyze browsing habits and deduce personal interests and preferences. The use of cookies and tracking software is controversial because companies can collect information about consumers without their explicit permission. Outside of the Web environment, marketing firms employ similarly controversial means to collect information about people and their buying habits. Each time a consumer uses a credit card, redeems frequent flyer points, fills out a warranty card, answers a phone survey, buys groceries using a store loyalty card, orders from a mail-order catalog, or registers a car with the DMV, the data is added to a storehouse of personal information about that consumer. In none of these cases do consumers explicitly consent to submitting their information to a marketing organization.

Marketing firms aggregate the information they gather about consumers to build databases that contain a huge amount of consumer behavioral data. They want to know as much as they can about consumers—who they are, what they like, how they behave, and what motivates them to buy. The marketing firms provide this data to companies so they can tailor their products and services to individual consumer preferences. Advertisers use the data to more effectively target and attract customers to their messages. Ideally, this means that buyers should be able to shop more efficiently and find products that are well suited for them. Sellers should be better able to tailor their products and services to meet their customers' desires and to increase sales. However, concerns about how all this data is actually used is a major reason why many potential Web shoppers have not yet made online purchases.

Large-scale marketing organizations such as DoubleClick employ advertising networks to serve ads to thousands of Web sites. When someone clicks on an ad at a company's Web site, tracking information about the person is gathered and forwarded to DoubleClick, where it is stored in a large database. This data includes a record of what the person clicked and what the person bought. A group of Web sites served by a single advertising network is called a collection of **affiliated Web sites**.

Marketers use cookies to recognize return visitors to their sites and to store useful information about them. The goal is to provide customized service for each consumer. When someone visits a Web site, the site "asks" that person's computer if it can store a cookie on the hard drive. If the computer agrees, it is assigned a unique identifier and a cookie with this identification number is placed on its hard drive. During a Web-surfing session, three types of data are gathered. First, as one browses the Web, "GET" data is collected. **GET data** reveals, for example, that the consumer visited an affiliated book site and requested information about the latest Dean Koontz book. Second, "POST" data is captured. **POST data** is entered into blank fields on an affiliated Web page when a consumer signs up for a service, such as the Travelocity service that sends an e-mail when airplane fares change for flights to favorite destinations. Third, the marketer monitors the consumer's surfing throughout any affiliated Web sites, keeping track of the information the user sought and viewed. This is known as **click-stream data**. Thus, as a single person surfs the Web, a tremendous amount of data is generated for marketers and sellers.

After cookies have been stored on your computer, they work behind the scenes whenever you surf the Web later, searching for information about you in network advertising databases. If a match is found, the information stored there about you can be used to tailor the ads and promotions presented as you browse Web sites. The marketer also knows what ads have been seen most recently and makes sure that they aren't seen again, unless the advertiser has decided to market using repetition. The marketer also tracks what sites are visited and uses the data to make educated guesses about the kinds of ads that would be most interesting to you.

There are four ways to limit or even stop the deposit of cookies on a hard drive: browsers can be set to limit or stop cookies, users can manually delete them from a hard drive, consumers can download and install a cookie-management program for free, or they can use anonymous browsing programs that don't accept cookies. (For example, *www. anonymizer.com* offers anonymous surfing services; by switching on the Anonymizer privacy button or going through its Web site, you can hide your identity from nosy Web sites.) However, an increasing number of Web sites lock visitors out unless they allow cookies to be deposited on their hard drives.

In addition to using cookies to track consumer data, marketers use **personalization software** to optimize the number, frequency, and mixture of their ad placements. It also is used to evaluate how visitors react to new ads. The goal is to turn first-time visitors to a site into paying customers and to facilitate greater cross-selling activities.

There are several types of personalization software. For example, **rules-based** personalization software uses business rules that are tied to customer-provided preferences or online behavior to determine the most appropriate page views and product information to display. For instance, if you use a Web site to book airline tickets to a popular vacation spot, rules-based software might ensure that you are shown ads for rental cars. **Collaborative filtering** offers consumer recommendations based on the types of products purchased by other people with similar buying habits. For example, if you bought a book by Dean Koontz, a company might recommend Stephen King books to you, based on the fact that a significant percentage of other customers have bought books by both authors.

Demographic filtering is another form of personalization software. It augments clickstream data and user-supplied data with demographic information associated with user zip codes to make product suggestions. Microsoft has captured age, sex, and location information for years through its various Web sites, including MSN and Hotmail. It has accumulated a vast database on tens of millions of people, each assigned a global user ID. Microsoft has also developed a technology based on this database that enables marketers to target one ad to men and another to women. Additional information such as age and location can be used as selection criteria.[37]

Yet another form of personalization software, **contextual commerce**, associates product promotions and other e-commerce offerings with specific content a user may receive in a news story online. For example, as you read a story about white-water rafting, you may be offered deals to buy rafting gear or a promotion for a vacation in West Virginia, a state with many rivers suited for white-water rafting. Instead of simply bombarding customers at every turn with standard sales promotions that get tiny response rates, marketers are getting smarter about where and how they use personalization. They are also taking great care to measure whether personalization is paying off. The intended result is that effective personalization increases online sales and improves consumer relationships.

Online marketers cannot capture personal information, such as names, addresses, and Social Security numbers, unless people provide them. Without this information, companies can't contact individual Web surfers who visit their sites. Data gathered about a user's Web browsing through the use of cookies is anonymous, as long as the network advertiser doesn't link the data with personal information. However, if a Web site visitor volunteers personal information, a Web site operator can use it to find additional personal information that the visitor may not want to disclose. For example, a name and address can be used to find a corresponding phone number, which can then lead to obtaining even more personal data. All this information becomes extremely valuable to the Web site operator, who is trying to build a relationship with Web site visitors and turn them into customers. The operator can use this data to initiate contact or sell it to other organizations with which they have marketing agreements.

Consumer data privacy has grown into a major marketing issue. Companies that can't protect or don't respect customer information have lost business and become defendants in class actions stemming from privacy violations. For example, privacy groups spoke out vigorously to protest the proposed merger of the Web ad server DoubleClick and the consumer database Abacus. The groups were concerned that the information stored in cookies would be combined with data from mailing lists, thus revealing the Web users' identities. This would enable a network advertiser to identify and track the habits of unsuspecting consumers. Public outrage and the threat of lawsuits forced DoubleClick to back off this plan.

Opponents of consumer profiling are also concerned that personal data is being gathered and sold to other companies without the permission of consumers who provide the data. After the data has been collected, consumers have no way of knowing how it is used or who is using it. For example, when Toysmart.com went bankrupt in 2000, it planned to sell customer information from its Web site to the highest bidder, to earn cash to pay its employees and creditors. This data included names, addresses, and ages of customers and their children. TRUSTe had licensed Toysmart.com to put the TRUSTe privacy seal on its Web site, provided that Toysmart.com never divulged customer information to a third party. Because Toysmart.com was planning to violate that agreement, TRUSTe submitted a legal brief asking the bankruptcy court to withhold its approval for the proposed sale. TRUSTe officials also registered a complaint with the FTC, which launched an investigation and then filed suit to stop Toysmart.com from selling its customer list and related information, in violation of the privacy policy that appeared on the company's Web site. Finally, Walt Disney Company, which owned 60 percent of Toysmart.com, bought the list and "retired it" to protect customers' privacy and to put an end to the controversy.

One potential solution to consumer privacy concerns is a screening technology called the **Platform for Privacy Preferences (P3P)**, which shields users from sites that don't provide the level of privacy protection they desire. Instead of forcing users to find and read through the privacy policy for each site they visit, P3P software in a computer's browser will download the privacy policy from each site, scan it, and notify users if the policy does not match their preferences. (Of course, unethical marketers can post a privacy policy that does not accurately reflect how data is treated.) The World Wide Web Consortium, an international industry group whose members include Apple, Commerce One, Ericsson, and Microsoft, created P3P and is supporting its development.

Treating Consumer Data Responsibly

When dealing with consumer data, strong measures are required to avoid customer relationship problems. The most widely accepted approach to treating consumer data responsibly is for a company to adopt the Code of Fair Information Practices and the 1980 OECD privacy guidelines. Under these guidelines, an organization collects only personal information that is necessary to deliver its product or service. The company ensures that the information is carefully protected and accessible only by those with a need to know, and that consumers can review their own data and make corrections. The company informs customers if it intends to use customer information for research or marketing, and it provides a means for them to opt out.

An increasing number of companies are also appointing executives to oversee their data privacy policies and initiatives. As a result of this increased focus, companies recognize the need to establish corporate data privacy policies. Some companies are appointing a **chief privacy officer (CPO)**, while others are assigning these duties to other senior managers. An effective CPO can avoid violating government regulations and reassure customers that their privacy will be protected. This requires the organization to give the CPO power to stop or modify major company marketing initiatives, if necessary. The CPO's general duties include training employees about privacy, checking the company's privacy policies for potential risks, figuring out if gaps exist and how to fill them, and developing and managing a process for customer privacy disputes.

The CPO should be briefed on *planned* marketing programs, information systems, or databases that involve the collection or dissemination of consumer data. The rationale for early involvement in such initiatives is to ensure that potential problems can be identified in the earliest stages, when it is easier and cheaper to fix them. Some organizations fail to address privacy issues early on, and it takes a negative experience to make them appoint a CPO. For example, United States Bancorp, a bank with more than $86 billion in assets, appointed a CPO, but only after spending $3 million to settle a lawsuit that accused the bank of selling confidential customer financial information to telemarketers. Table 4-3 provides useful guidance for ensuring that your organization treats consumer data responsibly. The preferred answer to each question is *yes*.

TABLE 4-3   Manager's checklist for treating consumer data responsibly

Questions	Yes	No
Do you have a written data privacy policy that is followed?	——	——
Can consumers easily view your data privacy policy?	——	——
Are consumers given an opportunity to opt in or opt out of your data policy?	——	——
Do you collect only the personal information needed to deliver your product or service?	——	——
Do you ensure that the information is carefully protected and accessible only by those with a need to know?	——	——
Do you provide a process for consumers to review their own data and make corrections?	——	——
Do you inform your customers if you intend to use their information for research or marketing and provide a means for them to opt out?	——	——
Have you identified a person who has full responsibility for implementing your data policy and dealing with consumer data issues?	——	——

Workplace Monitoring

As discussed in Chapter 2, many organizations have developed a policy on the use of IT to protect against employee abuses that reduce worker productivity or that could expose the employer to harassment lawsuits. The institution and communication of an IT usage policy establishes boundaries of acceptable behavior and enables management to take action against violators.

The potential for decreased productivity, coupled with increased legal liabilities from computer users, have led employers to monitor workers to ensure that the corporate IT usage policy is followed. More than 80 percent of major U.S. firms find it necessary to record and review employee communications and activities on the job, including phone calls, e-mail, Internet connections, and computer files. Some are even videotaping employees on the job. In addition, some companies employ random drug testing and psychological testing. With few exceptions, these increasingly common (and many would say intrusive) practices are perfectly legal.

The Fourth Amendment of the Constitution protects citizens from unreasonable government searches and is often invoked to protect the privacy of government employees. Public-sector workers can appeal directly to the "reasonable expectation of privacy" standard established by the Supreme Court ruling in *Katz v. United States* in 1998.[38]

However, the Fourth Amendment cannot be used to limit how a private employer treats its employees, because such actions are not taken by the government. As a result, public-sector employees have far greater privacy rights than those in private industry. Although private-sector employees can seek legal protection against an invasive employer under various state statutes, the degree of protection varies widely by state. Furthermore, state privacy statutes tend to favor employers over employees. For example, to sue successfully, employees must prove that they were in a work environment where they had a reasonable expectation of privacy. As a result, courts typically rule against employees who file privacy claims for being monitored while using company equipment. A company can defeat a privacy claim simply by proving that an employee had been given explicit notice that e-mail, Internet use, and files on company computers were not private and that their use might be monitored. When an employer engages in workplace monitoring, though, it must ensure that it treats all types of workers equally. For example, a company could get into legal trouble for punishing an hourly-paid employee more seriously for visiting inappropriate Web sites than it punished a monthly-paid employee.

Society is struggling to define the extent to which employers should be able to monitor the work-related activities of employees. On one hand, employers must be able to guarantee a work environment that is conducive to all workers, ensure a high level of worker productivity, and avoid the costs of defending against "nuisance" lawsuits. On the other hand, privacy advocates want federal legislation that keeps employers from infringing upon the privacy rights of employees. Such legislation would require prior notification to all employees of the existence and location of all electronic monitoring devices. Advocates also want restrictions on the types of information collected and the extent to which an employer may use electronic monitoring. As a result, many bills are being introduced and debated at both the state and federal level. As the laws governing employee privacy and monitoring continue to evolve, business managers must keep informed to avoid enforcing outdated usage policies. Organizations with global operations face an even bigger challenge, because the legislative bodies of other countries also debate these issues.

Spamming

Spamming is the transmission of the same e-mail message to a large number of people. Some people consider spam a nuisance and an invasion of privacy, but most of it is commercial advertising, sometimes for questionable products such as pornography and phony get-rich-quick schemes. For example, within the Usenet discussion system, in which Internet users can read and post messages to a number of distributed newsgroup categories, spammers frequently send the same message to dozens of newsgroups, some of which contain more than 1000 members. Spammers also target individual users with direct e-mail messages, building their mailing list by scanning Usenet postings, buying mailing lists, or searching the Web for addresses. Based on analysis by the message-filtering vendor CipherTrust, it blocked 86 percent of all spam that originated in the United States for its 1000-plus clients during May, June, and July 2004. China and Hong Kong, where about 23 percent of all spamming IP addresses reside, accounted for less than 3 percent of all spam.[39]

Spam is actually an extremely inexpensive method of marketing that is used by many legitimate organizations. For example, some companies might send e-mail to a broad cross-section of potential customers to announce the release of a new product in an attempt to increase initial sales. The cost of creating an e-mail campaign for a product or service is close to $1000, compared to tens of thousands of dollars for direct-mail campaigns. In addition, e-mail campaigns take only three weeks to develop, compared with three months for direct mail, and the turnaround time for feedback averages 48 hours for e-mail, as opposed to three weeks for direct mail. However, the benefits of spam to companies can be largely offset by the public's generally negative perception for receiving unsolicited ads.

Spam forces sometimes unwanted and objectionable materials into e-mail boxes, detracts from the ability of Internet users to communicate effectively because of full mailboxes and relevant e-mails hidden among many unsolicited messages, and costs Internet users and service providers millions of dollars annually. Usenet members can be overwhelmed with a barrage of spam and become frustrated that their Usenet group seems to have lost focus. It takes users time to scan and delete spam e-mail, a cost that can add up if they pay for Internet connection charges on an hourly basis. It also costs money for ISPs and online services to transmit spam, which are reflected in the rates charged to subscribers.

The Controlling the Assault of Non-Solicited Pornography and Marketing (CAN-SPAM) Act went into effect in January 2004. The act says it is legal to spam, provided the messages meet a few basic requirements—spammers cannot disguise their identity by using false return addresses, there must be a label in the message specifying that it is an ad or solicitation, and such e-mails must include a way for recipients to indicate they do not want future mass mailings. Not only has the CAN-SPAM Act failed to slow the flow of junk e-mail, but some believe that it actually has increased the flow of spam, because it legalizes the sending of unsolicited e-mail.[40]

For example, White Buffalo Ventures operates LonghornSingles.com, an online dating service. When the University of Texas started blocking unsolicited e-mail messages from the firm, the two organizations went to court. The University of Texas argued that White Buffalo was part of a larger spam problem that had crashed the university's computer system, and that it was responding to complaints from students and faculty. White Buffalo complained that the university violated its constitutional rights by filtering out the e-mails. It also claimed that its e-mail conformed to the requirements of the CAN-SPAM Act, which allows certain e-mails and superseded the university's antispam policy. The Fifth U.S. Circuit Court of Appeals ruled that the university did not violate the constitutional rights of the online dating service by blocking the e-mails. It further ruled that the CAN-SPAM Act does not preempt the university's policy and that the policy is permissible under the First Amendment.[41]

In August 2005, Scott Richter, the self-proclaimed "king of spam," agreed to pay Microsoft $7 million to settle a two-year lawsuit over billions of unsolicited e-mails with misleading subject lines and forged sender addresses in violation of state and federal law. Richter and his affiliates also agreed to comply with federal antispam laws and to stop sending spam to anyone who did not opt in to receive marketing e-mails.[42]

Advanced Surveillance Technology

A number of advances in information technology, such as surveillance cameras, facial recognition software, and systems that can pinpoint a person's position, provide exciting new data-gathering capabilities. However, these advances can also diminish individual privacy and complicate the issue of how much information should be captured about people's private lives.

Advocates of advanced surveillance technology argue that people have no legitimate expectation of privacy in a public place. Critics raise concerns about the use of surveillance to secretly store images of people, creating a new potential for abuse, such as intimidation of political dissenters or blackmail of people caught with the "wrong" person or in the "wrong" place. Critics also raise the possibility that such technology may not identity people accurately.

Camera Surveillance

Although it did not prevent terrorist bombings in July 2005, London has one of the world's largest public surveillance systems—the average person there might be photographed by 300 cameras in the course of a day. A number of U.S. cities plan to expand their surveillance systems accordingly. Chicago, which has the largest public video surveillance system in the United States, is expanding its 2000-camera network and encouraging businesses to provide the city live feeds from their surveillance cameras.[43] The New York City Metropolitan Transportation Authority awarded a $212 million contract to place 1000 video cameras and 3000 motion sensors in its subways, in an effort to implement enough obstacles to deter potential attackers of the nation's largest transit system[44] (see Figure 4-4).

A "smart surveillance system," which singles out people who are acting suspiciously, is under development in Australia. The system is based on computers that learn what "normal" behavior is and then look for patterns of behavior outside the norm. When the system detects unusual behavior, it alerts authorities so they can take preemptive action.[45]

Facial Recognition Software

There have been numerous experiments with facial recognition software to help identify criminal suspects and other undesirable characters. These experiments have yielded mixed results. The Rampart Division of the Los Angeles Police Department tested the software over a two-month period in 2004, resulting in 19 arrests and clearing one man the officers had suspected of being someone else. Officials in Tampa, Florida, stopped using it in 2003 because it didn't result in arrests. And at Boston's Logan International Airport in 2002, two systems failed 96 times to identify people who volunteered to help test it. The technology correctly identified 153 other volunteers.[46]

FIGURE 4-4   A crowd in a public place

GPS Chips

Global Positioning System (GPS) chips are being placed in many devices, from automobiles to cell phones, to precisely locate users. The FCC has asked cell-phone companies to implement methods for locating users so police, fire, and medical professionals can be accurately dispatched to 911 callers. Similar location-tracking technology is also available for personal digital assistants, laptop computers, trucks, and boats. Parents place these chips in their children's cars and use software that allows them to track the cars' whereabouts.

As a consequence, banks, retailers, and airlines are eager to gain real-time access to consumer location data, and have already devised a number of new services they want to provide—sending digital coupons for stores that particular consumers are near, providing the location of the nearest ATM, and updating travelers on flight and hotel information. Airlines are considering the use of wireless devices to enable passengers to check in for flights when they are close enough to the gate and then to monitor when the person passes through the gate.

Businesses claim that they will respect the privacy of wireless users and allow them to opt in or opt out of marketing programs that are based on their location data. Wireless spamming is a distinct possibility—a user might continuously receive wireless ads, notices for local restaurants, and shopping advice while walking down the street. Another concern is that the data could be used to track people down at any time or to figure out where they had been at some particular instant. The potential to reveal one's location when using a cell phone will cause some people to reconsider using one in the future.

Summary

1. What is the right of privacy, and what is the basis for protecting personal privacy under the law?

    As a legal concept, the right to privacy has four aspects: protection from unreasonable intrusion upon one's isolation, protection from appropriation of one's name or likeness, protection from unreasonable publicity given to one's private life, and protection from publicity that unreasonably places one in a false light before the public.

    A number of laws have been enacted over the past 40 years that affect a person's privacy; most of these actions address invasion of privacy by the government rather than by private industry. In addition, there is no overarching national data privacy policy, so current legislation is sometimes inconsistent and even conflicting. This legislation includes the Communications Act of 1934, the Freedom of Information Act of 1966, the Privacy Act of 1974, the Electronic Communications Privacy Act of 1986, the 1994 Communications Assistance for Law Enforcement Act, the Children's Online Protection Act of 1998, the European Community Directive 95/46/EC of 1998, the 1998 Gramm-Leach-Bliley Act, and the 2001 Patriot Act.

2. What are some of the laws that authorize electronic surveillance by the government, and what are the associated ethical issues?

    The Federal Wiretap Act outlines processes to obtain court authorization for surveillance of all kinds of electronic communication, including e-mail, fax, Internet, and voice, in criminal investigations. The Electronic Communications Privacy Act of 1986 set standards for access to stored e-mail and other electronic communications and records. The Foreign Intelligence Surveillance Act of 1978 allows wiretapping of aliens and citizens in the United States if probable cause is found that the target is a member of a foreign terrorist group or an agent of a foreign power. Executive Order 12333 is the legal authority for electronic surveillance outside the United States; it permits intelligence agencies to intercept communications outside the country without a court order. The Communications Assistance for Law Enforcement Act requires the telecommunications industry to build tools into its products that federal investigators can use—after getting court approval—to eavesdrop on conversations. The 2001 Patriot Act gave sweeping new powers to both domestic law enforcement and international intelligence agencies. Each of these acts was intended to authorize activities by law enforcement and intelligence agencies to better protect U.S. citizens. However, each act also raised considerable ethical issues of protecting individual privacy and the potential for misuse.

3. What are the two fundamental forms of data encryption, and how does each work?

    Data encryption is an essential tool for ensuring confidentiality, integrity, and authenticity of messages and business transactions. A public key encryption system uses two keys to encode and decode messages. One key of the pair, the message receiver's public key, is readily available to the public; anyone can use the key to send the person encrypted messages. The second key, the message receiver's private key, is kept secret and is known only by the message receiver. Its owner uses the private key to decrypt messages. A private key encryption system uses a single key both to encode and decode messages. Both the sender and receiver must know the key to communicate. It is critical that no one else learns the key, or all messages between the two can be decoded by others.

The development of faster computers has increased the importance of developing stronger encryption methods that cannot be cracked.

4. What is identity theft, and what techniques do identity thieves use?

    Identity theft occurs when someone steals key pieces of personal information to gain access to a person's financial accounts, apply for new credit or financial accounts, apply for loans or Social Security benefits, and create other obligations in someone else's name. The number of incidents of identity theft are increasing at a rapid pace. Identity thieves often resort to hacking databases, phishing, and using spyware to obtain identity information.

5. What are the various strategies for consumer profiling and the associated ethical issues?

    Marketing firms capture data from numerous sources to build databases that detail a large amount of consumer behavior data. Online marketers can place an electronic cookie on a user's hard drive so they can recognize consumers when they return to a Web site. Cookies can also be used to store information about consumers. Marketers employ personalization software to analyze a user's browsing habits and deduce their personal interests and preferences. The use of cookies and personalization software is controversial because it enables companies to discover information about consumers and potentially to share it with other companies without consumers' explicit permission. On the positive side, the use of such data enables organizations to customize the site visitor's experience and make it more convenient and effective.

6. What must organizations do to treat consumer data responsibly?

    A widely accepted approach to treating consumer data responsibly is the adoption of the Code of Fair Information Practices and the 1980 OECD privacy guidelines. Under these guidelines, an organization collects only the personal information that is necessary to deliver its product or service. The company ensures that the information is carefully protected and is accessible only by those with a need to know, and that it provides a process for consumers to review their own data and make corrections. In addition to adopting the OECD guidelines, many companies are appointing CPOs or other senior managers to oversee their data privacy policies and initiatives.

7. Why and how are employers increasingly using workplace monitoring?

    Employers record and review employee communications and activities on the job, including monitoring workers' phone calls, e-mail, Internet connections, and computer files. Society is struggling to define the extent to which employers should be able to monitor employees. On one hand, employers must be able to guarantee a work environment that is conducive to all workers, ensure a high level of worker productivity, and avoid the costs of defending against "nuisance" lawsuits. On the other hand, privacy advocates want federal legislation that keeps employers from infringing upon the privacy rights of employees.

8. What is spamming, and what ethical issues are associated with its use?

    Spamming is the transmission of the same e-mail message to a large number of people. Some people consider spam a nuisance and an invasion of privacy, but it is actually an extremely inexpensive method of marketing used by many legitimate organizations. On the other hand, spam forces unwanted and sometimes objectionable material into e-mail boxes, detracts from an Internet user's ability to communicate effectively, and costs Internet users and service providers millions of dollars annually.

9. What are the capabilities of advanced surveillance technologies, and what ethical issues do they raise?

A number of advances in information technology, such as surveillance cameras, facial recognition software, and GPS systems that can pinpoint a person's position, provide exciting new security capabilities. However, these advances also can diminish individual privacy and complicate the issue of how much information should be captured about people's private lives—where they are, who they are, and what they do behind closed doors.

Self-Assessment Questions

1. People are entitled to their privacy:
   a. at all times and under all conditions
   b. when there is a reasonable expectation of privacy
   c. while surfing the Internet
   d. at work

2. Legislation that protects people from data privacy abuses by private industry are almost nonexistent. True or False?

3. The _____ Act provides the public with the means to gain access to certain government records.

4. Web sites that cater to children must conform to a different set of data collection practices from those that cater strictly to adults. True or False?

5. The _____ philosophy of addressing privacy concerns with strict government regulation differs greatly from the _____ philosophy of having no federal policy and only self-regulation.

6. Judges frequently deny government requests for Title III wiretaps. True or False?

7. Which of the following identifies the numbers dialed for outgoing calls?
   a. pen register
   b. wiretap
   c. trap and trace
   d. gag and bound

8. _____ is the science of encoding messages so that only the sender and the intended receiver can understand them.

9. Which of the following is *not* a technique frequently employed by identity thieves?
   a. hacking databases
   b. spyware
   c. phishing
   d. trap and trace

10. In addition to using cookies to track consumer data, _____ is used by marketers to optimize the number, frequency, and mixture of their ad placements.

11. More than 80 percent of major U.S. firms record and review employee communications and activities on the job. True or False?

12. CAN-SPAM legalizes the sending of unsolicited e-mail. True or False?

Review Questions

1. Which Amendment of the U.S. Constitution addresses the right of privacy?

2. What is spyware? How is it used?

3. What is an opt-in policy? What is an opt-out policy?

4. Why do marketers engage in consumer profiling?

5. What benefits can consumer profiling provide to the consumer?

6. What is the difference between a pen register and trap and trace? What act allows these?

7. Identify and briefly describe three types of personalization software.

8. What is phishing? What is its intent?

9. Why is it said that public workers enjoy greater privacy protection than private-industry workers?

10. What are some legitimate business benefits of using spam to reach potential customers?

11. What is the FOIA? Why do whistle-blowers frequently use it?

12. What are the privacy guidelines set by the OECD? Why are they considered significant?

13. What is data encryption? What is the AES algorithm?

14. What is the role of a CPO?

15. Briefly distinguish between private key and public key encryption.

Discussion Questions

1. What laws and legal rulings provide the basis for the right of privacy?

2. Compare and contrast the European Union and United States' philosophies on data privacy. Which approach do you think is better? Why?

3. What benefits can consumer profiling provide to you as a consumer? Do these benefits outweigh the loss of your privacy?

4. How much effort should a Web site operator take to prevent preteen visitors who lie about their age from visiting their adult-oriented Web site? Should Web site operators be prosecuted under the COPA Act if preteens lie about their age and provide personal information on such sites?

5. A FOIA exemption prevents disclosure of records if it would invade someone's personal privacy. Develop a hypothetical example in which a person's privacy interests are clearly outweighed by competing public interests. Develop a second hypothetical example in which a person's privacy interests are not outweighed by competing public interests.

6. Summarize the Fourth Amendment of the U.S. Constitution. Does it apply to the actions of private industry?

7. Do you feel that information systems to fight terrorism should be developed and used even if they infringe on privacy rights or violate the Privacy Act and other such statutes?

8. Research the Web to find the latest developments on the Platform for Privacy Preferences (P3P). Write a short report summarizing your findings.

9. Why do employers monitor workers? Do you think they should be able to do so? Why or why not?

10. Do you think that law enforcement agencies should be able to use advanced surveillance technology, such as surveillance cameras combined with facial recognition software? Why or why not?

What Would You Do?

1. Your friend is considering using an online service to identify people with compatible personalities and attractive physical features who would be interesting to date. First, your friend must submit some basic personal information, then complete a five-page personality survey, and finally provide a recent photo. Would you advise your friend to do this? Why or why not?

2. As the information systems manager for a small manufacturing plant, you are responsible for all aspects of the use of information technology. A new inventory control system is being implemented to track the quantity and movement of all finished products stored in a local warehouse. Each time forklift operators move a case of product, they must first scan the UPC code on the case. The product information is captured, as well as the day, time, and forklift operator identification. This data is transmitted over a LAN to the inventory control computer, which then displays information about the case and where it should be placed in the warehouse.

   The warehouse manager is excited about using case movement data to monitor worker productivity. He will be able to tell how many cases per shift each operator moves, and he plans to use this data to provide performance feedback that could result in pay increases or termination. He has asked you if there are any potential problems with using the data in this manner, and, if so, what should be done to avoid them. How would you respond?

3. You are a writer for a tabloid magazine and want to get some headline-grabbing news about the stars of a popular TV show. You decide to file a separate Freedom of Information Act request for each of the show's stars with the FBI. Would you consider this an ethical approach to getting the information you want? (As a writer for a gossip magazine, would you care?) Do you think that the FBI would honor your request?

4. You are a new brand manager for Coach purses. You are considering the use of spam to promote the latest line of purses, which are targeted to young, affluent adults. List the advantages and disadvantages of such a marketing strategy. Would you recommend this means of promotion in this instance? Why or why not?

5. You are the CPO of a midsized manufacturing company, with sales of more than $250 million per year and almost $50 million from Internet-based sales. You have been challenged by the vice president of sales to change the company's Web site data privacy policy from an opt-in policy to an opt-out policy and to allow the sale of customer data to other companies. The vice president has estimated that this change would bring in at least $5 million per year in added revenue with little additional expense. How would you respond to this request?

Cases

1. Privacy Concerns Curtail Crime-Solving Database

In January 2002, the federal government earmarked $12 million for the development of the Multistate Anti-Terrorism Information Exchange (Matrix), a database that would collect and analyze information on potential terrorists and criminals. A number of states agreed to participate, and by the fall, authorities credited the Matrix database with helping to catch the Washington D.C. sniper. Other states quickly signed on, but not without attracting the attention of privacy groups, including the powerful American Civil Liberties Union (ACLU).

On October 30, 2003, the ACLU filed Freedom of Information Act requests in Ohio, Michigan, Connecticut, Pennsylvania, and New York, pertaining to the states' participation in the Matrix project. Dubbing Matrix a "surveillance program," the ACLU released an issue brief claiming that the database was an Orwellian project designed to achieve the same "Big Brother" aims as the Pentagon's Total Information Awareness (TIA) system. Congress had shelved the TIA system in September 2003 in response to intense pressure by privacy advocates. Their chief argument against TIA and Matrix is that data mining is dangerous.

Both systems gather data from existing government and commercial databases that store personal information about individual citizens. Although law enforcement officials insist that they use Matrix to provide fast access to data on potential criminals, Seisint, the company that created the database, originally boasted that Matrix would be able to root out terrorists.

Matrix can perform mathematical analyses of data to produce a "high terrorism factor" scoring system. In fact, Seisint handed over a list of 120,000 of the top-scoring names to the federal government—of the top 80 names, five were hijackers who participated in the September 11 attacks, and 45 were "potential terrorists" already under investigation.

This was no mean accomplishment, but Matrix has privacy advocates up in arms. After the CIA was accused of spying on civil rights activists in the 1970s, Congress prohibited the agency from monitoring innocent civilians. Matrix, the privacy advocates claim, is in clear violation of this policy. With one keystroke, they fear, this powerful program will create false positives: bad data or program errors can suddenly make an innocent person guilty.

However, Matrix supporters argue that law enforcement must have effective tools to catch criminals if they are to protect the public. In a world of increasing terrorist threats, the general public might well side with the system's supporters, especially considering that Matrix has already achieved considerable success.

So, why had all but four states pulled out of the project by mid-2005? The answer is simple. No official in IT or law enforcement could guarantee the security of the data. A virus attack, a break-in, or an unscrupulous employee could cause sensitive data to fall into the wrong hands. Furthermore, Matrix is operated by a private company and hence lacks government oversight or public accountability. For example, officials might request a person's dossier that is created and transmitted by Seisint employees, who themselves have not been cleared. Matrix had been up and running for more than a year before background checks were run on its employees.

Yet, Florida, Connecticut, Pennsylvania, and Ohio continue the effort to keep Matrix alive. As these state governments iron out the security kinks in the system, other states may join them, walking the narrow line between privacy concerns and the need for protection.

Questions:

1. Matrix uses information that has already been collected and is being used for marketing and other purposes. Why did privacy advocates suddenly sound the alarm when the same data used by marketing groups was accessed by law enforcement officials?

2. How have the Freedom of Information Act and the Privacy Act affected the development and use of Matrix?

3. What kind of oversight and security, if any, could make Matrix a viable crime prevention tool?

2. RFID Elicits Fears of "Big Brother" Technology

Radio frequency identification (RFID) tags have been around for 30 years, but have only recently become inexpensive enough to cause alarm among privacy advocates. The tag, a computer chip attached to an antenna, can broadcast its location and can store more data than a UPC code or bar code, including information as complex as how to construct a car. Because RFID tags do not need to be swiped or even placed in the line of sight of a reader, their data can be collected at greater distances and more efficiently.

For government and big business, RFID means big savings. Since January 2005, for example, Wal-Mart's top 100 suppliers have been required to provide RFID tags on cases and pallets of products so the company can reduce inventory costs.

RFID tags have many other potentially beneficial uses. Pet owners are tagging their cats and dogs so they can be found more easily if they get lost. Alzheimer's patients wear RFID bracelets in case they go missing. Livestock is tagged so government can stop the spread of illnesses such as mad cow disease. EasyPass RFID cards allow motorists to speed through toll booths at 55 mph.

Privacy advocates, however, are concerned. "How would you like it if, for instance, one day you realized your underwear was reporting on your whereabouts?" said California State Senator Debra Bowen.

Companies could conceivably tag individual products on shelves, rather than just pallets and cases in the warehouses, without notifying consumers. Once the customer purchases the item with a credit card, the company database could link the customer's location with data from the credit card, including the customer's name and home address. These companies could then track customers' movements and determine their shopping preferences. Worse yet, privacy advocates warn, the police or FBI could track customers, scanning a political rally, for example, to discover with whom they are associating.

There are a number of holes in this argument. First, RFID technology is probably six to 15 years away from being able to tag low-cost individual products. The cheapest tags currently cost about 40 cents apiece. Second, the FBI would somehow have to access a company's database to identify people who have tagged products.

Well-placed concerns, however, center on RFID identification documents. For example, Chase Bank and American Express have recently come out with "contactless" credit cards. Consumers don't even need to remove their credit cards from their wallets for RFID readers to scan these cards. Although these cards can only be scanned from a distance of a few centimeters, a second reader could potentially eavesdrop on this transaction and steal the information on the card.

Credit card companies claim that they have implemented security measures to foil any such attempts. Each contactless card possesses a unique number, separate from the credit card number, so that a thief who obtained the information through eavesdropping could not use the credit card number online. Because these cards are authenticated by the manufacturer through the scanning process, criminals who created fake cards would be identified immediately. In addition, RFID transmission cannot penetrate metal, so consumers can easily prevent a hidden reader from scanning their card by wrapping the card in aluminum foil.

Industry attempts to thwart RFID privacy legislation and the lack of full disclosure regarding RFID testing, however, have created an atmosphere of mistrust. Add to this the use of RFID cards to track students' whereabouts in New York and California high schools, and it is no wonder that "big brother" concerns abound. Both consumer education and full disclosure of RFID testing and research will be necessary if the adaptation of RFID technology is to continue unimpeded for the greater good.

Questions:

1. The federal government has considered using RFID technology on passports. What privacy issues might this raise?

2. How does current privacy legislation affect RFID use?

3. Companies such as Wal-Mart and Procter and Gamble would eventually like to tag individual products on store shelves to help keep them in stock and prevent shoplifting. What privacy laws could be enacted to help protect consumers in this situation?

3. Echelon—Top-Secret Intelligence System

Echelon is a top-secret electronic eavesdropping system that is managed by the U.S. National Security Agency (NSA) and known to be used by the intelligence agencies of England, Canada, Australia, and New Zealand. It can intercept and decrypt almost any electronic message sent anywhere in the world via satellite, microwave, cellular, or fiber-optic telecommunications, including radio and TV broadcasts, phone calls, computer-to-computer data transmission, faxes, and e-mail. It may have been in operation since the 1970s, but it wasn't until the 1990s that journalists, using the FOIA, were able to confirm its existence and gain insight into its capabilities. Although Echelon is the world's largest and most sophisticated surveillance network, it is by no means the only one. Russia, China, Denmark, France, the Netherlands, and Switzerland operate Echelon-like systems to obtain and process intelligence by listening in on electronic communications.

Which electronic transmissions are captured and what Echelon can do with the messages is subject to much conjecture. Even if all electronic messages worldwide were unencrypted, finding the messages that warranted further attention would be an enormous, computer-intensive task. As a result, Echelon probably targets communications to and from specific people and organizations rather than trying to assimilate all electronic messages. Thus, some subset of all possible messages is forwarded to the massive U.S. intelligence operations at Fort Meade, Virginia, where powerful computers look for code words or key phrases. Intelligence analysts peruse any conversation or document flagged by the system, and significant messages are then forwarded to the agency that requested the information.

A number of intelligence satellites in orbit are used to detect signals that normally dissipate into space—radio signals, mobile phone conversations, and microwave transmissions. In addition, at least six ground-based stations throughout the world monitor the communication satellites of Intelsat, the world's largest service provider of commercial satellite communications.

Computer processing speeds and the science of speech recognition probably are not advanced enough for a real-time global listening system capable of transcribing the hundreds of thousands of calls that are happening at any instant in time. However, Echelon is capable of voice-pattern matching and can identify speakers if their voice patterns are stored in its database. Also, it employs recording systems that can automatically trigger tape recordings based on "hearing" key words.

Echelon's special software and speech recognition technology can convert any audio communications into formatted, searchable text. A half-hour broadcast can be processed and stored in searchable format in 10 minutes. Currently, the software understands only American English, but the CIA is enhancing it to handle Chinese and Arabic. Other Echelon software alerts intelligence analysts any time a new page goes up on a Web site of interest. CIA personnel use special software to perform searches in English of Web sites developed in Chinese, Japanese, Russian, and eight other languages. The software then translates the text of the Web site into English.

This immense, highly sophisticated surveillance system apparently operates with little oversight, and the various agencies that run Echelon have provided few details about legal guidelines governing the project. In fact, the governments of the countries believed to be involved have not officially acknowledged Echelon's existence. Because of this, there is no way of knowing its true capabilities and exactly how it is used.

Echelon intercepts both sensitive government data and corporate information. It also provides the opportunity to illegally spy on private citizens. It is no wonder that privacy advocates are upset with the secrecy surrounding the system and its great potential for misuse. They feel that Echelon can be directed against virtually any citizen in the world with the full knowledge and cooperation of their government.

In the U.K., Echelon has already been accused of spying on organizations such as Amnesty International—an international agency that seeks to ensure fair and prompt trials for political prisoners and that opposes human rights abuses. In addition, in September 1999, the European Union released a report that accused Echelon of intercepting confidential company information and divulging it to favored competitors to help win contracts. The report alleged that Airbus Industries of France lost valuable contracts because information intercepted by Echelon was forwarded to the Boeing Company to help it obtain a competitive advantage.

In the United States, the ACLU and others are concerned that Echelon may be used without a court order to intercept communications involving Americans. The Foreign Intelligence Surveillance Act prohibits interception of certain communications for intelligence purposes without a court order, unless the Attorney General certifies that certain conditions are met. These conditions include a limitation that "there is no substantial likelihood that the surveillance will acquire the contents of any communication to which a United States person is a party."

Echelon supporters know that communications surveillance is successful in gathering enemy intelligence and was a key to the success of the allied military effort in World War II. They also argue that tragedies such as the attacks on September 11, 2001, and the Oklahoma City bombing are proof that surveillance systems are necessary to warn authorities and combat terrorism.

Within that context, the United States agreed to share highly classified material from Echelon with the Spanish government to aid in its battle against the Basque separatist group ETA. As a result, the Spanish are now receiving decoded intercepts relating to the ETA's plans for terrorist operations.

Questions:

1. Are you for or against the use of Echelon for eavesdropping on electronic communications? Why or why not? Is your opinion affected by the terrorist attacks of September 11?

2. Develop a set of plausible conditions under which the directors of Echelon would authorize using the system to listen to specific electronic communications.

3. What sort of expanded or new capabilities might Echelon develop in the next 10 years as information technology continues to improve? What additional privacy issues might be raised by these new capabilities?

End Notes

[1] Swipe Web site, www.we-swipe.us/about.html.

[2] Schmitz, Rob, "'Swipe' Combines ID with Art," *All Things Considered*, NPR Web site, www.npr.org/templates/story/story.php?storyId=1806119, April 1, 2004.

[3] Zetter, Kim, "Great Taste, Less Privacy," *Wired*, www.wired.com/news/privacy/0,1848,62182,00.html?tw=wn_polihead_4, February 6, 2004.

[4] BBBOnLine, www.bbbonline.org/privacy/.

[5] McCandlish, Stanton, "EFF's Top 12 Ways to Protect Your Online Privacy," version 2.0, www.eff.org/Privacy/?f=eff_privacy_top_12.html, April 10, 2002.

[6] *Personal Privacy in an Information Society: The Report of the Privacy Protection Study Commission*, transmitted to President Jimmy Carter on July 12, 1977, aspe.hhs.gov/datacncl/1977privacy/toc.htm.

[7] *The Privacy Journal*, www.privacyjournal.net/index.htm, September 19, 2005.

[8] Boatright, John R., *Ethics and the Conduct of Business*, Third Edition, 2000, Prentice Hall, Upper Saddle River, NJ, pages 166–168.

[9] Sternstein, Aliya, "TSA Launches Secure Flight," *FCW.com*, www.fcw.com, August 27, 2004.

[10] Weiss, Todd, R., "GAO: Secure Flight Antiterror Program Violates Privacy Laws," *Computerworld*, www.computerworld.com, July 26, 2005.

[11] Singel, Ryan, "Feds Push Flier Background Checks," *Wired News*, www.wired.com, August 15, 2005.

[12] OECD Web site, www.oecd.org, September 19, 2005.

[13] Betts, Mitch, "The Almanac: Privacy," *Computerworld*, www.computerworld.com, March 15, 2004.

[14] U.S. Department of Commerce, Safe Harbor Overview, www.export.gov/safeharbor/sh_overview.html, August 29 2005.

[15] Kerr, Jennifer C., "FCC Rule on Internet Class Said to Encourage Hacking," *InformationWeek*, www.informationweek.com, August 11, 2005.

16 Schwartz, John, "U.S. Selects a New Encryption Technique," *The New York Times*, www.nytimes.com, October 3, 2000.

17 Roberts, Paul, "NIST Says Data Encryption Standard Now 'Inadequate'," *Computerworld*, www.computerworld.com, July 29, 2004.

18 Schwartz, John, "U.S. Selects a New Encryption Technique," *The New York Times*, www.nytimes.com, October 3, 2000.

19 Vijayan, Jaikumar, "Data Security Breaches Reveal Encryption Need," *Computerworld*, www.computerworld.com, January 6, 2005.

20 Hayes, Frank, "Invisible Encryption," *Computerworld*, www.computerworld.com, June 13, 2005.

21 "National and State Trends in Fraud and Identity Theft," January–December 2004, Federal Trade Commission, www.consumer.gov/sentinel/pubs/Top10Fraud2004.pdf, February 1, 2005.

22 Levy, Steven and Stone, Brad, "Grand Theft Identity," *Newsweek*, pages 38–47, July 4, 2005.

23 Konrad, Rachel, "Consumers Battle Identity Theft Ring," *Cincinnati Enquirer*, page 5, February 25, 2005.

24 Holland, Jesse J., "Senate Plans Hearings on Data Brokers," *Cincinnati Enquirer*, page D5, February 25, 2005.

25 Gross, Grant, "Data Brokers Didn't Notify Consumers of Past Breaches," *Computerworld*, www.computerworld.com, April 13, 2005.

26 Gross, Grant, "Data Brokers Didn't Notify Consumers of Past Breaches," *Computerworld*, www.computerworld.com, April 13, 2005.

27 Mearian, Lucas, "Data Snafus Spur Action: Tape Mishap Prompts Call for Network Backup," *Computerworld*, www.computerworld.com, March 7, 2005.

28 Vijayan, Jaikumar and Weiss, Todd, "CardSystems Breach Renews Focus on Data Security," *Computerworld*, June 20, 2005.

29 CitiFinancial Statement on Lost Data Tapes, CitiGroup Press Room, www.citigroup.com/citigroup/press/2005/050606a.htm, June 6, 2005.

30 Gross, Grant, "Data Brokers Didn't Notify Consumers of Past Breaches," *Computerworld*, www.computerworld.com, April 13, 2005.

31 Perez, Juan Carlos, "Security Concerns to Stunt E-commerce Growth," *Computerworld*, www.computerworld.com, June 24, 2005.

32 Reuters, "Online Scammers Pose as Execs in 'Spear-Phishing'," *Computerworld*, www.computerworld.com, August 18, 2005.

33 Perez, Juan Carlos, "Security Concerns to Stunt E-commerce Growth," *Computerworld*, www.computerworld.com, June 24, 2005.

34 Martens, China, "Industry Looks to Unite Again to Tackle Spyware," *Computerworld*, www.computerworld.com, July 12, 2005.

35 Perez, Juan Carlos, "Security Concerns to Stunt E-commerce Growth," *Computerworld*, www.computerworld.com, June 24, 2005.

[36] Levy, Steven and Stone, Brad, "Grand Theft Identity," *Newsweek*, pages 38-47, July 4, 2005.

[37] "Microsoft Intros Search Ads with Demographic Filtering," *Marketing VOX*, www.marketingvox.com/archives/2005/03/16/, March 16, 2005.

[38] Kovach, Kenneth A., "Electronic Communication in the Workplace—Something's Got to Give," *Business Horizons*, July 2000.

[39] Keizer, Gregg, "Spam: Born in the USA," *InformationWeek*, www.informationweek.com, August 12, 2004.

[40] Claburn, Thomas, "Does CAN-SPAM Act Lead to More Spam?" *InformationWeek*, www.informationweek.com, June 3, 2004.

[41] Associated Press, "Court OKs Blocking Spam," *InformationWeek*, www.informationweek.com, August 4, 2005.

[42] Niccoli, James, "Microsoft Settles with 'Spam King' for $7 Million," *Computerworld*, www.computerworld.com, August 9, 2005.

[43] Dorning, Michael, "U.S. Cities Focus on Spy Cameras," *Chicago Tribune*, www.chicagotribune.com, August 8, 2005.

[44] Chan, Sewell, "Lockheed Martin IS Hired to Bolster Transit Security in N.Y.," *New York Times*, www.nytimes.com, August 23, 2005.

[45] Skatssoon, Judy, "News in Science – Smart Surveillance Has Alarm Bells Ringing," www.abc.net.au/science/news/stories/s1428126.htm, August 2, 2005.

[46] Associated Press, "LAPD: We Know That Mug," *Wired News*, www.wired.com, December 26, 2004.

Sources for Case 1

Hunker, Jeffrey, "A Dual-Edged Sword: Providing Information, Stealing Privacy," Security Pipeline, www.securitypipeline.com/showArticle.jhtml;jsessionid=AAZ4HXP0B4RVYQSNDBGCKHSCJUM EK JVN?articleId=18400971&pgno=2, March 15, 2004.

Ramasastry, Anita, "Why We Should Fear the Matrix," CNN Web site, www.cnn.com/2003/LAW/11/06/findlaw.analysis.ramasastry.matrix/, November 6, 2003.

Royse, David, "Matrix Crime-Solving Database Lives On," *InformationWeek*, www.informationweek.com/showArticle.jhtml;jsessionid=NWBAO0VOR3COUQSNDBGCKHS CJUMEKJVN?articleID=165701675, July 12, 2005.

Sources for Case 2

"Credit Cards that Don't Swipe," *Talk of the Nation*, NPR Web site, www.npr.org/templates/story/story.php?storyId=4664479, May 24, 2005.

Gilbert, Alorie, "Privacy Advocates Call for RFID Regulation," CNet News.com, news.com.com/2100-1029_3-5065388.html?tag=fd_top, September 18, 2003.

Gross, Grant, "RFID Policy Panel Raises Privacy Concerns," *Computerworld*, www.computerworld.com/softwaretopics/erp/story/0,10801,100904,00.html, April 7, 2005.

Sources for Case 3

"Echelon Surveillance System Threatens U.S. Privacy," *Online Newsletter*, September 2000, www.findarticles.com.

Schwartz, Mathew, "Intercepting Messages," *Computerworld*, pages 48–49, August 28, 2000.

Wilkinson, Isambard, "US Wins Spain's Favour With Offer to Share Spy Network Material," www.smh.com.au, June 18, 2001.

Lettice, John, "French Echelon Report Says Europe Should Lock Out US Snoops," *The Register*, www.theregister.co.uk, October 13, 2000.

Loeb, Vernon, "Making Sense of the Deluge of Data," *The Washington Post*, page A23, March 26, 2001.

CHAPTER 5

FREEDOM OF EXPRESSION

VIGNETTE

China Stifles Online Dissent

In 1995, the government of the People's Republic of China established the country's first Internet service provider (ISP). The decision reflected two conflicting needs. On the one hand, China's drive toward economic globalization requires the adoption of Western technologies that allow Chinese companies to market themselves to the West. On the other hand, the Communist Party's hold over the country rests on suppressing freedom of the press and freedom of expression. The party thus decided to introduce a government-controlled Internet, often referred to as the "Great Firewall of China."

The firewall blocks citizens from accessing Western news Web sites such as CNN, the New York Times, and Reuters. Search engines do not post results for certain terms, and words like *democracy*, *freedom*, and *independence* are banned from news sites and instant messaging services. E-mails disappear. Information posted by the outlawed religious group Falun Gong is routinely pulled off bulletin boards. In addition, the government requires commercial Web sites to register, which allows it to trace subversive content back to the source. Dozens of people have been jailed for posting seditious content.

To prevent Internet users from bribing cybercafé owners to avoid logging on using their state ID cards, the government has also assembled a cyberspace police force, estimated to number 30,000.

As the number of Chinese Internet users climbed toward 100 million in 2005, the number of personal Web sites exploded. To gain control of the content posted by an estimated 700,000 personal opinion blogs and online diaries, the government set June 30, 2005, as a deadline by which all private Web sites would have to register. Then, in September 2005, the government instituted laws prohibiting all sites, public or private, from providing news reports and commentary without government authorization. Violators face fines of up to $3700 and closure of their sites.

Although some technically savvy users may find ways around the new regulations, these measures will no doubt squelch the growing exchange of ideas that had been emerging online among China's urban elite.[1,2]

LEARNING OBJECTIVES

As you read this chapter, consider the following questions:

1. What is the legal basis for the protection of freedom of speech in the United States, and what types of speech are not protected under the law?

2. In what ways does the Internet present new challenges in the area of freedom of expression?

3. What key free-speech issues relate to the use of information technology?

FIRST AMENDMENT RIGHTS

The Internet enables a worldwide exchange of news, ideas, opinions, rumors, and information. Its broad accessibility, open-minded discussions, and anonymity make the Internet an ideal communications medium. It provides an easy and inexpensive way for a speaker to send a message indiscriminately to a large audience, potentially thousands of people worldwide. In addition, given the right e-mail addresses, a speaker can aim a message with laser accuracy at a select subset of powerful and influential people.

People must often make ethical decisions about how to use such remarkable freedom and power. Organizations and governments have attempted to establish policies and laws to help guide people as well as to protect their own interests. Businesses, in particular, have sought to conserve corporate network capacity, avoid legal liability, and improve worker productivity by limiting the nonbusiness use of IT resources.

The right to freedom of expression is one of the most important rights for free people everywhere. The First Amendment to the U.S. Constitution was adopted to guarantee this right and others. Over the years, a number of federal, state, and local laws have been found unconstitutional because they violated one of the tenets of this amendment. The First Amendment reads:

> Congress shall make no law respecting an establishment of religion, or prohibiting the free exercise thereof; or abridging the freedom of speech, or of the press; or the right of the people peaceably to assemble, and to petition the government for a redress of grievances.

In other words, the First Amendment protects the right to freedom of religion and freedom of expression from government interference. This amendment has been interpreted by the Supreme Court to apply to the entire federal government, even though it only expressly applies to Congress.

Numerous court decisions have broadened the definition of speech to include nonverbal, visual, and symbolic forms of expression, such as burning the American flag, dance movements, and hand gestures. Sometimes the speech at issue is unpopular or highly offensive to the majority of people; however, the Bill of Rights provides protection for minority views. The Supreme Court has also ruled that the First Amendment protects the right to speak anonymously as part of the guarantee of free speech.

However, the Supreme Court has held that the following types of speech are not protected by the First Amendment and may be forbidden by the government: obscene speech, defamation, incitement of panic, incitement to crime, "fighting words," and sedition (incitement of discontent or rebellion against a government). Two of these types of speech, obscene speech and defamation, are particularly relevant to information technology.

Obscene Speech

Miller v. California is the Supreme Court case that established a test to determine if material is obscene and therefore not protected by the First Amendment. Miller, after conducting a mass mailing campaign to advertise the sale of adult material, was convicted of violating a California statute prohibiting the distribution of obscene material. Some unwilling recipients of Miller's brochures complained to the police, initiating the legal proceedings. Although the brochures contained some descriptive printed material, they primarily consisted of pictures and drawings that explicitly depicted men and women engaging in sexual activity. In its 1973 ruling in *Miller v. California*, the Supreme Court determined that speech can be considered obscene and not protected under the First Amendment, based on the following three questions:

- Would the average person, applying contemporary community standards, find that the work, taken as a whole, appeals to the prurient interest?
- Does the work depict or describe, in a patently offensive way, sexual conduct specifically defined by the applicable state law?
- Does the work, taken as a whole, lack serious literary, artistic, political, or scientific value?

These three tests have become the U.S. standard for determining if something is obscene. The requirement that a work be assessed by its impact on an average adult in a community has raised many issues—who is an average adult, what are contemporary community standards, and, in the case of potentially obscene material displayed worldwide on the Internet, what is the community?

Defamation

The right to freedom of expression is restricted when the expressions, whether spoken or written, are untrue and cause harm to another person. The publication of a statement of alleged fact that is false and that harms another person is **defamation**. The harm is often of a financial nature, in that it reduces a person's ability to earn a living, work in a profession, or run for an elected office. An oral defamatory statement is **slander**, and a written defamatory statement is **libel**. Because defamation is defined as an untrue statement of fact, truth is an absolute defense against a charge of defamation. Although people have the right to express opinions, they must exercise care in their Internet communications to avoid possible charges of defamation. Organizations must also be on guard and prepared to take action in the event of libelous attacks against them.

FREEDOM OF EXPRESSION: KEY ISSUES

Information technology has provided amazing new ways to communicate with people around the world. With these new methods come responsibilities and new ethical problems. This section discusses a number of key issues related to freedom of expression, including controlling access to information on the Internet, anonymity, defamation, hate speech, and pornography.

Controlling Access to Information on the Internet

Although there are clear and convincing arguments to support freedom of speech on the Internet, the issue is complicated by the ease with which children can access the Internet. Even some advocates of free speech acknowledge the need to restrict children's access, but it is difficult to restrict Internet access by children without also restricting access by adults. In an attempt to address this issue, the U.S. government has passed laws, and software manufacturers have invented special software to block access to objectionable material. The following sections summarize these approaches.

The Communications Decency Act (CDA)

The Telecommunications Deregulation and Reform Act became law in 1996. Its purpose was to allow freer competition among phone, cable, and TV companies. Embedded in the Telecommunications Act was the Communications Decency Act (CDA), aimed at protecting children from online pornography. The CDA imposed $250,000 fines and prison terms of up to two years for the transmission of "indecent" material over the Internet.

The *Reno v. ACLU* suit, filed in February 1996, challenged the criminalizing of so-called indecency on the Internet. The government appealed the case to the Supreme Court after a three-judge federal panel ruled unanimously that the law unconstitutionally restricted free speech. Examples of indecency that were identified as potentially criminal by government witnesses included Internet postings of a photo of actress Demi Moore, naked and pregnant on the cover of *Vanity Fair*, and any use online of the infamous "seven dirty words." The plaintiffs included the American Civil Liberties Union (ACLU), Planned Parenthood, Stop Prisoner Rape, Human Rights Watch, and the Critical Path AIDS Project. Many of these organizations feared that much of their online material could be classified as indecent, because examples cited by government witnesses included speech about abortion, prisoner rape, safe sex practices, and many other sexually related topics. The plaintiffs argued that such information was important to both minors and adults.[3]

The problem with the CDA was its broad language and vague definition of indecency, a standard that was left to individual communities to decide. In June 1997, the Supreme Court ruled the law unconstitutional and declared that the Internet must be afforded "the highest protection available under the First Amendment."[4] The Supreme Court said in its ruling that "the interest in encouraging freedom of expression in a democratic society outweighs any theoretical but unproven benefit of censorship." The ruling also said that "the growth of the Internet has been and continues to be phenomenal. As a matter of constitutional tradition, and in the absence of evidence to the contrary, we presume government regulation of the content of speech is more likely to interfere with the free exchange of ideas than to encourage it."[5]

If the CDA had been judged constitutional, it would have opened all aspects of Internet content to legal scrutiny. Many current Web sites probably would either not exist or would look much different today had the law not been overturned. In addition, Web sites that might have been deemed indecent under the CDA would operate under an extreme risk of liability.

The Child Online Protection Act (COPA)

In October 1998, the Child Online Protection Act (COPA) was signed into law. The law states that "whoever knowingly and with knowledge of the character of the material, in interstate or foreign commerce by means of the World Wide Web, makes any communication for commercial purposes that is available to any minor and that includes any material that is harmful to minors shall be fined not more than $50,000, imprisoned not more than 6 months, or both."[6] (Subsequent sections of the act allow for penalties of up to $150,000 for each day of violation.)

The law became a significant battleground for proponents of free speech. Not only could it affect sellers of explicit material online and their potential customers, it could ultimately set standards for Internet free speech. Supporters of COPA (primarily the Department of Justice) argued that the act protected children from online pornography while preserving the rights of adults. However, privacy advocacy groups such as the Electronic Privacy Information Center, the ACLU, and the Electronic Frontier Foundation claimed that the language was overly vague and limited the ability of adults to access material protected under the First Amendment.

Following numerous hearings and appeals, the Supreme Court ruled in June 2004 that "there is a potential for extraordinary harm and a serious chill upon protected speech" if the law went into effect. The ruling made it clear that COPA was unconstitutional and could not be used to shelter children from online pornography.

Internet Filtering

An **Internet filter** is software that can be installed with a Web browser to block access to certain Web sites that contain inappropriate or offensive material. The best Internet filters use a combination of URL filtering, keyword filtering, and dynamic content filtering. With URL filtering, a particular URL or domain name is identified as an objectionable site and the user is not allowed access to it. Keyword filtering uses key words or phrases such as *sex*, *Satan*, and *gambling* to trigger the blocking of Web sites. With dynamic content filtering, each Web site's content is evaluated immediately before it is displayed, using such techniques as object analysis and image recognition.

Network administrators may choose to install filters on employees' computers to prevent them from viewing sites that contain pornography or other objectionable material. Employees who are unwillingly exposed to such material would have a strong case for sexual harassment. The use of filters can also ensure that employees do not waste their time viewing nonbusiness Web sites. Popular Internet filters include ContentProtect, CYBERsitter, NetNanny, and CyberPatrol. Another popular Internet filter, HateFilter, can be downloaded from the Anti-Defamation League's (ADL) Web site. If users try to access a blocked site advocating bigotry, hatred, or violence toward groups on the basis of their ethnicity, race, religion, or sexual orientation, HateFilter's redirect feature offers to link them to related ADL educational material.

Another filtering system is available through the Internet Content Rating Association (ICRA), a nonprofit organization whose members include Internet industry leaders such as AOL Europe, BellSouth, British Telecom, IBM, Microsoft, and Verizon. ICRA's mission is to enable the public to make informed decisions about electronic media through the open and objective labeling of content. Its goals are to protect children from potentially harmful material while safeguarding free speech on the Internet.

In the ICRA rating system, Web authors fill out an online questionnaire to describe the content of their site. The questionnaire covers such broad topics as chat rooms, the language used on the site, its nudity and sexual content, the violence depicted, and other content areas such as alcohol, drugs, gambling, and suicide. Within each broad category, the Web author is asked whether specific items or features are present or absent on the site. Based on the author's responses, ICRA then generates a content label (a short piece of computer code) that the author adds to the site. This label conforms to an Internet industry standard known as the Platform for Internet Content Selection (PICS). Internet users can then set their browsers to allow or disallow access to Web sites based on the information declared in the content label and their own preferences.

Note that ICRA does not rate Internet content—the content providers do. However, relying on Web site authors to do their own ratings has its weaknesses. For one, many hate sites and sexually explicit sites don't have ICRA ratings. They won't be blocked unless a browser is set to block all unrated sites, which can lead to blocking so many acceptable sites that Web surfing can become a useless activity. Site labeling also depends on the honesty with which Web site authors rate themselves. If authors lie when completing the ICRA questionnaire, their site can receive a content label that doesn't accurately reflect the content. For these reasons, site labeling is at best a complement to other filtering techniques.

Another approach to restricting access to Web sites is to subscribe to an Internet service provider (ISP) that performs the blocking itself. The blocking occurs at the ISP's server, as opposed to using software loaded on each user's computer. One ISP, ClearSail/Family.NET, prevents access to known Web sites that address such topics as bomb making, gambling, hacking, hate, illegal drugs, pornography, profanity, Satan, and suicide. ClearSail employees search the Internet each day to uncover new Web sites to add to ClearSail's block list. It blocks known URLs, known pornographic hosting services, keywords in URLs, and search keywords. There are millions of blocked Web pages as a result of ClearSail's filtering capability. Additionally, newsgroups are blocked because of the uncontrollable amount of pornography contained within them. The following Legal Overview explains the use of Internet filters in schools and libraries to block harmful material from minors.

LEGAL OVERVIEW

Children's Internet Protection Act (CIPA)

In another attempt to protect children from accessing pornography and other explicit material online, Congress passed the Children's Internet Protection Act (CIPA) in 2000. The act required federally financed schools and libraries to use some form of technological protection (such as an Internet filter) to block computer access to obscene material, pornography, and anything considered harmful to minors. Congress did not specifically define which content or Web sites should be forbidden and which measures should be used—these decisions were left to individual school districts and library systems. Any school or library that failed to comply with the law would no longer be eligible to receive federal money through the e-Pay program to help pay for the cost of Internet connections. The following points summarize CIPA:

- Under CIPA, schools and libraries subject to CIPA do not receive the discounts offered by the "E-Rate" program (discounts that make access to the Internet affordable to schools and libraries) unless they certify that they have certain Internet safety measures in place. These include measures to block or filter pictures that: (1) are obscene, (2) contain child pornography, or (3) when computers with Internet access are used by minors, are harmful to minors.

continued

151

- Schools subject to CIPA are required to adopt a policy to monitor online activities of minors.
- Schools and libraries subject to CIPA are required to adopt a policy addressing: (1) access by minors to inappropriate matter on the Internet and World Wide Web; (2) the safety and security of minors when using electronic mail, chat rooms, and other forms of direct electronic communications; (3) unauthorized access, including hacking and other unlawful activities by minors online; (4) unauthorized disclosure, use, and dissemination of personal information regarding minors; and (5) restricting minors' access to materials harmful to them. CIPA does not require the tracking of Internet use by minors or adults.[7]

Opponents of the law feared that it transferred power over education to private software companies who developed the Internet filters and defined which sites to block. Furthermore, opponents felt that the motives of these companies were unclear—for example, some of the filtering companies track students' Web surfing activities and sell the data to market research firms. Opponents also pointed out that some versions of these filters were ineffective. They blocked access to legitimate sites and allowed users to access objectionable sites. Yet another objection was that penalties associated with the act could cause schools and libraries to lose federal funds from the e-Pay program, which is intended to help bridge the so-called digital divide in America between the rich and poor and the urban and rural. Loss of federal funds would lead to a less capable version of the Internet for students at poorer schools, which have the fewest alternatives to federal aid.

CIPA's proponents contended that shielding children from drugs, hate, pornography, and other topics was a sufficient reason to justify filters. They argued that the Internet filters were highly flexible and customizable and that their critics exaggerated the limitations. Proponents pointed out that schools and libraries could still elect not to implement a children's Internet protection program; they just wouldn't receive federal money for Internet access.

Many school districts have implemented programs consistent with CIPA. Acceptance of an Internet filtering system is more meaningful if the system and its rationale are first discussed with parents, students, teachers, and administrators. Then the program can be refined, taking into account everyone's feedback. An essential element of a successful program is to require that students, parents, and employees sign an agreement outlining the school district's acceptable use policies for accessing the Internet. Controlling Internet access from one central school district network, rather than letting each school set up its own filtering system, reduces administrative effort and ensures consistency. Procedures must be defined to block new objectionable sites as well as remove blocks from Web sites that should be accessible.

continued

Implementing CIPA in libraries is much more difficult because their services are open to people of all ages, including adults who have First Amendment rights to access a broader range of Internet materials than are allowed under CIPA. One county library was sued for filtering, while another was sued for not filtering enough. At least one federal court has ruled that a local library board may not require the use of filtering software on all library Internet computers. A possible compromise for public libraries with multiple computers would be to allow unrestricted Internet use for adults, but to provide computers with only limited access for children.

The ACLU filed a suit to challenge CIPA. In May 2002, a three-judge panel in Eastern Pennsylvania held that "we are constrained to conclude that the library plaintiffs must prevail in their contention that CIPA requires them to violate the First Amendment rights of their patrons, and accordingly is facially invalid" under the First Amendment. The ruling instructed the government not to enforce the act. This ruling, however, was reversed in June 2003 by the U.S. Supreme Court in *U.S. v. American Library Association*. The Supreme Court, in a 6-3 decision, held that public libraries must purchase filtering software and comply with all portions of CIPA.

Rather than deal with all the technical and legal complications, some librarians say they wish they could simply train students and adults to use the Internet safely and wisely. After all, bad behavior in libraries did not start and will not end with the Internet.

Anonymity

The principle of **anonymous expression** allows people to state their opinions without revealing their identity. The freedom to express an opinion without fear of reprisal is an important right of a democratic society. Anonymity is even more important in countries that don't allow free speech. However, in the wrong hands, anonymous communication can be used as a tool to commit illegal or unethical activities.

Anonymous political expression played an important role in the early formation of the United States. Before and during the American Revolution, patriots who dissented against British rule often used anonymous pamphlets and leaflets to express their opinions. England had a variety of laws designed to restrict anonymous political commentary, and people found guilty of breaking these laws were subject to harsh punishment—from whippings to hangings. A famous case in 1735 involved a printer named John Zenger, who was prosecuted for seditious libel because he wouldn't reveal the names of anonymous authors whose writings he published. The authors were critical of the governor of New York. The British were outraged when the jurors refused to convict Zenger, in what is considered a defining moment in the history of freedom of the press.

Other democracy supporters often authored their writings anonymously or under pseudonyms. For example, Thomas Paine was an influential writer, philosopher, and statesman of the Revolutionary War era. He published a pamphlet called *Common Sense*, in which he criticized the British monarchy and urged the colonies to become independent by establishing a republican government of their own. Published anonymously in 1776, the pamphlet sold more than 500,000 copies when the population of the colonies was estimated to have been less than 4 million; it provided a stimulus to produce the Declaration of Independence six months later.

Despite the importance of anonymity in early America, it took nearly 200 years for the Supreme Court to render rulings that addressed anonymity as an aspect of the Bill of Rights. One of the first rulings was in the 1958 case of *NAACP v. Alabama*, in which the court ruled that the NAACP did not have to turn over its membership list to the state of Alabama. The court believed that members could be subjected to threats and retaliation if the list were disclosed, and that disclosure would restrict a member's right to freely associate in violation of the First Amendment.

Another landmark anonymity case involved a sailor threatened with discharge from the U.S. Navy because of information obtained from America Online. In 1998, following a tip, a Navy investigator asked America Online to identify the sailor, who used an Internet pseudonym to post information in a personal profile that suggested he might be gay. Thus, he could be discharged under the military's "don't ask, don't tell" policy on homosexuality. America Online admitted that its representative violated company policy in providing the information. A federal judge ruled that the Navy had overstepped its authority in investigating the sailor's sexual orientation and had also violated the Electronic Communications Privacy Act, which limits how government agencies can seek information from e-mail or other online data. The sailor received undisclosed monetary damages from America Online and, in a separate agreement, was allowed to retire from the Navy with full pension and benefits.[8]

Maintaining anonymity on the Internet is important to some computer users. They might be seeking help in an online support group, reporting defects about a manufacturer's goods or services, participating in frank discussions of sensitive topics, expressing a minority or antigovernment opinion in a hostile political environment, or participating in chat rooms. Other Internet users would like to ban Web anonymity because they think that its use increases the risks of defamation, fraud, libel, and exploitation of children.

Anonymous Remailers

Maintaining anonymity is a legitimate need for some Internet activities; however, the address in an e-mail message or Usenet newsgroup posting clearly identifies its author. Internet users who want to remain anonymous can send e-mail to an **anonymous remailer** service, where a computer program strips the originating address from the message. It then forwards the message to its intended recipient—an individual, chat room, or newsgroup—with either no address or a fictitious one. This ensures that no header information can identify the author. Some remailers use encryption and routing through multiple remailers to provide a virtually untraceable level of anonymity.

The use of a remailer keeps communications anonymous; what is communicated, and whether it is ethical or legal, is up to the sender. The use of remailers to enable people to commit unethical or even illegal acts in some states or countries has spurred controversy. Remailers are frequently used to send pornography, to illegally post copyrighted material to Usenet newsgroups, and to send unsolicited advertising to broad audiences (spamming). A corporate IT organization may want to employ filters or set the corporate firewall to prohibit employees from accessing remailers, or to send a warning message each time an employee communicates with a remailer.

John Doe Lawsuits

Businesses must protect against the public expression of opinions that might hurt their reputations and the public sharing of company confidential information. When anonymous employees reveal harmful information over the Internet, the potential for broad dissemination is enormous and can require great effort to identify the people involved and stop them.

In a **John Doe lawsuit**, the identity of the defendant is temporarily unknown. Such suits are common in Internet libel cases, where the defendant communicates using a pseudonym or anonymously. Corporations often file these lawsuits because they are upset by anonymous e-mail messages that criticize the company or reveal company secrets. For example, Raytheon filed a lawsuit in 1999 for $25,000 in damages against 21 John Does for allegedly revealing company financial results on a Yahoo! message board, along with other information that the company claimed hurt its reputation. Raytheon received a court order to subpoena Yahoo! and several ISPs for the identity of the 21 unnamed defendants. Eventually, Raytheon traced the identities of all 21 people who posted the alleged company secrets. Four employees voluntarily left the company and others received counseling about sharing confidential company information.[9]

America Online, Yahoo!, and other ISPs receive more than a thousand subpoenas a year to reveal the identity of John Does. Free-speech advocates argue that if someone charges libel, the anonymity of the Internet poster should be preserved until the libel is proved. Otherwise, the subpoena power can be used to silence anonymous, critical speech.

Proponents of such lawsuits point out that most John Doe cases are based on serious allegations of wrongdoing, such as libel or disclosure of confidential information. For example, stock-price manipulators can use chat rooms to affect the share price of a stock—especially those of very small companies with just a few outstanding shares. In addition, competitors of an organization might try to create a feeling that a company is a miserable place to work, which could discourage job candidates from applying, investors from buying stock, or consumers from buying company products. Proponents of John Doe lawsuits argue that perpetrators should not be able to hide behind anonymity to avoid responsibility for their actions.

Anonymity on the Internet is not guaranteed. By filing a lawsuit, companies gain immediate subpoena power, and many message board hosts release information right away, often without notifying the poster. Everyone who posts comments in a public place on the Internet must consider the consequences if their identities are exposed. Furthermore, everyone who reads anonymous postings on the Internet should think twice about believing what they read.

The California State Court in *Pre-Paid Legal v. Sturtz et al.* set a legal precedent that courts apply to subpoenas requesting the identity of anonymous Internet speakers. The case involved a subpoena issued by Pre-Paid Legal Services (PPLS), which requested the identity of eight anonymous posters on Yahoo!'s Pre-Paid message board. Attorneys for PPLS argued that it needed the posters' identities to determine whether they were subject to a voluntary injunction that prevented former sales associates from revealing PPLS's trade secrets.

The Electronic Frontier Foundation (EFF) represented two of the John Does whose identities were subpoenaed. EFF attorneys argued that the message board postings cited by PPLS revealed no company secrets, but merely indicated that the eight John Does were disparaging the company and its treatment of sales associates. They argued further that requiring the John Does to reveal their identities would let the company punish them for speaking out and set a dangerous precedent that would discourage other Internet users from voicing criticism. Without proper safeguards on John Doe subpoenas, a company could use the courts to uncover its critics.

EFF attorneys urged that the court apply a four-part test adopted by the federal courts in *Doe v. 2TheMart.com, Inc.* to determine whether a subpoena for the identity of Internet speakers should be upheld. In that case, the federal court ruled that a subpoena should be enforced only when the following occurs:

- The subpoena was issued in good faith and not for any improper purpose.
- The information sought related to a core claim or defense.
- The identifying information was directly and materially relevant to that claim or defense.
- Adequate information was unavailable from any other source.

In August 2001, a judge in Santa Clara County Superior Court invalidated the subpoena to Yahoo! requesting the posters' identities. He ruled that the messages were not obvious violations of the injunctions invoked by PPLS, and that the First Amendment protection of anonymous speech outweighed PPLS's interest in learning the identity of the speakers.

John Doe lawsuits are also used in the fight against spam. In April 2003, AOL, Earth-Link, Microsoft, and Yahoo! formed an antispam alliance. These companies have filed lawsuits under the federal CAN-SPAM law in an attempt to reduce the spam problem for Internet users worldwide. As of October 2004, Microsoft alone had filed more than 100 legal actions worldwide, including 75 in the United States. In many of these cases, the defendants are unknown and must be referred to as John Doe until further investigation can reveal their true identities.[10]

National Security Letters

A **National Security Letter (NSL)** requires financial institutions to turn over electronic records about the finances, telephone calls, e-mail, and other personal information of suspected terrorists or spies. Recent developments have expanded the scope and power of NSLs to the point that some believe they now represent a threat to freedom of speech.

Before the USA Patriot Act, the U.S. Attorney General or a Deputy Attorney General had to authorize each NSL. Under Section 505 of the act, however, the FBI is allowed to use an NSL to obtain records from banks and other financial institutions if they are sought for an intelligence or terrorism investigation. Section 505 also includes a gag provision prohibiting any firm that receives an NSL from even disclosing the fact that the FBI is seeking information. Thus, a firm cannot inform its customers that they are being investigated. The Intelligence Authorization Act for Fiscal Year 2004 modified Section 505 to give even greater powers to the FBI. The act expanded the scope of discovery beyond financial institutions to a host of other businesses, including ISPs. The act also dropped the requirement to obtain court approval and did not require any other independent check on the validity or scope of the inquiry. Many believe that the redefined NSL eliminates the potential for people to retain their anonymity and compromises the privacy rights of all Americans.

In April 2004, the ACLU and an anonymous ISP filed a lawsuit challenging the FBI's power to issue NSLs. In September 2004, Judge Victor Marrero of the Southern District of New York struck down the NSL statute and the associated gag provision, stating "Democracy abhors undue secrecy...An unlimited government warrant to conceal, effectively a form of secrecy per se, has no place in our open society." The U.S. Court of Appeals for the Second Circuit was expected to hear the government's appeal of this case after Congress finished debating the reauthorization of the Patriot Act in 2006.

Defamation and Hate Speech

In the United States, speech that is merely annoying, critical, demeaning, or offensive enjoys protection under the First Amendment. Legal recourse is possible only when hate speech turns into clear threats and intimidation against specific citizens. Persistent or malicious harassment aimed at a specific person can be prosecuted under the law, but general, broad statements expressing hatred of an ethnic, racial, or religious group cannot. A threatening private message sent over the Internet to a person, a public message displayed on a Web site describing intent to commit acts of hate-motivated violence, and libel directed at a particular person are all actions that can be prosecuted.

In addition to First Amendment protection, another difficulty in enforcing laws against defamation and hate speech is the ease of anonymous communication over the Internet. Anyone can send hateful e-mails and avoid easy identification by using a remailer service that guarantees complete anonymity. From time to time, America Online and other ISPs have voluntarily agreed to prohibit their subscribers from sending hate messages using their services. Because such prohibitions are included in the service contracts between a private ISP and its subscribers and do not involve the federal government, they do not violate the subscribers' First Amendment rights.

After an ISP implements such a prohibition, it must monitor the use of its service to ensure that the regulations are followed. When a violation occurs, the ISP must take action to prevent it from happening again. For example, if a subscriber who participates in a chat room engages in hate speech that violates the ISP's terms of service, the subscriber's account can be cancelled or the subscriber can be warned and forbidden from participating in the chat room. To aid enforcement of such prohibitions, the ISP can also encourage subscribers to report violators to the appropriate company representatives. Of course, ISP subscribers who lose an account for violating the ISP's regulations may resume their hate speech by simply opening a new account with some other, more permissive ISP.

Although they may implement a speech code, public schools and universities are legally considered agents of the government and therefore must follow the First Amendment's prohibition against speech restrictions based on content or viewpoint. Corporations, private schools, and private universities, on the other hand, are not part of the state or federal government. As a result, they may prohibit students, instructors, and other employees from engaging in offensive speech using corporate, school, or university computers, the Internet, or e-mail services.

Despite the protection of the First Amendment and the challenges posed by anonymous expression, there are instances of U.S. citizens being successfully sued or convicted of crimes relating to hate speech:

- A former student was sentenced to one year in prison for sending e-mail death threats to Asian-American students at the University of California, Irvine. His e-mail was signed "Asian hater," and his letters stated that he would make it his career to find and kill every Asian himself.
- A coalition of antiabortion groups was ordered to pay more than $100 million in damages after placing information on a Web site about doctors and clinic workers who perform abortions, including photos, home addresses, license plate numbers, and even the names of their spouses and children. Three of the doctors listed on the site were murdered and others were wounded. A jury found that the Web site provided information that resulted in a real threat of bodily harm and awarded damages. However, in March 2001, the U.S. Court of Appeals for the Ninth Circuit reversed the decision. The court ruled that the coalition made no statements mentioning violence and that publication of the personal information did not constitute a serious expression of intent to harm.[11]
- Varian Medical Systems won a $775,000 jury verdict in an Internet defamation and harassment lawsuit against former employees who posted thousands of messages that accused managers of being homophobic and of discriminating against pregnant women.[12]

Most other countries do not provide constitutional protection for hate speech. For example, promoting Nazi ideology is a crime in Germany, and denying the occurrence of the Holocaust is illegal in many European countries. Authorities in Britain, Canada, Denmark, France, and Germany have charged people for crimes involving hate speech on the Internet.

A U.S. citizen who posts material on the Internet that is illegal in a foreign country can be prosecuted if he subjects himself to the jurisdiction of that country—for example, by visiting there. As long as the person remains in the United States, he is safe from prosecution, because U.S. laws do not allow a person to be extradited for engaging in an activity protected by the U.S. Constitution, even if the activity violates the criminal laws of another country.

Pornography

Many adults and free-speech advocates believe that nothing is illegal or wrong about purchasing adult pornographic material made for and by consenting adults. They argue that the First Amendment protects such material. On the other hand, most parents, educators, and other child advocates are upset by the thought of children viewing pornography. They are deeply concerned about its impact on children and fear that increasingly easy access to pornography encourages pedophiles and sexual molesters.

Clearly, the Internet has been a boon to the pornography industry by providing fast, cheap, and convenient access to more than 60,000 Web sex sites. Perhaps most importantly, access via the Internet enables pornography consumers to avoid offending others or being embarrassed by others observing their purchases. There is no question that adult pornography on the Internet is big business and generates lots of traffic. About one in four regular Internet users (almost 21 million Americans) visits a Web sex site at least once a month—more than the number of visitors to Web sports sites. Forrester Research estimates that sex sites on the Web generate at least $1 billion a year in revenue.[13]

The CAN-SPAM Act can be a deterrent in fighting the dissemination of pornography. For example, three people were indicted by an Arizona grand jury in August 2005 for violating the CAN-SPAM Act by sending massive amounts of unsolicited e-mail that advertised pornographic Web sites. The e-mails allegedly contained pornographic images, so the grand jury also leveled felony obscenity charges for transmission of hard-core pornography images. The defendants face multiple counts of spamming and criminal conspiracy that carry a maximum sentence of five years each. The number of Internet users who received spam from the operation is estimated to be in the tens of millions.[14]

Spim, or instant messaging spam, is also becoming a problem; more than 30 percent of IM users receive unsolicited instant messages. Spim violators can also be charged under the CAN-SPAM Act. For example, a man was arrested in Los Angeles in February 2005 and charged with sending unsolicited instant messages that advertised refinancing services and pornography.[15]

U.S. organizations must be very careful when dealing with pornography in the workplace. It is against U.S. obscenity laws to publish pornography, and by providing computers, training, and Internet access, companies can be seen by the law as publishers because they have enabled employees to store pornographic material and retrieve it on demand.

Some companies believe that they have a duty to stop the viewing of pornography in the workplace. As long as they can show that they were taking reasonable steps and determined actions to prevent it, they have a valid defense if they become the subject of a sexual harassment lawsuit. If it can be shown that a company made just a half-hearted attempt at stopping pornography and gave up when things got a bit difficult, then its defense in court would be weak. Reasonable steps include establishing a computer usage policy that prohibits access to pornography sites, identifying those who violate the policy, and taking action against those users—no matter how embarrassing it is for the users or how harmful it might be for the company.

A few companies take an opposite viewpoint. Their legal and human resources departments believe that the company cannot be held liable if they don't know employees are viewing, downloading, and distributing pornography. Thus, they advise management to ignore a problem by never investigating it, thereby ensuring they can claim that they never knew it was happening. Many people would consider such an approach unethical and would view management as shirking an important responsibility to provide a work environment free of sexual harassment. Employees unwillingly exposed to pornography would have a strong case for sexual harassment because they could claim that pornographic material was available in the workplace and that the company took inadequate measures to control the situation.

In contrast to adult pornography, numerous federal laws address child pornography. Possession of child pornography is a federal offense punishable by up to five years in prison. The production and distribution of such materials carry harsher penalties—decades or even life in prison is not an unusual sentence. In addition to these federal statutes, all states have enacted laws against the production and distribution of child pornography, and all but a few states have outlawed the possession of child pornography. South Carolina even passed a law in July 2001 that requires computer technicians who discover child pornography on clients' computers to give their names and addresses to law enforcement officials.[16]

Table 5-1 is a manager's checklist for dealing with issues of freedom of expression in the workplace. In each case, the preferred answer is *yes*.

TABLE 5-1 Manager's checklist for handling freedom of expression in the workplace

Questions	Yes	No
Do you have a written data privacy policy that is followed?	___	___
Does your corporate IT usage policy discuss the need to conserve corporate network capacity, avoid legal liability, and improve worker productivity by limiting the nonbusiness use of information resources?	___	___
Have means been implemented to limit employee access to nonbusiness Web sites (for example, Internet filters, firewall configurations, or use of an ISP that blocks access to such sites)?	___	___
Does your corporate IT usage policy discuss the inappropriate use of anonymous remailers?	___	___
Has your corporate firewall been set to detect the use of anonymous remailers?	___	___
Has your company (in cooperation with legal counsel) formed a policy on the use of John Doe lawsuits to identify the authors of libelous, anonymous e-mail?	___	___
Does your corporate IT usage policy make it clear that defamation and hate speech have no place in the business setting?	___	___
Does your corporate IT usage policy prohibit the viewing or sending of pornography?	___	___
Is employee e-mail regularly monitored for defamatory, hateful, and pornographic material?	___	___
Does your corporate IT usage policy tell employees what to do if they receive hate mail or pornography?	___	___

Summary

1. What is the legal basis for the protection of freedom of speech in the United States, and what types of speech are not protected under the law?

   The First Amendment protects the right to freedom of religion and freedom of expression from government interference. Numerous court decisions have broadened the definition of speech to include nonverbal, visual, and symbolic forms of expression. Sometimes the speech at issue is unpopular or highly offensive to the majority of people; however, the Bill of Rights provides protection for minority views. The Supreme Court has also ruled that the First Amendment protects the right to speak anonymously as part of the guarantee of free speech. Obscene speech, defamation, incitement of panic, incitement to crime, "fighting words," and sedition are not protected by the First Amendment and may be forbidden by the government.

2. In what ways does the Internet present new challenges in the area of freedom of expression?

   The Internet enables a worldwide exchange of news, ideas, opinions, rumors, and information. Its broad accessibility, open-minded discussions, and anonymity make it an ideal communications medium. People must often make ethical decisions about how to use such remarkable freedom and power. Organizations and governments have attempted to establish policies and laws to help guide Internet use as well as protect their own interests. Businesses, in particular, have sought to conserve corporate network capacity, avoid legal liability, and improve worker productivity by limiting the nonbusiness use of IT resources.

3. What key free-speech issues relate to the use of information technology?

   Controlling access to Internet information is one key issue. Although there are clear and convincing arguments to support freedom of speech on the Internet, the issue is complicated by the ease with which children can access the Internet. The conundrum is that it is difficult to restrict Internet access by children without also restricting access by adults. The U.S. government has passed several laws in an attempt to address this issue, and software manufacturers have invented special applications whose goal is to block access to objectionable material.

   Anonymous communication is another key issue. Businesses must protect against the public expression of opinions that might hurt their reputations and avoid the public sharing of company confidential information. When anonymous employees reveal harmful information over the Internet, the potential for broad dissemination is enormous, and can require great effort to identify the people involved and stop them. Internet users who want to remain anonymous can send e-mail to an anonymous remailer service, but whether the communication is ethical is up to the person sending the e-mail.

   A National Security Letter (NSL) requires organizations to turn over electronic records about the finances, telephone calls, e-mail, and other personal information of suspected terrorists or spies. Use of NSLs is highly controversial and is undergoing legal challenge.

   The spread of defamation and hate speech is another important topic, especially for ISPs. In the United States, Internet speech that is merely annoying, critical, demeaning, or offensive enjoys protection under the First Amendment. Legal recourse is possible only when hate speech turns into clear threats and intimidation against specific citizens. From time to time, ISPs have voluntarily agreed to prohibit their subscribers from sending hate messages using their services. Because such prohibitions can be included in the service contracts between a private ISP and its subscribers and do not involve the federal government, they do not violate the subscribers' First Amendment rights. After an ISP implements such a prohibition, it must monitor the use of its service to ensure that the regulations are followed. When a violation occurs, the ISP must take action to prevent it from happening again.

Another issue involves the use of information technology to access, store, and distribute pornography. As long as companies can show that they were taking reasonable steps to prevent pornography, they have a valid defense if they are subject to a sexual harassment lawsuit. These steps include establishing a computer usage policy that prohibits access to pornography sites, identifying those who violate the policy, and taking action against those users—no matter how embarrassing it is for the users or how harmful it might be for the company.

Self-Assessment Questions

1. The _____ is the most basic legal guarantee to the right of freedom of expression in the United States.

   a. Bill of Rights

   b. Fourth Amendment

   c. First Amendment

   d. Constitution

2. The right to freedom of expression has been broadened to include nonverbal, visual, and symbolic forms of expression. True or False?

3. An important Supreme Court case that established a test to determine if material is obscene and therefore not protected speech was _____ .

4. An oral statement that is false and that harms another person is called _____ .

   a. defamation

   b. slander

   c. libel

   d. freedom of expression

5. The Communications Decency Act, which was passed in 1996 and was aimed at protecting children from online pornography, was eventually ruled unconstitutional. True or False?

6. The U.S. Supreme Court in *U.S. v. American Library Association* ruled that public libraries must install Internet filtering software to comply with all portions of _____ .

   a. the Children's Internet Protection Act

   b. the Child Online Protection Act

   c. the Communications Decency Act

   d. the Internet Filtering Act

7. Anonymous expression, which allows people to state opinions without revealing their identity, has been found unconstitutional. True or False?

8. A lawsuit in which the true identity of the defendant is temporarily unknown is called

    _____ .

9. *Doe v. 2TheMart.com* set legal guidelines for granting a subpoena to identify anonymous Internet speakers. True or False?

10. Under Section 505 of the _____ , the FBI is allowed to use an NSL to obtain records from banks and other financial institutions if the records are sought for an intelligence or terrorism investigation. Section 505 also includes a gag provision prohibiting any firm that receives an NSL from even disclosing the fact that the FBI is seeking information.

11. AOL and other ISPs often include a clause in their service contracts that prohibits subscribers from sending hate messages using their services. Such clauses have generally been ruled to violate the subscribers' First Amendment rights and so are unenforceable. True or False?

12. Which of the following statements about Internet pornography is true?

    a. The First Amendment is often used to protect distributors of adult pornography over the Internet.

    b. There are fewer than 60,000 Web sex sites.

    c. About one in six regular Internet users visits a Web sex site at least once per month.

    d. In contrast to adult pornography, few federal laws address child pornography.

Review Questions

1. State the First Amendment. To whom has the Supreme Court interpreted it to apply?

2. To what forms of speech have the courts applied the First Amendment? What types of speech are not protected?

3. *Miller v. California* defined which three conditions for judging whether speech is obscene?

4. Distinguish between defamation, slander, and libel.

5. What are the key elements of the Children's Internet Protection Act? To whom does it apply?

6. What are Internet filters? How do they work?

7. What is the role of an anonymous remailer?

8. Why can corporations and private universities prohibit students, instructors, and other employees from engaging in offensive speech using company equipment or company services, while public universities and institutions cannot?

9. What is a John Doe lawsuit? In what context is one frequently applied?

10. What is a National Security Letter?

11. What sort of hate speech can be prosecuted under the law? What forms cannot?

12. What are the two extreme viewpoints taken by companies when dealing with pornography in the workplace?

Discussion Questions

1. Outline a scenario in which you might be acting ethically but still want to remain anonymous while using the Internet. How might someone learn your identity even if you attempt to remain anonymous?

2. Do you think exceptions should be made to the First Amendment to make it easier to prosecute people who send hate e-mails and use the Internet to try to recruit people to their cause? Why or why not?

3. What can an ISP do to limit the distribution of hate e-mail? Why are such actions not considered a violation of the subscriber's First Amendment rights?

4. Go to the ICRA Web site at *www.icra.org*. After exploring this site, briefly describe how its rating process works. What are the advantages and disadvantages of this system?

5. What is a John Doe lawsuit? Do you think that corporations should be allowed to use a subpoena to identify a John Doe before proving that the person has defamed a company? Why or why not?

6. How has the USA Patriot Act changed the use of National Security Letters? Do you think the expansion of this tool is warranted in today's world climate? Why or why not?

7. Do you think further efforts are appropriate to limit the dissemination of pornography on the Internet? Why or why not?

What Would You Do?

1. A friend contacts you about joining his company, Anonymous Remailers Anonymous. He would like you to lead the technical staff at a 25 percent increase in salary and benefits over your current position. Your initial project would be to increase protection for users of the company's anonymous remailer service. In discussing the opportunity with your friend, you learn that some of the firm's customers are criminal types and purveyors of pornography and hate mail. Although your friend cannot be sure, he admits it is possible that terrorists may use his firm's services. Would you accept the generous job offer? Why or why not?

2. You are the vice president of human resources and are working with a committee to complete your company's computer usage policy. What advice would you offer the committee regarding Internet pornography? Would the policy be laissez-faire, or would it require strict enforcement of tough corporate guidelines? Why?

3. You are a member of your company's computer support group and have just helped a user to upgrade his computer. As you run tests after making the upgrade, you are surprised to find that the user has disabled the Internet filter software that is supposed to be standard on all corporate computers. What would you do?

4. Imagine that you receive a hate e-mail at your school or job. Does your school or workplace have a policy that covers such issues? What would you do?

5. You are the chairperson of the board of directors at your county's public library system. You plan to install Internet filtering software so that the library system will conform to the Children's Internet Protection Act and receive federal funding. Outline a plan to complete the installation that appropriately involves patrons and employees of the library and ensures minimal problems with the filters.

Cases

1. Employee Blogs Raise Freedom-of-Expression Issues

January 17, 2005, was Mark Jen's first day as associate product manager for Google. He was a little disappointed to be working in the advertising unit, as he had wanted to work on consumer products and "revolutionize the way people use computers and the Internet." On his first day, he showed up early, received his security badge, sat through a boring three-hour orientation, and filled out paperwork. He spent the rest of the day searching Google's corporate intranet, which to his surprise was very disorganized, recreating "the chaos of the Internet on its internal network." That evening, Mark returned home and started a new blog to chronicle his experiences at Google. He mused that maybe one day he would compile the material into a book, but—in the meantime—his entries might be of interest to his family and friends.

Little did Mark realize that his blog, called Ninetyninezeroes—one zero short of a mathematical googol—would attract the attention overnight of thousands of people curious about life inside Google. His entries frankly discussed his first day on the job, a global sales meeting, and Google's compensation packages. His comments evidently included information about Google's future products and economic performance. Within a couple of days, Mark's audience had grown into the tens of thousands.

The following week, Ninetyninezeroes was suddenly offline for a couple days. Then, on January 26, Mark revealed that he had been asked to take down sensitive information about the company. On January 27, the entries stopped altogether, and rumors were rampant as the number of visitors to his blog climbed toward 100,000 per day. On February 9, Mark finally checked back in to let his readers know that he had been fired on January 28.

Within the blogosphere, Mark became a cause celebre and Google's reputation suffered. The incident sent a shock wave through the IT industry, forcing companies to evaluate and establish their own blogging policies.

In sharp contrast to Google, Sun Microsystems encourages its employees to blog. Perhaps the most popular is "Jonathan's Blog," featuring the writings of Jonathan Schwartz, the company president, who candidly discusses Sun's strategies, IT trends, and criticisms of competitors. Other Sun bloggers have also attracted large followings.

Although companies are aware that employee blogging involves risks that employees might reveal company secrets or breach federal security disclosure laws, they also recognize that blogging is a new way to reach out to partners, customers, and employees and to reshape their corporate image.

Within a few months of Mark's dismissal, companies such as Sun, IBM, and Yahoo! began to formulate and publish employee blogging policies. Many guidelines suggested that employees use their common sense: Don't reveal company secrets. Don't insult people that you interact with for eight hours each day. Check with the Public Relations Department if you are not sure whether you should include specific information. Other guidelines provided suggestions that might improve the quality of the blog and make it more appealing to potential readers: Be interesting. Don't use pen names or ghost writers. Be authentic.

Both the IT industry and Mark Jen, who did not blame Google, learned something. By the spring of 2005, Mark had found a job at Plaxo, where he helped create its official blogging guidelines, among other responsibilities.

Questions:

1. Do you believe Mark Jen's First Amendment rights were violated by Google? Why or why not?

2. Employee bloggers sometimes use pseudonyms. Under what conditions might the blogging service be forced to provide the real identity of the blogger?

3. Following Mark Jen's dismissal, Google made it clear that it allows employees to maintain blogs and even to post information about problems with Google's blogging service. What reasons might Google have for releasing this information to the public?

2. U.S. Companies Aid and Abet Chinese Internet Censorship

In January 2003, Representative Christopher Cox (R-Calif.) introduced a bill called the Global Internet Freedom Act, which would provide funding for the development of technologies that would counteract Internet censorship and jamming. The introduction of the bill followed a congressional panel during which IT professionals had emphasized the urgency of the situation.

"The Chinese government is already on its third generation of firewall technology, and we haven't even started Version one of our counterstrategy yet," argued Paul Baranowski of Peekabooty, a project that developed a computer program to circumvent government censorship of the Internet. "If we do not do something soon, they may be able to close off the country completely."

Ironically, while the U.S. government was considering allocating $50 million to anticensorship technologies, U.S. companies were helping to build the censorship infrastructure. IT giants Cisco, Sun Microsystems, Yahoo!, Google, and Microsoft have all provided technology with which the Chinese government denies its citizens free speech.

For example, state-of-the-art Cisco routers can set up 750,000 filters that block users not only from accessing large domains, but specific areas within a domain. "Our perspective is that it's the user, not Cisco, that determines the functionality and uses to which the technology is put," said Cisco spokesperson Terry Alberstein.

Yahoo! has been under fire from human rights groups for years for restructuring its Chinese search engine to prevent access to Web sites that the communist government considers subversive or lewd. "It's just really important for us to have good relations and good partnerships with governments all over the world," said Yahoo! chief executive officer Terry Semel.

In the Chinese version of MSN Spaces, Microsoft's blogging site, phrases and words like *Dalai Lama* and *democracy* are banned. Google recently made what it termed "a difficult decision" to remove headlines linked to news stories that the communist government finds objectionable.

So, even if Congress hadn't tabled the Global Internet Freedom Act in committee, what chance would Peekabooty or some other small, anticensorship IT company have against the infrastructure these giants are building? A year ago, the answer may have been "slim." Then, in an action that concerned the free world and set a dangerous precedent, Yahoo! handed over the identify of a political dissident to the communist government.

In June 2004, Chinese journalist Shi Tao e-mailed a Communist Party document to a "foreign hostile element." The document warned of possible unrest during the anniversary of the Tiananmen Square massacre. In April 2005, Shi Tao was arrested and sentenced to 10 years in jail. In September 2005, the story broke across the world that Yahoo! had provided the Chinese government with the information needed to identify Shi Tao.

The huge Chinese market is a powerful temptation for the IT giants, who maintain that they must observe local laws. If they refuse to comply with the demands of the Chinese government, they will simply be replaced by their competition. Yet the IT giants might have to contend with another challenge: In May 2005, Boston Common Asset Management, an investment group that holds substantial shares in Cisco, filed a shareholder resolution demanding that Cisco implement a comprehensive human rights policy with respect to China and other countries that engage in Internet censorship.

Questions:

1. The Chinese government frequently emphasizes that its online censorship prevents citizens from accessing pornographic sites, deemphasizing its political censorship. How do you think these two types of censorship differ?

2. Why have human rights groups criticized Cisco and Yahoo! more than other companies that comply with the demands of the Chinese government?

3. If U.S. companies do not comply with the demands of China's communist government, they may lose business to their competitors. What do you think these companies should do?

3. The Electronic Frontier Foundation (EFF)

The EFF was founded in 1990 by John Perry Barlow, a lyricist for the Grateful Dead, and Mitch Kapor, a founder of the Lotus Development Corporation, which developed the Lotus 1-2-3 spreadsheet software. The EFF is a nonprofit, nonpartisan organization whose goal is to protect fundamental civil liberties related to technology, including privacy and freedom of expression on the Internet. It frequently undertakes court cases as an advocate of preserving individual rights.

The EFF's mission includes educating the press, policy makers, and the general public about civil liberties. Its Web site (*www.eff.org*) provides an extensive collection of information on issues such as censorship, free expression, digital surveillance, encryption, and privacy. The Web site gets more than 100,000 hits a day and is among the most visited sites on the Internet. Many visitors are key managers within companies who are responsible for making decisions about how to act ethically and legally in applying information technology.

Although more than 80 percent of its annual operating budget comes from individual donations from concerned citizens, the EFF also has corporate sponsors, including Intuit, Netscape, Oracle, Pacific Bell, Sun Microsystems, and law firms. Many of these sponsors hope to do a better job of minimizing the negative impacts of IT on society, help IT users become more responsible in fulfilling their role and responsibilities, and eventually build a more knowledgeable marketplace for their products. On the other hand, the EFF has developed many critics over the years for what some see as its bias against most forms of regulation.

Questions:

1. What reasons might a firm give for joining and supporting the EFF?

2. Visit the EFF Web site and develop a list of the current "hot issues." Research one EFF issue that interests you, and write a brief paper summarizing the EFF's position. Discuss whether you support this position and why.

3. The vice president of public affairs for your midsized telecommunications equipment company has suggested that the firm donate $100,000 to the EFF and become a corporate sponsor. The CFO has asked if you, the CIO, support this action. What would you say?

End Notes

[1] Joffe-Walt, Benjamin, "China's Leaders Launch Smokeless War Against Internet and Media Dissent," *The Guardian*, www.guardian.co.uk/china/story/0,7369,1578189,00.html, September 26, 2005.

[2] Kahn, Joseph, "China Tightens Its Restrictions for News Media on the Internet," *New York Times*, www.nytimes.com/2005/09/26/international/asia/26china.html?ex=1128744000&en =d94ff777a3ccb20d&ei=5070, September 26, 2005.

[3] "Supreme Court Rules: Cyberspace Will be Free! ACLU Hails Victory in Internet Censorship Challenge," American Civil Liberties Union Web site, www.aclu.org, June 27, 1996.

[4] Goff, Leslie Jaye, "Technology Flashback—1996 Internet Content Decency Debate," *Computerworld*, www.computerworld.com, November 29, 1999.

[5] "U.S. Supreme Court Decision, Reno et al. v. ACLU, et al.," www2.epc.org, June 27, 1997.

[6] Title XIV-Child Online Protection Act, Electronic Privacy Information Center Web site, www.epic.org/free_speech/censorship/copa.html, September 27, 2005.

[7] Children's Internet Protection Act, Federal Communications Commission Web site, www.fcc.gov/cgb/consumerfacts/cipa.html, September 29, 2005.

[8] Machilis, Sharon, "Sailor Settles with AOL, Navy," *Computerworld*, www.computerworld.com, June 12, 1998.

[9] Quistgaard, Kaitlin, "Raytheon Triumphs Over Yahoo Posters' Anonymity," www.salon.com, May 24, 1999.

[10] "Microsoft Files More Anti-Spam Lawsuits in Conjunction with Leading ISPs," Press Release at Microsoft Web site, www.microsoft.com/presspass/press/2004/oct04/10-28canspamfollowuppr.mspx, October 28, 2004.

[11] "Planned Parenthood of the Columbia/Williamette Inc. v. American Coalition of Life Activists," www.ce9.uscourts.gov, March 28, 2001.

[12] "Companies' Lawsuits Seek to Silence Employees," *ACLU News*, ACLU Web site, www.aclu.org, January 7, 2002.

[13] Egan, Timothy, "Wall Street Meets Pornography," *New York Times*, www.nytimes.com, October 25, 2000.

[14] Williams, Martyn, "Three Indicted in U.S. Spam Crackdown," *Computerworld*, www.computerworld.com, August 26, 2005.

[15] Roberts, Paul, "Arrest, But No Relief from IM Spam," *Computerworld*, www.computerworld.com, February 22, 2005.

[16] Swanson, Sandra, "You Be The Judge," *Information Week*, www.informationweek.com, pages 18-20, August 6, 2001.

Sources for Case 1

Jen, Mark, "Archives January 2005," Plaxoed! Blog, blog.plaxoed.com/2005/01/.

Perez, Juan Carlos, "Fired Google Blogger Reflects, Moves On," PC World, www.pcworld.idg.com.au/index.php/id;1106244715;fp;512;fpid;653243282, February 17, 2005.

Pimentel, Benjamin, "Writing the Codes on Blogs: Companies Figure out What's OK, What's Not in Online Realm," *San Francisco Chronicle*, www.sfgate.com/cgi-bin/article.cgi?file=/chronicle/archive/2005/06/13/BUGOMD64QH51.DTL&type=tech, June 13, 2005.

Sources for Case 2

Grebb, Michael, "China's Cyberwall Nearly Concrete," *Wired News*, www.wired.com/news/politics/0,1283,56195,00.html, November 5, 2002.

Johnson, Tim, "Critics Say U.S. Companies Enable Censorship," *Miami Herald*, www.miami.com/mld/miamiherald/news/12199310.htm, July 24, 2005.

Kirby, Carrie, "Chinese Internet vs. Free Speech: Hard Choices for U.S. Tech Giants," *San Francisco Chronicle*, www.sfgate.com/cgi-bin/article.cgi?file=/c/a/2005/09/18/MNGDUEPNLA1.DTL&type=tech, September 18, 2005.

Sources for Case 3

Kizilos, Peter, "(Interview with) Lori Fena," *Online*, www.findarticles.com, March-April 1998.

EFF Web site, www.eff.org, June 18, 2002.

CHAPTER **6**

INTELLECTUAL PROPERTY

VIGNETTE

Apple Trade Secrets Revealed?

Nicholas Ciarelli is an avid fan of Apple who has operated a popular Apple Web site (*www.thinksecret.com*) since he was 13. He is now a student at Harvard and editor of *The Crimson*, the school's newspaper. Ciarelli, whose site dishes out information and speculation about Apple, incurred the firm's wrath by heralding the arrival of a new Mac mini for $499 and the iLife software package, two weeks before they were officially announced at the Macworld Conference and Expo in San Francisco.

Apple, which is notorious for being excessively secretive about its business plans and products, reacted to the unauthorized announcement as an attack on its valuable trade secrets. Its attorneys filed a lawsuit claiming that Ciarelli "solicited information about unreleased products...from individuals who violated their confidentiality agreements with Apple by providing details that were later posted on the Internet."

Ciarelli was quoted as saying, "I employ the same legal newsgathering practices used by any other journalist. I talk to sources of information, investigate tips, follow up on leads, and corroborate details. I believe these practices are reflected in Think Secret's track record."

On one side of the dispute, some believe that the Uniform Trade Secrets Act prevents third parties from revealing information obtained from people bound by nondisclosure agreements. On the other hand, many believe that Ciarelli's actions were protected on the basis of First Amendment free-speech rights, and that Apple should have focused on identifying and punishing the employees who leaked the information.[1,2,3,4,5]

LEARNING OBJECTIVES

As you read this chapter, consider the following questions:

1. What does the term *intellectual property* encompass, and why are companies so concerned about protecting it?
2. What are the strengths and limitations of using copyrights, patents, and trade secret laws to protect intellectual property?
3. What is plagiarism, and what can be done to combat it?
4. What is reverse engineering, and what issues are associated with applying it to create a look-alike of a competitor's software program?
5. What is open source code, and what is the fundamental premise behind its use?
6. What is the essential difference between competitive intelligence and industrial espionage, and how is competitive intelligence gathered?
7. What is cybersquatting, and what strategy should be used to protect an organization from it?

WHAT IS INTELLECTUAL PROPERTY?

Intellectual property is a term used to describe works of the mind, such as art, books, films, formulas, inventions, music, and processes, that are distinct and "owned" or created by a single person or group.

Copyright law protects authored works such as art, books, film, and music. Patent laws protect inventions, and trade secret laws help safeguard information that is critical to an organization's success. Together, copyright, patent, and trade secret legislation forms a complex body of law that addresses the ownership of intellectual property. Such laws can also present potential ethical problems for IT companies and users—for example, some innovators believe that copyrights, patents, and trade secrets stifle creativity by making it harder to build on the ideas of others.

Defining and controlling the appropriate level of access to intellectual property are complex tasks. For example, protecting computer software has proven to be difficult because it has not been well categorized under the law. Software has sometimes been treated as the expression of an idea that can be protected under copyright law, but it also has been treated as a process for changing a computer's internal structure, making it eligible for protection under patent law. At one time, software was even judged to be a series of mental steps, making it inappropriate for ownership and ineligible for any form of protection.

Copyrights

Copyright and patent protection is established in the U.S. Constitution in Article I, section 8, clause 8, which specifies that Congress shall have the power "to promote the Progress of Science and useful Arts, by securing for limited Times to Authors and Inventors the exclusive Rights to their respective Writings and Discoveries."

A **copyright** grants the creators of "original works of authorship in any tangible medium of expression, now known or later developed, from which they can be perceived, reproduced, or otherwise communicated, either directly or with the aid of a machine or device, the exclusive right to distribute, display, perform, or reproduce the work in copies or to prepare derivative works based upon the work" (U.S. Code, Title 17, Section 102(a)). The author may grant this exclusive right to others. As new forms of expression develop, they can be awarded copyright protection. For example, audiovisual works were added and computer programs were assigned to the literary works category in the Copyright Act of 1976.

Copyright law guarantees developers the rights to their works for a certain amount of time. Since 1960, the term of copyright has been extended 11 times from its original limit of 28 years. The Sonny Bono Copyright Term Extension Act, signed into law in 1998, set the following time limits:

- For works created after January 1, 1978, copyright protection endures for the life of the author plus 70 years.
- For works created but not published or registered before January 1, 1978, the term endures for the life of the author plus 70 years, but in no case expires earlier than December 31, 2004.
- For works created before 1978 that are still in their original or renewable term of copyright, the total term was extended to 95 years from the date the copyright was originally secured.

These extensions were championed mainly by movie studios that wanted to retain the rights to their early films. Opponents argued that lengthening the copyright period made it more difficult for artists to build on the work of others, thus stifling creativity and innovation. The act was legally challenged by Eric Eldred, a bibliophile who wanted to put digitized editions of old books online. The case went all the way to the Supreme Court, which ruled the act constitutional in 2003.

The types of work that can be copyrighted include architecture, art, audiovisual works, choreography, drama, graphics, literature, motion pictures, music, pantomimes, pictures, sculptures, sound recordings, and other intellectual works, as described in Title 17 of the U.S. Code. To be eligible for a copyright, a work must fall within one of the preceding categories, and it must be original. Copyright law has proven to be extremely flexible in covering new technologies—software, video games, multimedia works, and Web pages can all be protected. However, evaluating the originality of a work can cause problems and lead to litigation.

Some works are not eligible for copyright protection, including those that have not been fixed in a tangible form of expression (such as an improvisational speech) and works that consist entirely of common information that contain no original authorship, such as a chart showing conversions between European and American units of measure.

Copyright law tries to strike a balance between protecting an author's rights and enabling public access to copyrighted works. To this end, the **fair use doctrine** (Title 17, section 107) established four factors for courts to consider whether a particular use of copyrighted property is fair and can be allowed without penalty:

- The purpose and character of the use (such as commercial use or nonprofit, educational purposes)
- The nature of the copyrighted work
- The portion of the copyrighted work used
- The effect of the use upon the value of the copyrighted work

The concept that an idea cannot be copyrighted, but that the expression of an idea can be, is key to understanding copyright protection. For example, an author cannot copy the exact words someone else used to describe his feelings during a World War II battle, but he can still convey the sense of horror the other person felt. Also, there is no copyright infringement if two parties independently develop a similar or even identical work. For example, if two writers happened to use the same phrase to describe a key historical figure, neither would be guilty of infringement. Of course, independent creation can be extremely difficult to prove or disprove.

Copyright infringement occurs when someone copies a substantial and material part of another's copyrighted work without permission. The courts have a wide range of discretion in awarding damages—from $200 for innocent infringement to $100,000 for willful infringement.

One area of copyright infringement involves the worldwide sale of counterfeit consumer supplies, such as toner and ink-jet cartridges.[6] Computer printer manufacturers profit through the sales of their own cartridges, of course, and argue that customers who use counterfeit supplies risk inferior print quality, toner leakage, high cartridge failure rates, and increased equipment downtime. Smart shoppers know, however, that refurbished cartridges sell at a considerable discount.

One company that objected to the use of refurbished cartridges was Lexmark International, a manufacturer and supplier of printers, associated supplies, and services. In 2002, Lexmark filed suit against Static Control Components (SCC), a producer of third-party components used to make refurbished printer cartridges. The suit alleged that SCC's

Smartek chips included Lexmark software in violation of copyright law and the Digital Millennium Copyright Act. In February 2003, Lexmark was granted an injunction that prevented SCC from selling the chips until the case could be resolved at trial. However, the ruling was overturned in October 2003 by the U.S. Court of Appeals for the Sixth Circuit, which said that copyright law should not be used to inhibit interoperability between the products of rival vendors. The appeals court upheld its own decision in February 2005.[7]

Software Copyright Protection

The use of copyrights to protect computer software raises complicated issues of interpretation. For example, a software manufacturer can observe the operation of a competitor's copyrighted program and then create a program that accomplishes the same result and performs in the same manner. To prove infringement, the copyright holders must show a striking resemblance between their software and the new software that could be explained only by copying. However, if the new software's manufacturer can establish that it developed the program on its own, without any knowledge of the existing program, there is no infringement. For example, two software manufacturers could conceivably develop separate programs for a simple game such as tic-tac-toe without infringing the other's copyright.

The WIPO Copyright Treaty

The World Intellectual Property Organization (WIPO), headquartered in Geneva, Switzerland, is an agency of the United Nations that was established in 1967. Its goal is the "maintenance and further development of the respect for intellectual property throughout the world."[8] It has 182 member nations and administers 23 international treaties. Since the 1990s, WIPO has strongly advocated for the interests of intellectual property owners.

The WIPO Copyright Treaty, adopted in 1996, provides additional copyright protections to address electronic media. The treaty ensures that computer programs are protected as literary works and that the arrangement and selection of material in databases is also protected. It provides authors with control over the rental and distribution of their work and prohibits circumvention of any technical measures put in place to protect the works. The WIPO Copyright Treaty is implemented in U.S. law by the Digital Millennium Copyright Act (DMCA).[9]

The Digital Millennium Copyright Act

The DMCA was signed into law in November 1998 and was written in compliance with the global copyright protection treaty from WIPO. The DMCA added new provisions, making it an offense to do the following:

- Circumvent a technical protection.
- Develop and provide tools that allow others to access a technologically protected work.
- Manufacture, import, provide, or traffic in tools that enable others to circumvent protection and copy a protected work.

Violations of these provisions carry both civil and criminal penalties, including up to five years in prison, a fine of up to $500,000 for each offense, or both. Unlike traditional copyright law, the statute does not govern copying, but the distribution of tools and software that can be used for copyright infringement, as well as for legitimate noninfringing use. Although the DMCA explicitly outlaws technologies that can defeat copyright protection devices, it does permit reverse engineering for encryption, interoperability, and computer security research.

Several cases brought under the DMCA dealt with the use of software to enable the copying of DVD movies. For example, motion picture companies supported the development and worldwide licensing of the Content Scramble System (CSS), which enables a DVD player or a computer drive to decrypt, unscramble, and play back, but not copy, motion pictures on DVDs. However, a software program called DeCSS can break the encryption code and enable users to copy DVDs. The posting of this software on the Web in January 2000 led to a lawsuit by major movie studios against its author. After a series of cases, courts finally ruled that the use of DeCSS violated the DMCA's anticircumvention provisions.

Opponents of DMCA say that it gives holders of intellectual property so much power that it actually restricts the free flow of information. For example, under DMCA, Internet service providers (ISPs) are required to remove access to Web sites that allegedly break copyright laws—even before the copyright infringement has been proven. Companies that provide Internet access to music and videos face legal action and the failure of their businesses if they do not gain approval to publish content from the music and movie industry.

Patents

A **patent** is a grant of a property right to inventors; it is issued by the U.S. Patent and Trademark Office (USPTO). A patent permits its owner to exclude the public from making, using, or selling a protected invention, and it allows legal action against violators. Not only does a patent prevent copying, it prevents independent creation, unlike a copyright. Even if someone else invents the same item independently and with no prior knowledge of the patent holder's invention, the second inventor is excluded from using the patented device without permission of the original patent holder. The rights of the patent extend only to the United States and its territories and possessions, not to foreign countries.

The value of patents to a company cannot be underestimated. IBM has more than 40,000 patents active worldwide, including 25,000 in the United States. In 2004, IBM obtained 3248 U.S. patents, the 12th consecutive year it has received more U.S. patents than any other company.[10] IBM's licensing of patents and technologies generates more than $15 billion in annual revenue.

To obtain a U.S. patent, an applicant must file with the USPTO according to strict requirements. As part of the application, the USPTO searches the prior art, starting with patents and published material that have already been issued in the same area. **Prior art** is the existing body of knowledge that is available to a person of ordinary skill in the art. The USPTO will not issue a patent for an invention whose professed improvements are already present in the prior art. Although the USPTO employs some 3000 examiners to research the originality of each patent application, it still takes an average of 25 months to process one. Such delays can be costly for companies that want to bring patented products to market quickly. People who are trained in the patent process, rather than the inventors themselves, prepare about 80 percent of all patent applications.

The main body of law that governs patents is in Title 35 of the U.S. Code. According to sections 103–105 of Title 35, an invention must pass the following four tests to be eligible for a patent:

- It must fall into one of five statutory classes of items that can be patented: processes, machines, manufactures (such as objects made by humans or machines), compositions of matter (such as chemical compounds), and new uses in any of the previous four classes.
- It must be useful.
- It must be novel.
- It must not be obvious to a person having ordinary skill in the same field.

The U.S. Supreme Court has ruled that three classes of items cannot be patented: abstract ideas, laws of nature, and natural phenomena. Mathematical subject matter, standing alone, is also not entitled to patent protection. Thus, Pythagoras could not have patented his formula for the length of the hypotenuse of a right triangle ($c^2 = a^2 + b^2$).

Patent infringement occurs when someone makes unauthorized use of another's patent. Unlike copyright infringement, there is no specified limit to the monetary penalty if patent infringement is found. In fact, if a court determines that the infringement is intentional, it can award up to three times the amount of the damages claimed by the patent holder. The most common defense against patent infringement is a counterattack on the claims of infringement and the validity of the patent itself. Even if the patent is valid, the plaintiff must still prove every element of at least one claim and that the infringement caused some sort of damage.[11]

Software Patents

A software patent "claims as all or substantially all of its invention some feature, function, or process embodied in instructions that are executed on a computer."[12] Prior to 1981, the courts regularly turned down requests for such patents, giving the impression that software could not be patented. In the 1981 *Diamond v. Diehr* case, the Supreme Court denied a patent to Diehr, who had developed a process control computer and sensors to monitor the temperature inside a rubber mold. However, the USPTO interpreted the court's reasoning to mean that just because an invention used software, it did not mean that the invention could not be patented. Based on this ruling, courts have slowly broadened the scope of protection for software-related inventions.

The creation of a new Court of Appeals for the Federal Circuit in 1982 further improved the environment for the use of patents in software-related inventions. This court is charged with hearing all patent appeals and is generally viewed as providing stronger enforcement of patents and more effective punishment, including triple damages for willful infringement.

Since the early 1980s, the USPTO has granted as many as 20,000 software-related patents per year. Applications software, business software, expert systems, and system software have been patented, as well as software processes such as compilation routines, editing and control functions, and operating system techniques. Even electronic font types and icons have been patented.

Before obtaining a software patent, a developer should do a patent search, which can be lengthy and expensive. However, even a thorough search may not identify all potential infringements, because the USPTO's classification system is complex and software patents may be classified in several categories, making them difficult to find. If a patent search misses something, there is some risk of an expensive patent infringement lawsuit. The Software Patent Institute is building a database of information to document known patented software and assist the USPTO and others in researching prior art in the software arena.

Some software experts think that too many software patents are being granted and that they inhibit new software development. For example, in September 1999, Amazon.com obtained a patent for "one-click shopping" based on the use of a "shopping cart" purchase system for electronic commerce. In October 1999, Amazon.com sued Barnes & Noble for allegedly infringing this patent with its Express Lane feature. The filing of the suit prompted complaints about business method patents, which many critics deride as overly broad and unoriginal concepts that do not merit patents and stifle innovation. Some critics considered one-click shopping little more than a simple combination of existing Web technologies. Following preliminary court hearings and the filing of injunctions, and the discovery that others had used the one-click technology before Amazon.com even began business, Amazon.com and Barnes & Noble settled out of court in March 2002.

Software engineers rarely take the time or effort to search patent databases for new inventions that could benefit their projects, partly because software patents are described in obscure language and because engineers risk paying triple damages for knowingly infringing one. As a result, many software patent infringements are for independent inventions—the same invention discovered by two different parties without knowledge of the other's work.

Most large software companies have cross-licensing agreements in which each agrees not to sue the other over patent infringements. For example, Microsoft has agreements with IBM, Sun Microsystems, SAP, Hewlett-Packard, Siemens, Cisco, and AutoDesk. Microsoft is pursuing further cross-licensing agreements and wants to have deals in place with as many as 40 companies by 2010. This strategy to obtain the rights to technologies that it might use in its products provides a tremendous amount of development freedom to Microsoft without risk of expensive litigation.[13]

Major IT firms have little interest in cross-licensing with smaller firms, so small businesses have no choice but to license patents if they use them. As a result, small businesses must pay an additional cost from which larger companies are exempt. Furthermore, small businesses are generally unsuccessful in enforcing their patents against larger companies. Should a small business bring a patent infringement suit against a large firm, it can overwhelm the small business with multiple patent suits, whether they have merit or not. Considering that a software patent lawsuit can cost millions of dollars, the small firm simply cannot afford to fight; instead, it must settle and license its patents to the large company.

IBM announced in 2005 that it would donate 500 patents for free use by developers to help them innovate and build new hardware and software. The announcement represented a major shift in IBM's intellectual property strategy and was meant to encourage other patent holders to donate their own intellectual property. It also placed IBM in direct opposition to Microsoft and other major software developers. IBM's strategy may also have offended zealous patent defenders, such as major pharmaceutical and media companies, who are important IBM customers.[14]

Inventors sometimes employ a tactic called defensive publishing as an alternative to filing for patents. Under this approach, a company publishes a description of its innovation in a bulletin, Web site, conference paper, or trade journal. This establishes an idea's legal existence as prior art, which obviously provides competitors with access to the innovation. However, because the idea becomes part of the prior art, competitors cannot patent the idea or charge licensing fees to other users of the technology or technique. This approach costs mere hundreds of dollars, requires no lawyers, and is fast.

One company, Intellectual Ventures, prefers to obtain and enforce software patent rights rather than build and market usable software systems. Intellectual Ventures has raised more than $350 million from large companies, including Apple, eBay, Google, Intel, Microsoft, Nokia, and Sony, to purchase more than 1000 patents. The firm's investors are protected from being sued over these patents, while other companies must pay royalties for using the patents or face legal action.

Submarine Patents and Patent Farming

A standard is "a definition or format that has been approved by a recognized standards organization or is accepted as a de facto standard by the industry. Standards exist for programming languages, operating systems, data formats, communications protocols, and electrical interfaces."[15] Standards are extremely useful because they enable hardware and software from different manufacturers to work together.

A technology, process, or principle that has been patented may be embedded—knowingly or unknowingly—within a standard. If so, the patent owner can demand a royalty payment from any party that implements the standard, or may refuse to permit certain parties from using the patent, thus effectively blocking them from using the standard. A patent that is hidden within a standard and does not surface until the standard is broadly adopted is called a **submarine patent**. A devious patent holder might influence a standards organization to make use of its patented item without revealing the existence of the patent. Then, later, the patent holder might demand royalties from all parties that use the standard. This strategy is known as **patent farming**.[16]

One possible example of a submarine patent could be Patent no. 5,838,906, which is owned by the University of California and licensed exclusively to a small software company called Eolas Technologies. The patent describes how a Web browser can use external applications. The patent holder did not make the patent known for years and then sued Microsoft for use of the principle. The university and Eolas received a $520 million award in August 2003 after a federal jury found that Microsoft's Internet Explorer browser infringed the patent. In November 2003, the patent office began a review of the patent based on a request from world-renowned Tim Berners-Lee, father of the World Wide Web and director of the World Wide Web Consortium. He argued that the 1998 patent should be invalidated because of the existence of prior art, or previous examples of the technology's use.[17] In January 2004, a federal judge upheld the original decision, requiring Microsoft to pay $520 million on grounds that Internet Explorer infringed the patent. The judge rejected Microsoft's request for a new trial and, adding insult to injury, ordered the payment of more than $45 million in interest.[18]

A possible example of patent farming involved Rambus, a designer and manufacturer of computer memory technology. Rambus allegedly influenced a standards organization to use its technology without disclosing that it had a patent application in process. Years later, Rambus was enmeshed in a series of lawsuits with three of the world's leading memory chip makers, including Infineon in Germany, Micron Technology in the United States, and Hynix Semiconductor in South Korea. Rambus claimed that rival producers of dynamic RAM chips infringed its patents. Collectively, the penalties for patent infringement could have been worth billions of dollars.[19] In January 2005, a U.S. Federal court ruled that Hynix had infringed several Rambus patents for memory products.

Trade Secret Laws

The Uniform Trade Secrets Act (UTSA) was drafted in the 1970s to bring uniformity to U.S. states in the area of trade secret law. The first state to enact the UTSA was Minnesota in 1981, followed by 39 more states and the District of Columbia. In Chapter 2, a trade secret was defined as business information that represents something of economic value, has required effort or cost to develop, has some degree of uniqueness or novelty, is generally unknown to the public, and is kept confidential. Similarly, the UTSA defines a trade secret as "information, including a formula, pattern, compilation, program, device, method, technique, or process, that:

- Derives independent economic value, actual or potential, from not being generally known to, and not being readily ascertainable by, persons who can obtain economic value from its disclosure or use.
- Is the subject of efforts that are reasonable under the circumstances to maintain its secrecy."

Under these terms, computer hardware and software can qualify for trade secret protection by the UTSA.

The Economic Espionage Act (EEA) of 1996 imposes penalties of up to $10 million and 15 years in prison for the theft of trade secrets. Before the EEA, there was no specific criminal statute to help pursue economic espionage; the FBI was investigating nearly 800 such cases in 23 countries when the EEA was enacted. Today, intellectual property loss from industrial and economic espionage costs U.S.-based businesses more than $300 billion annually, according to estimates by Guardsmark, a New York security services firm.[20]

As with the UTSA, information is considered a trade secret under the EEA only if companies take steps to protect it. Trade secret protection begins by identifying all the information that must be protected—from undisclosed patent applications to market research and business plans—and developing a comprehensive strategy for keeping the information secure.

Employees are the greatest threat to the loss of company trade secrets—they might accidentally disclose trade secrets or steal them for monetary gain. Organizations must educate employees about the importance of maintaining the secrecy of corporate information. Trade secret information should be labeled clearly as confidential and should only be accessible by a limited number of people. Most organizations have strict policies regarding nondisclosure of corporate information.

Because organizations can risk losing trade secrets when key employees leave, they often try to prohibit employees from revealing secrets by adding **nondisclosure clauses** to employment contracts. Thus, employees cannot take copies of computer programs or reveal the details of software owned by the firm, even when they leave. However, enforcing employment agreements can be difficult.

Another option for preserving trade secrets is to have an experienced member of the Human Resources Department conduct an exit interview with each departing employee. A key step in the interview is to review a checklist that deals with confidentiality issues. At the end of the interview, the departing employee is asked to sign an acknowledgment of responsibility not to divulge any trade secrets.

Employers can also use **noncompete agreements** to protect intellectual property from being used by competitors when key employees leave. Such agreements require employees not to work for any competitors for a period of time, perhaps one to two years. When courts are asked to settle disputes over noncompete agreements, they consider the reasonableness of the restrictions and how the restrictions protect the legitimate interests of the former employer. The courts also consider geographic area and the length of time of the restrictions in relation to the pace of the industry. A 2001 survey by TMP Executive Search revealed that 78 percent of telecommunications, software, and hardware companies surveyed ask at least some employees to sign noncompete agreements when they are hired. A standard agreement follows:

> "The employee agrees as a condition of employment that, in the event of termination for any reason, he/she will not engage in a similar or competitive business for a period of two years, nor will he/she contact or solicit any customer with whom Employer conducted business during his/her employment. This restrictive covenant shall be for a term of two years from termination, and shall encompass the geographic area within a 100-mile radius of Employer's place of business."

One dispute over noncompete agreements involved Compuware, which makes software to help people run high-end IBM mainframe computers. IBM began to sell its own version of such software years ago. Compuware sued IBM in March 2002, claiming that IBM had stolen trade secrets to copy its software by recruiting former Compuware employees to develop the software, despite nondisclosure agreements signed by the employees. Compuware also alleged that IBM used portions of its copyrighted software for IBM's new File Manager and Fault Analyzer products. Compuware charged that the IBM products had many features that were substantially similar to its own, including certain bugs that strongly indicated Compuware's source code was copied. In its opening court statements in February 2005, IBM denied Compuware's claims, stating that it merely used public information and its own expertise to create products that helped lower prices in a market Compuware had dominated.[21] After a five-week trial, IBM agreed to license at least $400 million of Compuware software and services over four years. The companies also established a cross-licensing agreement to cover patents related to both companies.[22]

The increasing use of corporate networks and the Internet has heightened the risk of losing trade secrets stored on computers. A company's IT organization must implement firewalls and other safeguards to limit outside access to corporate computers, encrypt sensitive e-mail, restrict access to servers, and secure databases to ensure that electronic copies of the firm's trade secrets are not accidentally or deliberately released. An area of special concern is the use of remote computers by employees who travel or take work home on a regular basis. Such computers can be easily stolen and are often used carelessly.

For example, CIA director John Deutch used his home computer in 1996 to work on top-secret material. The computer was not configured with the special hardware and software needed for classified work and was also used to connect to the Internet, creating an opportunity for hackers around the world to steal secret information. After being called before the Senate Armed Services Committee, Deutch received a reprimand and lost his security clearance. He was subjected to a damaging investigation and later received a controversial pardon from departing President Bill Clinton.

The following legal overview explains some of the nuances of what constitutes a trade secret and what does not.

LEGAL OVERVIEW

The Battle Over Customer Lists

Losing customer information to competitors is a growing concern in industries in which companies compete for many of the same clients. There are numerous cases of employees making unauthorized use of their employer's customer list. For example, Harrah's Entertainment charged that a former employee copied a customer list of 450 wealthy patrons, including names, contact information, and credit and account histories, from a Harrah's database before leaving to work at a competing casino.

Also, a hacker accessed AOL's subscriber list and used the e-mail addresses to promote his Internet gambling business. Information on AOL's 30 million subscribers is maintained in the company's data warehouse, and access is limited to a few employees. The hacker, who had worked at AOL for more than five years, didn't have access to the data, but impersonated another employee to reach it. The hacker got screen names, zip codes, credit card types, and telephone numbers, but not credit card numbers because AOL stores them separately.

So, does a firm's customer list represent a trade secret that is protected by intellectual property law? Legally, a customer list is not automatically considered a trade secret. If a company doesn't treat the list as valuable, confidential information internally, neither will the court. The courts must consider two main factors in making this determination. First, did the firm take prudent steps to keep the list secret by taking the following actions?

- Labeling it as confidential
- Storing it in a locked facility or in a password-protected computer file
- Limiting access to a small number of people on a need-to-know basis
- Requiring employees to sign a nondisclosure agreement that specifically mentions customer lists

Second, did the firm expend money or effort to develop the customer list? The more the firm invested to build its customer list and the more that the list provides the firm with a competitive advantage, the more likely the courts are to accept the list as a trade secret. A customer list that can be easily recreated by accessing publicly available sources, such as trade journals or even the telephone book, will not qualify as a trade secret.

continued

Many people wrongly assume that customer lists are public information, so workers should be educated about the confidentiality of such lists. At some companies, employees are briefed on court cases involving stolen customer lists and threatened with prosecution for stealing trade secrets.[23,24,25]

Trade secret law has several key advantages over the use of patents and copyrights in protecting companies from losing control of their intellectual property:

- There are no time limitations on the protection of trade secrets, as there are with patents and copyrights.
- There is no need to file any application, make disclosures to any person or agency, or disclose a trade secret to outsiders to gain protection. After the USPTO issues a patent, competitors can obtain a detailed description of it.
- Patents can be ruled invalid by the courts, meaning that the affected inventions no longer have patent protection; this risk does not exist for trade secrets.
- No filing or application fees are required to protect a trade secret.

Because of these advantages, more technology worldwide is protected as trade secrets rather than by patent.

Trade secret law protects only against the *misappropriation* of trade secrets. If competitors come up with the same idea on their own, it is not misappropriation; the law doesn't prevent someone from using the same idea if it was developed independently. For example, developing a software program from scratch that duplicates the functionality of another program would be legal, but stealing the program's source code would be illegal. Also, trade secret law does not prevent someone from analyzing an end product to figure out the trade secret behind it.

Trade secret protection and patent law differ greatly from country to country. For example, the Philippines provides no legal protection for trade secrets at all. Pharmaceuticals, methods of medical diagnosis and treatment, and information technology cannot be patented in some European countries. Many Asian countries require foreign corporations operating there to transfer rights to their technology to locally controlled enterprises. (Coca-Cola reopened its operations in India in 1993 after halting sales for 16 years to protect the "secret formula" for its soft drink, even though India's vast population represented a huge potential market.) American businesses that seek to operate in foreign jurisdictions or enter international markets must take these differences into account.

The WTO and the WTO TRIPs Agreement

The World Trade Organization (WTO) was established in 1995 to deal with the rules of international trade. These rules are based on WTO agreements that are negotiated and signed by members of the world's trading nations and ratified in their parliaments. The WTO is headquartered in Geneva, Switzerland, and had 148 member nations as of 2004. Its goal is to help producers of goods and services, exporters, and importers to conduct their business.

Nations recognize that intellectual property has become increasingly important in world trade, and yet the extent of protection and enforcement of intellectual property rights vary around the world. As a result, the WTO developed the Agreement on Trade-Related Aspects of Intellectual Property Rights, or the **TRIPs Agreement**, to establish minimum levels of protection that each government must provide to the intellectual property of all WTO members. This binding agreement requires member governments to ensure that intellectual property rights can be enforced under their laws and that penalties for infringement are tough enough to deter further violations. Table 6-1 provides a brief summary of copyright, patent, and trade secrets protection under the TRIPs Agreement.

TABLE 6-1   Summary of the WTO TRIPs Agreement

Form of intellectual property	Key terms of agreement
Copyright	Computer programs are protected as literary works. Authors of computer programs and producers of sound recordings have the right to prohibit the commercial rental of their works to the public.
Patent	Patent protection must be available for inventions for at least 20 years and for both products and processes in almost all fields of technology. (Controversy has arisen over whether this protection applies to computer software.)
Trade secrets	Trade secrets and other types of undisclosed information that have commercial value must be protected against breach of confidence and other acts that are contrary to honest commercial practices. However, reasonable steps must have been taken to keep the information secret.

KEY INTELLECTUAL PROPERTY ISSUES

This section discusses several issues that apply to intellectual property and information technology, including plagiarism, reverse engineering, open source code, competitive intelligence, and cybersquatting.

Plagiarism

Plagiarism is the theft and passing off of someone's ideas or words as one's own. The explosion of electronic content and the growth of the Internet have made it easy to cut and paste paragraphs into term papers and documents without proper citation or quotation marks. To compound the problem, hundreds of Internet "paper mills" enable users to download entire term papers. Although some paper mills post warnings that their services should be used for research purposes only, many users pay scant heed. As a result, plagiarism has become an issue from elementary schools to the highest levels of academia.

Plagiarism is also common outside academia. Popular literary authors, playwrights, musicians, journalists, and even software developers have been accused of it:

- Best-selling author and TV pundit Stephen Ambrose was accused of plagiarism in six of his books.

- Jayson Blair quit the New York Times in May 2004 after he was accused of plagiarism and faking quotes and other information in news stories. Executive Editor Howell Raines and Managing Editor Gerald Boyd resigned in the fallout from the scandal.
- Former Beatles member George Harrison became entangled for decades in litigation over similarities between his hit "My Sweet Lord," released in 1970, and "He's So Fine," which was composed by Ronald Mack and recorded by the Chiffons in 1962.
- The SCO Group, owner of the UNIX operating system, sued IBM for $1 billion, alleging that IBM used proprietary code from a UNIX program that SCO owns to help create the Linux operating system. (See Case 2 at the end of the chapter for more details.)

Despite school codes of ethics that clearly define plagiarism and prescribe penalties that range from no credit on a paper to expulsion, many students still do not understand what constitutes plagiarism and believe that all electronic content is in the public domain. Other students knowingly commit plagiarism because they feel pressure to achieve a high GPA or because they are too lazy or pressed for time to do original work.

Some instructors say that being familiar with a student's style of writing, grammar, and vocabulary enables them to determine if the student actually wrote a paper. In addition, plagiarism detection systems (see Table 6-2) allow teachers, corporations, law firms, and publishers to check the originality of documents and manuscripts. These systems work by checking submitted material against one or more of the following databases of electronic content:

- More than 4.5 billion pages of publicly accessible electronic content on the Internet
- Millions of works published in electronic form, including newspapers, magazines, journals, and electronic books
- A database of all papers submitted to the plagiarism detection service from one institution
- A database of all papers submitted to the service from all participating institutions

TABLE 6-2 Partial list of plagiarism detection services and software

Name of service	Web site	Provider
iThenticate	www.ithenticate.com/	iParadigms
Turnitin	www.turnitin.com/	iParadigms
MyDropBox	www.mydropbox.com/	MyDropBox LLC
Glatt Plagiarism Services	www.plagiarism.com/	Glatt Plagiarism Services
EVE Plagiarism Detection	www.canexus.com/eve/	CaNexus

Here are some practical steps to combat student plagiarism:[26]

- Help students understand what constitutes plagiarism and why they need to properly cite sources.
- Show students how to document Web pages and materials from online databases.
- Schedule a major writing assignment so that portions of it are due over the course of the term, thus reducing the likelihood that students will get into a time crunch.
- Tell students that you know about Internet paper mills.
- Educate students about plagiarism detection services, and tell students that you know how to use these services.

Reverse Engineering

Reverse engineering is the process of taking something apart in order to understand it, build a copy of it, or improve it. Reverse engineering was originally applied to computer hardware, but is now commonly applied to software as well. Reverse engineering of software involves analyzing it to create a new representation of the system in a different form or at a higher level of abstraction. Often, reverse engineering begins by using program code from which you can extract design-stage details, which is a higher abstraction level in the life cycle. In other words, design-level details about an information system are more conceptual and less defined than the program code of the same system.

One frequent use of reverse engineering for software is to convert an application that ran on one vendor's database to run on another's (for example, from dBASE to Access or from DB2 to Oracle). Database management systems use their own programming language for application development. As a result, organizations that want to change database vendors are faced with rewriting their existing applications using the new vendor's database programming language. The cost and length of time required for this redevelopment can deter an organization from changing vendors and deprive it of the possible benefits of converting to an improved database technology.

Reverse engineering can use the code of the current database programming language to recover the design of the information system application. Next, code-generation tools can be used to take the design and produce code (forward engineer) in the new database programming language. This reverse engineering and code-generating process greatly reduces the time and cost needed to migrate the organization's applications to the new database management system. No one challenges the right to use this process to convert in-house developed applications to a new database management system—after all, the application was developed and is "owned" by the company using it. It is quite another matter, though, to use this process on a purchased software application developed by outside parties. Most IT managers would consider this action unethical, because the software user does not actually own the right to the software. In addition, a number of intellectual property issues would be raised, depending on whether the software was licensed, copyrighted, or patented.

Other reverse engineering issues involve tools called compilers and decompilers. A **compiler** is a language translator that converts computer program statements expressed in a source language (such as COBOL, Pascal, or C) into machine language (a series of binary codes of 0s and 1s) that the computer can execute. When a software manufacturer provides a customer with its software, it usually provides the software in machine language form. Tools called reverse engineering compilers or **decompilers** can read the machine language and produce the source code. For example, REC is a decompiler that reads an executable, machine-language file and produces a C-like representation of the code used to build the program.

Decompilers and other reverse engineering techniques can be used to analyze a competitor's program by examining its coding and operation to develop a new program that either duplicates the original or that will interface with the program. Thus, reverse engineering provides a way to gain access to information that another organization may have copyrighted or classified as a trade secret.

The courts have ruled in favor of using reverse engineering to enable interoperability. In the early 1990s, video game maker Sega developed a computerized lock so that only Sega video cartridges would work on its entertainment systems. This essentially shut out competitors from making software for the Sega systems. *Sega Enterprises Ltd. v. Accolade, Inc.* dealt with rival game maker Accolade's use of a decompiler to read the Sega software source code. With the code, Accolade could create new software that circumvented the lock and ran on Sega machines. After an initial lawsuit and appeals, the courts finally ruled that if someone lacks access to the unprotected elements of an original work, and has a "legitimate reason" for gaining access to those elements, disassembly of a copyrighted work is considered to be a fair use under section 107 of the Copyright Act. The unprotected element in this case was the code necessary to enable software to interoperate with the Sega equipment. The court reasoned that to refuse someone the opportunity to create an interoperable product would allow existing manufacturers to monopolize the market, making it impossible for others to compete. This ruling had a major impact on the video game industry, allowing video game makers to create software that would run on multiple machines.

Software license agreements increasingly forbid reverse engineering. As a result of the increased legislation affecting reverse engineering, some software developers are moving their reverse-engineering projects offshore to avoid U.S. rules. An example of legislation that involved reverse engineering was the Semiconductor Chip Protection Act (SCPA) of 1984, which established a new type of intellectual property protection for mask works—two- or three-dimensional layouts of an integrated circuit used to make semiconductor chips.[27] The act also "permits reverse engineering of a mask work solely for the purposes of teaching, analyzing, or evaluating the concepts or techniques embodied in the mask work or in the circuitry, logic flow, or organization of components used in the mask work. The person who performs legitimate reverse engineering may incorporate the results in an original mask work which is made to be distributed."[28]

The ethics of using reverse engineering can be debated. Some argue that its use is fair if it enables a company to create software that interoperates with another company's software or hardware and provides a useful function. This is especially true if the software's creator refuses to cooperate by providing documentation to help create interoperable software. From the consumer's standpoint, such stifling of competition can increase costs and reduce business options. Reverse engineering can also be a useful tool in detecting software bugs and security holes.

Others argue strongly against the use of reverse engineering, and say it can uncover the design of software that someone else has developed at great cost and taken care to protect. This unfairly robs the creator of future earnings, opponents say. Without a pay-off, what is the business incentive for software development?

Open Source Code

Historically, the makers of proprietary software have not made their source code available, but some developers do not share that philosophy. **Open source code** refers to any program whose source code is made available for use or modification as users or other developers see fit. The basic premise behind open source code is that when programmers can read, redistribute, and modify the code, the software improves, can be adapted to meet new needs, and bugs can be rapidly identified and fixed. Open source code advocates believe that this process produces better software than the traditional closed model. A considerable amount of open source code is available, including the Linux operating system; the MySQL AB, Ingres, and Sybase database management systems; the Apache Web server; BIND (the software behind the domain name system for the Internet); and the Mozilla Web browser.

One open source developer is Db4objects, which provides a popular object database called db4o to thousands of Java and .NET application developers. Db4objects chose to employ a dual-licensing strategy, offering both open source and commercial options. The open source license is for firms that use db4o in-house or develop and distribute their own derivative work as free software. Such open source users may elect to purchase support for the software. A commercial license is required if a firm wants to embed db4o in a commercial product or to gain access to premium services and support. Companies that develop and ship proprietary products also can receive support and indemnification (protection from possible copyright infringement claims made by third parties).[29]

Why would firms or individual developers create open source code if they do not receive money for it? Here are several reasons:

- Some people share code to earn respect for solving a common problem in an elegant way.
- Some people have used open source code that was developed by others and feel the need to pay back.
- A firm may be required to develop software as part of an agreement to address a client's problem. The firm is not paid directly for the software, but it is paid for the employees' time spent to develop the software. The firm may then decide to license the code as open source and use it to promote its prowess or as an incentive to attract other potential clients with a similar problem.
- A firm may develop open source code in anticipation of earning software maintenance fees when end users need changes and turn to the original developer to implement the changes.
- A firm may develop useful code but be reluctant to license and market it, so the firm might "donate" the code to the general public.

Many organizations cite the high reliability of open source code as their reason for adopting it. They argue that such software is peer reviewed and that mature, open source code is more reliable than closed, proprietary software. Reasoning, a firm that provides code inspection services, examined the Apache open source Web server Version 2.1 and the TCP/IP protocol stack implementation in Version 2.4.19 of the open source Linux kernel.

It found that the two products were on par with the quality of commercial software—on the order of 0.51 defects per 1000 lines of code.[30]

There are various definitions of what constitutes open source code, each with its own idiosyncrasies. For example, the GNU General Public License (GPL) was a precursor to the open source code defined by the Open Source Initiative (OSI). The GPL is intended to protect GNU software from being made proprietary, and it lists terms and conditions for copying, modifying, and distributing free software. The OSI is a nonprofit organization that advocates for open source and certifies open source licenses.[31] Its certification mark, "OSI Certified," may be applied only to software distributed under an open source license that meets OSI criteria, as described at its Web site, *www.opensource.com*.

A software developer could attempt to make its program open source simply by putting it into the public domain with no copyright. This would allow people to share the program and their improvements, but it would also allow others to revise the original code and then distribute the resulting software as their own proprietary product. Users who received the program in the modified form would no longer have the freedoms associated with the original software. Use of an open source license avoids this scenario.

In some cases, for-profit developers define their own version of an open source software license to meet their own business needs. For the OpenSolaris operating system, Sun defined its Common Development and Distribution License (CDDL). Under this license, CDDL code can be combined with non-CDDL code, including proprietary code. Sun's CDDL OpenSolaris code has been the subject of 1670 patents, but they can *only* be used with Sun's code. If a developer changes any of the code, Sun loses the right to use the 1670 patents. These features of Sun's CDDL go against the original spirit and intent of open source and free software. From a practical standpoint, the features mean that no one can distribute software that combines Linux and OpenSolaris code.[32]

Competitive Intelligence

Competitive intelligence is the gathering of legally obtainable information to help a company gain an advantage over its rivals. For example, some companies have employees who monitor the public announcements of property transfers to detect any plant or store expansions of a competitor. An effective competitive intelligence operation requires the continual gathering, analysis, and evaluation of data with controlled dissemination of useful information to decision makers. Competitive intelligence is often integrated into a company's strategic plans and decision making. Many companies, such as Eastman Kodak, Monsanto, and United Technologies, have established formal competitive intelligence departments. Some companies have even employed former CIA analysts to assist them.

Competitive intelligence is not **industrial espionage**, which employs illegal means to obtain business information that is not available to the general public. In the United States, industrial espionage is a serious crime that carries heavy penalties. Almost all the data needed for competitive intelligence can be collected from examining published information or interviews, as shown in the following list. By coupling this data with analytical tools and industry expertise, an experienced analyst can make deductions that lead to significant information.

- 10-K or annual report
- An SC 13D acquisition—a filing by shareholders who report owning more than 5 percent of common stock in a public company

- 10-Q or quarterly reports
- Press releases
- Promotional materials
- Web sites
- Analyses by the investment community, such as a Standard and Poor's stock report
- Dun & Bradstreet credit reports
- Interviews with suppliers, customers, and former employees
- Calls to the competitor's customer service group
- Articles in the trade press
- Environmental impact statements and other filings associated with a plant expansion or construction
- Patents

Competitive intelligence gathering has become enough of a science that nearly 25 colleges and universities offer courses or entire programs, including the institutions in the following list:

- American Military University
- Brigham Young University
- California Institute of Technology Industrial Relations Center
- Drexel University
- Idaho State University
- The Anderson School at UCLA
- Indiana University
- Rutgers University
- University of Pittsburgh
- Macquarie University, Graduate School of Management, Australia
- University of South Australia, School of Communication, Information, and New Media, Australia
- De Montfort University, United Kingdom
- THESEUS Institute, International Management Institute, France
- Lund University, Sweden
- University of Stockholm, Sweden

Also, the Society of Competitive Intelligence Professionals (*www.scip.org*) offers ongoing training, programs, and conferences.

Without proper management safeguards, the process of gathering competitive intelligence can cross over to industrial espionage and "dirty tricks." One frequent trick is to enter a bar near a competitor's plant or headquarters, strike up a conversation, and ply people for information after their inhibitions have been weakened by alcohol. Competitive intelligence analysts must avoid such unethical or illegal actions, including lying, misrepresentation, theft, bribery, or eavesdropping with illegal devices. See Table 6-3 for useful guidelines. The preferred answer to each question in the checklist is *yes*.

TABLE 6-3 A manager's checklist for running an ethical competitive intelligence operation

Questions	Yes	No
Has the competitive intelligence organization developed a mission statement, objectives, goals, and a code of ethics?	___	___
Has the company's legal department approved the mission statement, objectives, goals, and code of ethics?	___	___
Do analysts understand the need to abide by their organization's code of ethics and corporate policies?	___	___
Is there a rigorous training and certification process for analysts?	___	___
Do analysts understand all applicable laws, domestic and international, including the Uniform Trade Secrets Act and the Economic Espionage Act, and do they understand the critical importance of abiding by them?	___	___
Do analysts disclose their true identity and organization prior to any interviews?	___	___
Do analysts understand that everything their firm learns about the competition must be obtained legally?	___	___
Do analysts respect all requests for anonymity and confidentiality of information?	___	___
Has the company's legal department approved the processes for gathering data?	___	___
Do analysts provide honest recommendations and conclusions?	___	___
Is the use of third parties to gather competitive intelligence carefully reviewed and managed?	___	___

Failure to act prudently with competitive intelligence can get analysts and companies into serious trouble. For example, the Procter & Gamble Company (P&G) admitted publicly in 2001 that it unethically gained information about Unilever, its competitor in the multibillion-dollar hair-care business. Competitive intelligence managers at P&G had hired a contractor, who in turn hired several subcontractors to spy on P&G's competitors. Unilever was the primary target; it markets brands such as Salon Selectives, Finesse, and Thermasilk, while P&G manufactures Pantene, Head & Shoulders, and Pert.

In at least one instance, the espionage included going through dumpsters on public property outside Unilever corporate offices in Chicago. In addition, competitive intelligence operatives were alleged to have misrepresented themselves to Unilever employees, suggesting that they were market analysts. (P&G confirms the dumpster diving, but it denies that misrepresentation took place.) The operatives captured critical information about Unilever's brands, including new-product rollouts, selling prices, and operating margins.

When senior P&G officials discovered that a firm hired by the company was operating unethically, P&G immediately stopped the campaign and fired the three managers responsible for hiring the firm. P&G then did something unusual—it blew the whistle on itself, confessed to Unilever, returned stolen documents to Unilever, and started negotiations with them to set things straight. P&G's chairman of the board was personally involved in ensuring that none of the information obtained would ever be used in P&G business plans.[33] Several weeks of high-level negotiations between P&G and Unilever executives led to a secret agreement between the two companies. P&G is believed to have paid tens of millions of dollars to Unilever. In addition, several hair-care product executives were transferred to other units within P&G.[34]

Experts in competitive intelligence agree that the firm hired by P&G crossed the line of ethical business practices by sorting through Unilever's garbage. However, they also give P&G credit for going to Unilever quickly after it discovered the damage. Such prompt action was seen as the best approach. If no settlement had been reached, of course, Unilever could have taken P&G to court, where embarrassing details would have been revealed, causing more bad publicity for a company that is generally perceived as highly ethical. Unilever also stood to lose from a public trial. The trade secrets at the heart of the case may have been disclosed in depositions and other documents during a trial, which could have devalued proprietary data.

Cybersquatting

A **trademark** is anything that enables a consumer to differentiate one company's products from another's. A trademark may be a logo, package design, phrase, sound, or word. Consumers often cannot examine goods or services to determine their quality or source, so instead they rely on the labels attached to the products. Trademark law gives the trademark's owner the right to prevent others from using the same mark or a confusingly similar mark.

The United States has a federal system that stores trademark information; merchants can consult this information to avoid adopting marks that have already been taken. Merchants seeking trademark protection apply to the USPTO if they are using the mark in interstate commerce or if they can demonstrate a true intent to do so. Trademark protection lasts as long as a mark is in use.

Companies that want to establish an Internet presence know that the best way to capitalize on the strength of their brand names is to make the names part of the domain names for their Web sites. When Web sites were first established, there was no procedure for validating the legitimacy of requests for Web site names, which were given out on a first-come, first-served basis. **Cybersquatters** registered domain names for famous trademarks or company names to which they had no connection, with the hope that the trademark's owner would buy the domain name for a large sum of money.

The main tactic organizations use to circumvent cybersquatting is to protect a trademark by registering numerous domain names and variations as soon as the organization knows it wants to develop a Web presence (for example, UVXYZ.com, UVXYZ.org, and UVXYZ.info). In addition, trademark owners who rely on non-English-speaking customers should register their names in multilingual form. Registering additional domain names is far less expensive than attempting to force cybersquatters to change or abandon their domain names.

Other tactics can also curb cybersquatting. For example, the Internet Corporation for Assigned Names and Numbers (ICANN) is a nonprofit corporation responsible for managing the Internet's domain name system. ICANN is in the process of adding seven new top-level domains (.aero, .biz, .coop, .info, .museum, .name, and .pro) to the system. Current trademark holders will be given time to assert rights to their trademarks in the new top-level domains before registrations are opened to the general public. ICANN also has a Uniform Domain Name Dispute Resolution Policy, under which most types of trademark-based domain name disputes must be resolved by agreement, court action, or arbitration before a registrar will cancel, suspend, or transfer a domain name. The ICANN policy is designed to provide for the fast, relatively inexpensive arbitration of a trademark owner's complaint that a domain name was registered or used in bad faith.

The Anti-cybersquatting Consumer Protection Act (ACPA), enacted in 1999, allows trademark owners to challenge foreign cybersquatters who might otherwise be beyond the jurisdiction of U.S. courts. Under this act, trademark holders can seek civil damages of up to $100,000 from cybersquatters that register their trade names or similar-sounding names as domain names. It also helps trademark owners challenge the registration of their trademark as a domain name even if the trademark owner has not created an actual Web site.

Summary

1. What does the term *intellectual property* encompass, and why are companies so concerned about protecting it?

    Intellectual property is a term used to describe works of the mind, such as art, books, films, formulas, inventions, music, and processes, that are distinct and "owned" or created by a person or group. Copyrights, patents, trademarks, and trade secrets provide a complex body of law regarding the ownership of intellectual property, which represents a large and valuable asset to most companies. If these assets are not protected, other companies can copy or steal them, resulting in significant loss of revenue and competitive advantage.

2. What are the strengths and limitations of using copyrights, patents, and trade secret laws to protect intellectual property?

    A copyright grants the author of an original work the exclusive right to distribute, display, perform, or reproduce the work in copies, to prepare derivative works based upon the work, and to grant these exclusive rights to others. Copyright law has proven to be extremely flexible in covering new technologies, including software, video games, multimedia works, and Web pages. However, evaluating the originality of a work can be difficult and has led to litigation. Copyrights provide less protection for software than patents; software that produces the same result in a slightly different way does not infringe a copyright because no copying occurred.

    A patent enables an inventor to sue people who manufacture, use, or sell the invention without permission while the patent is in force. Not only does a patent prevent copying, it prevents independent creation (an allowable defense to a copyright infringement claim). Unlike copyright infringement, for which monetary penalties are limited, if the court determines that patent infringement exists and is intentional, it can award up to triple the amount of the damages claimed by the patent holder.

    To qualify as a trade secret, information must have economic value and must not be readily ascertainable. In addition, the trade secret's owner must have taken steps to maintain its secrecy. Trade secret law doesn't prevent someone from using the same idea if it was developed independently or from analyzing an end product to figure out the trade secret behind it. However, trade secret law has three key advantages over the use of patents and copyrights in protecting companies from losing control of their intellectual property: there are no time limitations on the protection of trade secrets, unlike patents and copyrights; there is no need to file any application or otherwise disclose a trade secret to outsiders to gain protection; and there is no risk that a trade secret might be found invalid in court.

3. What is plagiarism, and what can be done to combat it?

    To plagiarize means to steal and pass off the ideas and words of another as one's own. Students need to understand what constitutes plagiarism and how to properly cite sources, especially information from electronic sources and databases. Plagiarism detection systems enable people to check the originality of documents and manuscripts.

4. What is reverse engineering, and what issues are associated with applying it to create a look-alike of a competitor's software program?

Reverse engineering is the process of breaking something down in order to understand it, build a copy of it, or improve it. Reverse engineering was originally applied to computer hardware, but is now commonly applied to software. In some situations, reverse engineering can be considered unethical because it enables access to information that another organization may have copyrighted or classified as a trade secret. Recent court rulings, shrinkwrap license agreements for software that forbid reverse engineering, and restrictions and exceptions in the DMCA have made reverse engineering a riskier proposition in the United States.

5. What is open source code, and what is the fundamental premise behind its use?

Open source code refers to any program whose source code is made available for use or modification as users or other developers see fit. The basic premise behind open source code is that when programmers can read, redistribute, and modify it, the software improves. It can be adapted to meet new needs, and bugs can be rapidly identified and fixed.

6. What is the essential difference between competitive intelligence and industrial espionage, and how is competitive intelligence gathered?

Competitive intelligence is not industrial espionage, which employs illegal means to obtain business information that is not readily available to the general public. In the United States, industrial espionage is a serious crime that carries heavy penalties. Almost all the data needed for competitive intelligence can be collected from carefully examining published information or through interviews. Competitive intelligence analysts must take care to avoid unethical or illegal behavior, including lying, misrepresentation, theft, bribery, or eaves-dropping with illegal devices.

7. What is cybersquatting, and what strategy should be used to protect an organization from it?

The main tactic organizations use to circumvent cybersquatting is to protect a trademark by registering numerous domain names and variations as soon as they know they want to develop a Web presence. In addition, trademark owners who rely on non-English-speaking customers should also register their names in multilingual form.

Self-Assessment Questions

1. Which of the following is an example of intellectual property?

    a. a business process for the efficient handling of medical insurance claims

    b. a computer program

    c. the design for a computer wafer

    d. all of the above

2. The term of a copyright, originally 28 years, has been extended many times and now can be as long as the life of the author plus 70 years. True or False?

3. The _____ was signed into law in 1998 and was written in compliance with the global copyright protection treaty from WIPO; the law makes it illegal to circumvent a technical protection of copyrighted materials.

4. Not only does a _____ prevent copying, it prevents independent creation, unlike a copyright.

5. Unlike copyright infringement, there is no specified limit to the monetary penalty if patent infringement is found. True or False?

6. A _____ is a form of protection for intellectual property that does not require any disclosures or filing of an application.

   a. copyright

   b. patent

   c. trade secret

   d. trademark

7. Plagiarism is an issue only in academia. True or False?

8. The courts have consistently ruled that copyright law should not inhibit interoperability between the products of rival vendors. True or False?

9. Which of the following statements is true about open source code?

   a. There is only one definition of open source code.

   b. Research shows that popular open source code is on par with the quality of commercial software.

   c. A very limited amount of open source code is in use today.

   d. Putting source code into the public domain with no copyright is equivalent to creating open source code.

10. Almost all the data needed for competitive intelligence can be collected from carefully examining published information or through interviews. True or False?

11. The main tactic used to circumvent cybersquatting is to register numerous domain name variations as soon as an organization knows it wants to develop a Web presence. True or False?

Review Questions

1. Identify three frequent means of protecting computer software as a form of intellectual property.

2. To whom does a copyright grant the exclusive right to distribute, display, perform, or reproduce the work?

3. What is the fair use doctrine? What four factors do courts consider in determining whether a particular use of copyrighted property is fair and can be allowed without penalty?

4. What is a copyright infringement claim? What defense can a defendant adopt in a copyright infringement case?

5. What is WIPO, and what are its goals?

6. What is the DMCA? Why do its opponents say that it restricts the free flow of information?

7. What rights does a patent extend to the inventor?

8. What is independent creation? Can it be used as a defense in a patent infringement claim?

9. What is meant by prior art? How does it factor into the requirements for a patent?

10. Do software developers typically perform a thorough patent search to identify potential infringements? Why or why not?

11. What is meant by a submarine patent? What is patent farming?

12. What is a trade secret? What must an organization do to file a successful claim under the Economic Espionage Act?

13. What represents the greatest threat to the loss of a firm's trade secrets?

14. What is a nondisclosure clause? What is a noncompete agreement?

15. What is plagiarism?

16. What is reverse engineering as it applies to computer software?

17. What is open source code?

18. What is the difference between competitive intelligence and industrial espionage?

19. Under what laws might someone be punished for industrial espionage?

20. What is a trademark? What is its purpose?

Discussion Questions

1. Explain the concept that an idea cannot be copyrighted, but the expression of an idea can be, and why this distinction is a key to understanding copyright protection.

2. What is a cross-licensing agreement? How do large software companies use them? Do you think their use is fair to small software development firms? Why or why not?

3. Identify at least five actions a firm should take to protect its trade secrets.

4. What is the best way for a software manufacturer to protect new software? Why?

5. Discuss whether a company's customer list is a trade secret.

6. Identify and briefly discuss three key advantages that trade secret law has over the use of patents and copyrights in protecting intellectual property.

7. What is the WTO, and what is the scope and intent of the TRIPs Agreement?

8. Outline an approach that a university might take to successfully combat plagiarism.

9. Does the DMCA prohibit or permit the use of reverse engineering? Explain your answer.

10. Why might an organization elect to use open source code instead of proprietary software?

11. How might a corporation use reverse engineering to convert to a new database management system? How might it use reverse engineering to uncover the trade secrets behind a competitor's software?

12. What recent developments have challenged the reverse engineering of software in the United States? Is this development good or bad for the software industry? Why?

13. Compare and contrast the key issues in the *Sega v. Accolade* and *Lexmark v. Static Control Components* reverse engineering lawsuits.

What Would You Do?

1. Your friend has e-mailed you the URL of a new Web site that permits free music downloading. The site maintains a central database of thousands of the most current and popular music tracks. You vaguely remember that the Recording Industry Association of America (RIAA) issued thousands of subpoenas to Internet service providers across the United States in an attempt to get the names of people offering music on file-sharing networks, such as KaZaA and Grokster. But this is different, isn't it? You are simply downloading from a central database; you're not sharing your music with others. Are there any legal or ethical issues that would keep you from accessing this Web site?

2. Because of the expense, your company's CFO had to approve a $500,000 purchase order for hardware and software needed to upgrade the servers used to store data for the Product Development Department. Everyone in the department had expected an automatic approval, and were disappointed when the purchase order request was turned down. Management said that the business benefits of the expenditure were not clear. Realizing that she needs to develop a more solid business case for the order, the vice president of Product Development has come to you for help. Can you help her identify arguments related to protecting intellectual property that might strengthen the business case for this expenditure?

3. You have been asked to lead your company's new competitive intelligence organization. What would you do to ensure that members obey applicable laws and the company's own ethical policies?

4. You are the vice president for Software Development at a small, private firm. Sales of your firm's products have been strong, but you recently detected a patent infringement by one of your larger competitors. Your in-house legal staff has identified three options: (1) ignore the infringement out of fear that your larger competitor will file numerous countersuits; (2) threaten to file suit, but negotiate for an out-of-court settlement of $2 million; or (3) point out the infringement and negotiate strongly for a cross-licensing agreement with the competitor, which has numerous patents you had considered licensing. Which option would you pursue and why?

5. You are the human resources manager for a large software developer. You have the opportunity to hire three experienced but dissatisfied members of a competitor's sales force. Through informal means, you discover that all three signed nondisclosure agreements with their current employer. What issues might you face if the salespeople accept job offers from your firm? How can you minimize the potential problems? What additional steps would you take before authorizing your firm to extend a job offer?

Cases

1. RealNetworks vs. Apple

The RealPlayer Music Store from RealNetworks is a music purchase program that allows users to access more than 350,000 songs. Users can find and buy individual music tracks or albums and access account information using the RealPlayer software. Because users own the music they buy, they can record or transfer it to music-playing devices without extra fees or charges. The purchased music is stored in the RealAudio (.rax) format. Hundreds of millions of tracks have been downloaded by RealPlayer users around the world.

Many music-playing devices include proprietary copy-protection schemes that limit what music can be played on them, depending on where the music was purchased. Apple Computer, for example, uses its FairPlay copy-protection technology for formatting music played on its highly successful iPod. Because Apple refuses to license FairPlay, Apple's iTunes service was the sole source of music for the iPod.

In a move to increase the number of devices that can play music from its Internet store, RealNetworks developed digital rights-translation software called Harmony. The software allows consumers to play songs downloaded from the RealPlayer Music Store on dozens of music devices, including those from Apple, Creative, iRiver, and Rio. Many consumers welcomed the arrival of Harmony, believing they (not music device manufacturers) should choose what music they can hear.

Apple issued a highly critical statement of RealNetworks, accusing it of using a hacker's tactics and ethics to break into the iPod. Apple also said it was investigating the legality of the actions, including violation of the Digital Millennium Copyright Act (DMCA). Apple further pointed out that because iPod software is updated from time to time, the Harmony technology would probably not continue to work.

According to a RealNetworks press release, "Harmony follows in a well-established tradition of fully legal, independently developed paths to achieve compatibility. There is ample and clear precedent for this activity—for instance, the first IBM-compatible PCs from Compaq. Harmony's method of locking content from Real's music store is compatible with the iPod, Windows Media DRM devices, and Helix DRM devices. Harmony technology does not remove or disable any digital rights management system. Apple has suggested that new laws such as the DMCA are relevant to this dispute. In fact, the DMCA is not designed to prevent the creation of new methods of locking content and explicitly allows the creation of interoperable software."

Several questions about the development of Harmony are crucial to understanding the potential legal implications:

- Did RealNetworks break the copy protection used by iTunes and violate the DMCA?
- If so, does the Harmony software fall under exceptions of the DMCA that allow for interoperability?
- Did RealNetworks create Harmony by reverse engineering the original iTunes software? Under the DMCA, you must be very careful about reverse engineering to try to overcome mechanisms placed in the software to prevent disclosure.

Many people believe that the DMCA gives more protection than copyright law and that it allows companies to skew the market with a form of protectionism. They believe it allows technological protectionism and prevents people from marketing cheaper compatible products.

Questions:

1. Research the Internet on the status of the RealNetworks vs. Apple dispute. Write a one-page summary of your findings.
2. Under what conditions might a court find that RealNetworks violated the DMCA and must stop selling its Harmony software products?
3. Under what conditions might a court find that RealNetworks did not violate the DMCA and can continue to sell and distribute its Harmony products?
4. Do you think the DMCA should be changed? If so, how should it be changed? If not, why not?

2. The SCO Group vs. IBM

UNIX is a computer operating system that was developed by AT&T Bell Laboratories in 1969. At the time, this interactive, multiuser, multitasking system was considered state-of-the-art. After several years of successful use in-house and in academia, AT&T licensed UNIX as a commercial product. IBM, Hewlett-Packard, Sun Microsystems, and Silicon Graphics became the principal U.S.-based UNIX licensees. Each of these companies modified the original AT&T version of UNIX to operate on their proprietary CPU processor chips. These various versions of UNIX came to be known as UNIX "flavors," and many of them are still in use today.

The SCO Group is a software firm that sells and services its UNIX operating system and related products in 82 countries worldwide. SCO, formerly known as Caldera, is a NASDAQ-traded firm based in Utah. Novell acquired all rights, title, and interest to the UNIX software code from AT&T in 1992 for $750 million in Novell stock. SCO then acquired the UNIX assets from Novell in 1995, including the UNIX System V source code and UnixWare, a version of UNIX designed for high-power computing tasks.

SCO spent considerable resources modifying the UnixWare product, and by 1998 it could run on Intel-based processors. Because the earlier Intel processors were considered to have inadequate processing power for more demanding enterprise applications, no other UNIX vendor developed an Intel version of UNIX. With the rapid growth of Intel's performance capabilities and popularity in the marketplace, SCO found itself in a highly desirable market position. UnixWare soon had several significant customers, including DaimlerChrysler, Goodyear, Lucent, and NASDAQ.

Linux is a UNIX-like operating system that runs on many different CPU chips. Its name is a combination of Linus (after Linus Torvalds, its inventor) and UNIX. The initial market positioning of Linux was to create a free operating system for developers and computer hobbyists. As such, Linux posed little, if any, commercial threat to UNIX. In an interesting development, some Linux users took it upon themselves to modify SCO's UNIX software to enable UNIX applications to run on Linux. Technically, this was illegal for anyone but SCO to do under its UNIX license.

President and CEO Darl McBride joined SCO in 2002 and recognized that the UNIX System V source code was a valuable asset. In February 2003, SCO allegedly found UNIX System V code and UNIX derivative code in Linux kernel version 2.4. The implications, at least to SCO, were clear. If Linux contained SCO's proprietary code, then it was violating SCO's UNIX license. SCO could challenge the legality of the Linux license, and thousands of corporate and individual Linux users would potentially owe SCO millions of dollars in licensing fees. Such action could radically affect how Linux was developed and sold and undermine its viability as a corporate platform.

SCO filed legal action against IBM for misappropriation of trade secrets, tortious interference, unfair competition, and breach of contract. SCO charged that in the original AT&T/IBM UNIX licensing agreements, the parties agreed to a set of provisions that required IBM to protect the UNIX code against unrestricted disclosure, unauthorized transfer or disposition, and unauthorized use by others. The complaint alleged that IBM attempted to improperly destroy the economic value of UNIX on Intel processors to benefit IBM's new Linux services business. SCO requested damages of at least $1 billion and demanded that IBM cease anticompetitive practices based on specific requirements sent in a notification letter to IBM. If these requirements were not met, SCO would have the authority to revoke IBM's AIX (the IBM flavor of UNIX) license.

In its response to SCO's charges, IBM stated "that contrary to Caldera's allegations, by its lawsuit, Caldera seeks to hold up the open source community (and development of Linux in particular) by improperly seeking to assert proprietary rights over important, widely used technology and impeding the use of that technology by the open source community."

To win its case against IBM, SCO must convince a jury that IBM knowingly mixed a derivative of SCO's code into its Linux products. If so, then SCO's property is being given away illegally under Linux's General Public License, which mandates free distribution of source code.

In Spring 2003, Novell challenged SCO's assertion that it owns the copyrights and patents to UNIX System V, pointing out that the asset purchase agreement between Novell and SCO in 1995 did not transfer these rights to SCO. In addition, Novell sought evidence for SCO's assertion that certain UNIX System V code had been copied into Linux. In early 2004, SCO responded by suing Novell for "slander of title," a claim sometimes invoked in disputes over ownership of real estate. Applying that idea to UNIX copyrights, SCO alleged that Novell publicly claimed that SCO didn't own the copyrights, thus discouraging Linux users from paying royalties to SCO.

In June 2003, SCO announced that it had terminated IBM's right to use or distribute any software modified from or based on UNIX System V source code. In the amended complaint, SCO sought additional damages from IBM's multibillion-dollar, AIX-related businesses. The following month, SCO announced it would offer UnixWare licenses tailored to support run-time, binary use of Linux for all commercial users of Linux based on kernel version 2.4.x and later. SCO also announced that it would not sue Linux customers who purchased a UnixWare license for past copyright violations.

SCO further upset the open source community and Linux users in March 2004, when it filed lawsuits against AutoZone, the auto parts store, and automobile maker DaimlerChrysler. SCO sued AutoZone for using Linux and failing to pay UNIX licensing fees. SCO sued Daimler-Chrysler for contractual violations, not copyright violations, because DaimlerChrysler is a UNIX licensee and AutoZone is not. SCO claimed it gave DaimlerChrysler 30 days to certify its compliance with SCO's licensing terms.

AutoZone asked in April 2004 that its case be stayed until other lawsuits involving SCO were resolved. The company further claimed that SCO had failed to identify which, if any, of SCO's copyrights it violated. The judge granted the stay in July 2004. Also that month, a Michigan judge dismissed the SCO Group's lawsuit against DaimlerChrysler.

In August 2003, IBM answered SCO's lawsuit with 10 counterclaims along the following lines: SCO is maliciously attacking major IBM products, Linux and UNIX, which is damaging IBM's business; SCO itself used and sold Linux products under the GPL, thus violating the very agreement it was accusing others of violating; and SCO has violated four IBM patents.

Meanwhile, SCO's business has not gone well. Its 2004 revenue was $43 million, down 46 percent from the previous year. SCO cited competitive pressures and a decrease in its UNIX licensing revenue, which dropped from $10 million in the fourth quarter of 2003 to just $1 million in the fourth quarter of 2004.

Questions:

1. Research the Internet to learn the status of the SCO-IBM dispute. Write a one-page summary for your instructor.

2. What does this case teach you about the sale and purchase of rights to software code?

3. Some critics feel that SCO's business tactics are despicable: acquire the rights to software you did not write, claim that lines of the code are embedded in a competing and more popular program, and then demand a licensing fee from anyone who uses the rival software. Others think that SCO's tactics are perfectly legal. What is your opinion and why?

4. Discuss the pros and cons of SCO's decision to sue corporate Linux users.

3. Lotus vs. Borland

By the early 1990s, conflicting court decisions over the previous decade had confused software developers, especially regarding the "look and feel" of software. Must every software product have its own unique look and feel? Could a company improve its products by incorporating features that existed in competing products? Was it legal to build and market a clone—a product externally identical to another product? The *Lotus v. Borland* lawsuit and associated appeals lasted more than five years, and went all the way to the Supreme Court. The case set a precedent that clarified the limits of software copyright protection.

The first electronic spreadsheet on the market was Daniel Bricklin's VisiCalc, introduced in 1979. The program enabled users to easily prepare budgets, forecast profits, analyze investments, and summarize tax data. VisiCalc was a huge success, selling more than 100,000 copies in its first year on the market, at a time when only a few million people owned PCs. For the first time, users found an application compelling enough to make them want to buy a computer. VisiCalc and other early spreadsheet programs are credited with being catalysts for the PC revolution, but neither the software nor the concept of a spreadsheet program was patented or copyrighted.

Mitch Kapor and Jonathan Sachs founded Lotus Development Corporation in 1982. In 1983, Lotus released its flagship spreadsheet product, Lotus 1-2-3, which, like VisiCalc, enabled users to perform accounting functions on a computer. Lotus 1-2-3 was highly successful, generating sales of $53 million in its first year and $156 million the next year.

Users manipulated and controlled the Lotus 1-2-3 program with a series of menu commands, such as Copy, File, Print, and Quit. Users could select these commands either by highlighting them on the screen or typing their first letter. Lotus 1-2-3 also allowed users to write a macro, a series of commands activated by a single macro keystroke. The program could then recall and perform the designated commands automatically.

Another company, Borland, was founded in 1983 with the objective of developing a program that was superior to Lotus 1-2-3. In 1987, Borland released its Quattro spreadsheet program to the public. Quattro gave users a choice for how they could communicate with the spreadsheet program—by using menu commands designed by Borland, or by using the commands and command structure found in Lotus 1-2-3. The latter capability helped attract Lotus 1-2-3 users to the Borland software. To provide this dual functionality, Borland copied only the words and structure of Lotus's menu command hierarchy, but not any of its competitor's computer code.

Lotus sued Borland in 1990, charging that the execution of commands in Quattro copied the look and feel of the Lotus 1-2-3 interface. In August 1992, a U.S. District Court found that Borland included in its spreadsheet "a virtually identical copy of the entire Lotus 1-2-3 menu tree" and infringed the copyright of Lotus. The judge held that the menu command hierarchy—the selection of menu items and their particular arrangement in an inverted tree structure—was a protectable element of the program that Borland had infringed. The judge also ruled that Borland's "key reader"—a program feature that enabled Quattro to interpret and execute Lotus 1-2-3 macros—was infringing because it employed a table that reproduced the entire Lotus 1-2-3 menu command hierarchy, with the first letter of each command substituted for the full command name.

The court subsequently entered an injunction against Borland that prohibited further sales or distribution of its spreadsheet products. In response, Borland appealed, and meanwhile shipped new versions of its spreadsheet that did not include the infringing features. The district court reaffirmed its decision in July 1993.

Borland decided to get out of the spreadsheet business, and sold its Quattro spreadsheet software to Novell Inc. in March 1994. One year later, the U.S. Court of Appeals for the First Circuit reversed the district court ruling that Borland had infringed the copyright of Lotus 1-2-3. The Court found that the Lotus menu command hierarchy was a "method of operation" that was excluded from copyright protection because the command hierarchy "provides the means by which users control and operate Lotus 1-2-3." The ruling was welcomed by Borland, which would have found it extremely difficult to pay the estimated $100 million in damages sought by Lotus.

The case was appealed to the Supreme Court, and in January 1996, five years after the suit began, the court affirmed the ruling for Borland by the First Circuit Court of Appeals. The case was significant for the software industry, which had been riddled with infringement lawsuits due to ambiguities in copyright law. The ruling made it clear that software copyrights could be successfully challenged, which further discouraged the use of copyright to protect software innovations. As a result, developers had to go through the more difficult and expensive patent process to protect their software products. These additional costs would either be absorbed by developers or passed on to users.

Borland is still in business, providing products and services targeted to software developers, including a Borland version of the C++ programming language and the database software packages Visual dBASE, Paradox, and InterBase. Recent annual sales were in the neighborhood of $300 million. Meanwhile, IBM purchased Lotus in 1995 for $3.5 billion. Lotus 1-2-3 is still widely used, although it is not as popular as the Microsoft Excel spreadsheet program.

Questions:

1. Go to your school's computer lab or a PC software store and experiment with current versions of any two of the Quattro, Excel, and Lotus 1-2-3 spreadsheet programs. Write a brief paragraph summarizing the similarities and differences in the "look and feel" of these two programs.

2. The courts took several years to reverse their initial decision and rule in favor of Borland. What impact did this delay have on the software industry? How might things have been different if Borland had received an initial favorable ruling?

3. Assume that you are the manager of Borland's software development. With the benefit of hindsight, what different decisions would you have made about Quattro?

End Notes

1 Gardner, David W., "Lawyer for Harvard Student Battling Apple Will Seek to Dismiss Lawsuit," *InformationWeek,* www.informationweek.com, February 3, 2005.

2 Gillmor, Dan, "Apple Suit is Wrong Kind of Different," *Computerworld*, www.computerworld.com, February 7, 2005.

3 Gline, Matthew A., "Introducing the iLawsuit," *The Crimson*, January 19, 2005.

4 Tartakoff, Joseph M., "Apple Sues Students," *The Crimson*, January 12, 2005.

5 "Apple Targets Harvard Student for Product 'Leaks,'" *InformationWeek*, www.informationweek.com, January 13, 2005.

6 "To Protect Its Customers' Printing Experience, Lexmark Continues Pursuing Supplies Counterfeiters," PR Newswire, March 3, 2005.

7 Niccoli, James, "Court Rules Against Lexmark in Printer Case," *Computerworld*, www.computerworld.com, February 22, 2005.

8 "World Intellectual Property Organization," Wikipedia (The Free Online Encyclopedia), en.wikipedia.org, February 1, 2005.

9 "Medium Term Plan for WIPO Program Activities—Vision and Strategic Direction of WIPO," WIPO Web site, www.wipo.int, February 1, 2005.

10 IBM Web site, www.IBM.com/ibm/licensing/patents, March 3, 2004.

11 "Software Patents," Wikipedia (The Free Online Encyclopedia), en.wikipedia.org/, February 1, 2005.

12 Syrowik, David R. and Cole, Roland R., "The Challenge of Software Patents," The Software Patent Institute Database of Software Technologies, www.spi.org/primsrpa.htm.

13 Evris, Joris, "Microsoft, Autodesk in Patent Licensing Deal," *ARNnet*, www.arn.ing.com, December 17, 2004.

14 Wrolstad, Jay, "IBM Releases 500 Software Patents," *CIO Today,* www.ciotoday.com, January 5, 2005.

15 "Standard," Webopedia, the Online Encyclopedia, www.webopedia.com.

16 Perens, Bruce, "Patent Farming," perens.com/Articles/PatentFarming.html, February 3, 2005.

17 McMillian, Robert, "Eolas Patent Rejected, But Decision Not Seen as Final," *Computerworld*, www.computerworld.com, March 8, 2004.

18 Associated Press, "Federal Judge Upholds $520 Million Verdict Against Microsoft," *InformationWeek*, www.informationweek.com, January 15, 2004.

19 Blau, John, "Infineon Challenges Rambus Patent Infringement Suit," *InformationWeek*, www.informationweek.com, December 22, 2004.

20 Guardsmark Web site, www.guardsmark.com/issues, March 3, 2005.

21 Karush, Sarah, "Compuware Tells Jury 'IBM Tried to Destroy Us,'" *InformationWeek*, www.informationweek.com, February 16, 2005.

22 Gonsalves, Antone, "Update: IBM Settles, Compuware Talks About Moving Forward," *InformationWeek*, www.informationweek.com, March 22, 2005.

23 Betts, Mitch, "Sidebar: Legal Issues: Gray Areas," *Computerworld*, www.computerworld.com, July 19, 2004.

24 Sullivan, Bob, "AOL Customer List Stolen, Sold to Spammer," MSNBC News, www.msnbc.msn.com, June 24, 2004.

25 Violino, Bob, "Protecting the Data Jewels: Valuable Customer Lists," *Computerworld*, www.computerworld.com, July 19, 2004.

26 "Cheating 101: Paper Mills and You," Teaching Effectiveness Seminar, March 5, 1999, Coastal Carolina University, www2.sjsu.edu/ugs/curriculum/cheating.htm, January 8, 2005.

27 U.S. Copyright Office, "Circular 100, Federal Statutory Protection for Mask Works," www.copyright.gov/circs/circ100.html#introduction, March 6, 2005.

28 U.S. Copyright Office, "Circular 100, Federal Statutory Protection for Mask Works," www.copyright.gov/circs/circ100.html#introduction, March 6, 2005.

29 Krill, Paul, "Object Database Goes Open Source," *Computerworld*, www.computerworld.com, December 3, 2004.

30 Krill, Paul, "Open Source Code Quality Endorsed," *InfoWorld*, www.infoworld.com, July 1, 2003.

31 Open Source Initiative Web site, www.opensource.org, January 29, 2005.

32 Hayes, Frank, "Open-Source Foes," *Computerworld*, www.computerworld.com, January 31, 2005.

33 Procter & Gamble press release, www.pg.com, September 6, 2001.

34 Sewrer, Andy, "P&G's Covert Operation," *Fortune*, www.fortune.com, September 17, 2001.

Sources for Case 1

Gonsalves, Antone, "RealNetworks Slashes Prices for Music Downloads," *InformationWeek*, www.informationweek.com, August 17, 2004.

Gonsalves, Antone, "RealNetworks Bypasses Apple With Songs For The iPod," *InformationWeek*, www.informationweek.com, July 26, 2004.

Hulme, George V., "Apple Threatens RealNetworks," *InformationWeek*, www.informationweek.com, July 29, 2004.

"RealNetworks Statement About Harmony Technology and Creating Consumer Choice," www.RealNetworks.com, July 29, 2004.

Sources for Case 2

"April 30, 2003, IBM Response to SCO Group Complaint," SCO v. IBM Web site at www.caldera.com/ibmlawsuit/ibm_response_to_sco-group_complaint_on_april30_2003.pdf, January 11, 2005.

GNU General Public License, Version 2, 1991, at www.gnu.org/copyleft/gpl.html, January 26, 2005.

"SCO Files Lawsuit Against IBM," SCO Investor Relations press release, www.sco.com, March 7, 2004.

Berinato, Scott, "The SCO Slugfest," *CIO Magazine*, August 1, 2004.

Hayes, Frank, "Weirder and Weirder," *Computerworld*, www.computerworld.com, January 26, 2004.

Jayson, Seth, "The Perils of Prosecuting the Penguin," *Business 2.0*, www.business2.com, April 26, 2004.

Koch, Richard, "Open Source Reality Check," *CIO*, www.cio.com, December 28, 2004.

McMillian, Robert, "SCO Shows Disputed Code to IBM," *Computerworld*, January 14, 2004.

Milliard, Elizabeth, "SCO Dented by Dismissal of Daimler Chrysler Suit," *LinuxInsider*, July 7, 2004.

Weiss, Todd R., "Defiant IBM Calls UNIX Indemnification Unnecessary," *Computerworld*, January 26, 2004.

Sources for Case 3

Brandel, William, "'Look and Feel' Reversal Re-ignites Copyright Fight," *Computerworld*, March 13, 1995.

Goetz, Martin A., "Copycats or Criminals?" *Computerworld*, June 12, 1995.

"Borland Prevails in Lotus Copyright Suit," Borland press release, www.borland.com, January 16, 1996.

SOFTWARE DEVELOPMENT

VIGNETTE

Antivirus Bug Brings Computers to a Standstill

At 7:30 a.m. Tokyo time on Saturday, April 23, 2005, Trend Micro released antivirus update Official Pattern Release (OPR) 2.594.00. At 9:02 a.m., it was pulled off Active Update. During those 92 minutes, hundreds of thousands of computer processors across the world came grinding to a halt.

The update referenced a data file that identified known viruses. This file contained a bug, a loop that gradually sapped nearly 100 percent of the processing power of PCs in Japan, Europe, and the Middle East. By 2 p.m. on Saturday, more than 300,000 Trend Micro customers had called to report problems. "The problem was caused by our virus labs in Manila during the checking process when a part of the test wasn't performed," explained Trend Micro spokeswoman Naomi Ikenomoto.

The company created the update to handle Rbot, a prolific worm that gained notoriety when an attack on the Italian senate caused computer monitors there to display pornographic photographs. The worm allowed hackers to steal information, and some variants of Rbot observed computer users via their own Webcams.

In the rush to protect customers from Rbot, the company failed to test the update on PCs that use Windows XP Service Pack 2, and possibly a few other systems. The company cited human error and responded by channeling more resources into quality assurance.

In the meantime, the mistake cost Trend Micro $8 million. The company lost an estimated 700 business licenses and 28,300 customer licenses. The mistake also caused the company to reduce its second-quarter revenue forecast by 16.7 percent. However, because corporations cannot quickly shift antivirus vendors, much of the impact was likely to be felt in 2006.[1,2,3]

LEARNING OBJECTIVES

As you read this chapter, consider the following questions:

1. Why do companies require high-quality software in business systems, industrial process control systems, and consumer products?

2. What ethical issues do software manufacturers face in making trade-offs between project schedules, project costs, and software quality?

3. What are the four most common types of software product liability claims, and what actions must plaintiffs and defendants take to be successful?

4. What are the essential components of a software development methodology, and what are its benefits?

5. How can Capability Maturity Model Integration improve an organization's software development process?

6. What is a safety-critical system, and what actions are required during its development?

STRATEGIES TO ENGINEER QUALITY SOFTWARE

High-quality software systems are easy to learn and easy to use. They perform quickly and efficiently to meet their users' needs. They operate safely and dependably and have a high degree of availability so that system downtime is kept to a minimum. Such software has long been required to support the fields of air traffic control, nuclear power, automobile safety, health care, military and defense, and space exploration. Now that computers and software have become an integral part of our lives, more and more users are demanding high quality in their software. They cannot afford system crashes, lost work, or lower productivity, nor can they tolerate security holes through which intruders can spread viruses, steal data, or shut down Web sites. Software manufacturers are struggling with economic, ethical, and organizational issues associated with improving the quality of their software. This chapter covers many of these issues.

A **software defect** is any error that, if not removed, could cause a software system to fail to meet its users' needs. The impact of these defects can be trivial; for example, a computerized sensor in a refrigerator's ice cube maker might fail to recognize that the tray is full and continue to make ice. Other defects could lead to tragedy—the control system for an automobile's antilock brakes could malfunction and send the car into an uncontrollable spin. The defect may be subtle and undetectable, such as a tax preparation package that makes a minor miscalculation, or the defect might be glaringly obvious, such as a payroll program that generates checks with no deductions for Social Security or other taxes.

For example, the following defects were uncovered in widely used software packages in the fall of 2005:

- Apple Computer had to create patches for 10 vulnerabilities in its Mac OS X operating systems 10.3.9 (dubbed Panther) and 10.4.2 (Tiger). Some of the bugs were serious enough that the Danish security company Secunia labeled the fixes as "highly critical." In the previous month, Apple set a firm record with patches for more than 40 bugs.[4]
- Versions of RealPlayer and Helix Player that work with the Linux operating system had to be patched against a vulnerability that enabled a hacker to execute commands remotely, provided he could convince a user to open certain files. The exploit code was even published on the Internet, leading some security firms to label the bug as critical.[5]
- Bugs in the Kaspersky Labs' Anti-Virus engine could be exploited by attackers to grab complete control of a PC protected by the company's Windows products. The Moscow-based security vendor had to implement a stopgap measure, building and releasing a package of signatures that detect possible exploits.[6]
- As of October 2005, Internet Explorer (IE) had six unpatched vulnerabilities, according to eEye Digital Security, which tracks the bugs it has submitted to Microsoft. During the same period, Danish vulnerability tracker Secunia said that IE 6.0 was afflicted with a dozen unpatched bugs.[7]

- Users of Symantec Corporation's AntiVirus Scan Engine versions 4.0 and 4.3 were advised to upgrade their software to eliminate a critical security bug. The flaw could theoretically allow an attacker to take control of an affected system.[8]
- According to security vendor Finjan Software, a bug in Google could have allowed attackers to grab a Google user's cookie. If the user was logged on to his Google account, the stolen cookie would have let the attacker view the user's saved searches or alerts, or even assume the user's identity in Google Groups.[9]

Sometimes, there are even errors in the patches implemented to correct bugs. For example, problems were reported with a Microsoft security update that was designed to fix a critical flaw in Windows. Under certain conditions, security update MS05-051 caused numerous problems, including blocking access to the Microsoft update Web site, displaying a blank logon screen without icons, and creating issues with Office applications.[10]

Software quality is the degree to which a software product meets the needs of its users. **Quality management** addresses how to define, measure, and refine the quality of the development process and the products developed during its various stages. These products, such as statements of requirements, flowcharts, and user documentation, are known as **deliverables**. The objective of quality management is to help developers deliver high-quality systems that meet the needs of their users. Unfortunately, the first release of any software rarely meets all its users' expectations. A software product usually does not work as well as its users would like until it has been used for a while, been found lacking in some ways, and then corrected or upgraded.

A primary cause for poor software quality is that developers do not know how to design quality into software from the very start, or do not take the time to do so. Software developers must define and follow a set of rigorous engineering principles and be committed to learn from past mistakes. In addition, they must understand the environment in which their systems will operate and design systems that are as immune to human error as possible.

Even if software is well-designed, programmers make mistakes in turning design specifications into lines of code. According to estimates, an experienced programmer unknowingly injects about one defect into every 10 lines of code. The programmers aren't incompetent or lazy—they're just human. Everyone makes mistakes, but in software, these mistakes result in defects.[11] For example, the Microsoft Windows XT operating system is composed of four hundred million lines of code bundled with products made by other Microsoft divisions. Even if 99.9 percent of the defects were identified and fixed before the product was released to the public, there would still be about one bug per 10,000 lines of code, or roughly 40,000 bugs in Windows XT. Thus, software that is used daily by workers worldwide probably contains thousands of bugs.

Another factor that causes poor-quality software is the extreme pressure that manufacturers feel to reduce the time to market of their products. They are driven by the need to beat the competition in delivering new functionality to users, to begin generating revenue to recover the cost of development, and to show a profit for shareholders. The resources and time budgeted to ensure quality are often cut under the intense pressure to ship the new product. When forced to choose between adding more user features or doing more testing, most software development managers decide in favor of more features. After all, they reason, defects can always be patched in the next release, which will give customers an automatic incentive to upgrade. The additional features will make the release more useful and therefore easier to sell to customers. Although customers are stakeholders who are key to the software's success, and they may benefit from new features, they also bear the burden of errors that aren't caught or fixed during testing. Thus, many customers challenge whether the decision to cut quality in favor of feature enhancement is ethical.

As a result of the lack of consistent quality in software, many organizations avoid buying the first release of a major software product or prohibit its use in critical systems. Their rationale is that the first release usually has many defects that cause user problems. Because of the many defects in the first two popular Microsoft operating systems (DOS and Windows) and their tendency to crash unexpectedly, many believe that Microsoft did not have a reasonably reliable operating system until its third major variation—Windows NT. Even software products that have been reliable over a long period can falter unexpectedly when the operating conditions change. For instance, the software in the Cincinnati Bell telephone switch had been thoroughly tested and had operated successfully for months after it was deployed in 1985. Later that same year, however, when the time changed from daylight saving time to standard time, the switch failed because it was overwhelmed by the number of calls to the local "official time" phone number from people who wanted to set their clocks. The volume of simultaneous calls to the same number was a change in operating conditions that no one had anticipated.

The Importance of Software Quality

Most people think of business information systems when they first think about software. **Business information systems** are a set of interrelated components that include hardware, software, databases, networks, people, and procedures that collect data, process it, and disseminate the output. One common business system captures and records business transactions. For example, a manufacturer's order-processing system captures order information, processes it to update inventory and accounts receivable, and ensures that the order is filled and shipped on time to the customer. Other examples include an electronic-funds transfer system that moves money among banks and an airline's online ticket reservation system. The accurate, thorough, and timely processing of business transactions is a key requirement for such systems. A software defect can be devastating, resulting in lost customers and reduced revenue. How many times would bank customers tolerate having their funds transferred to the wrong account before they stopped doing business with that bank?

Another type of business system is the **decision support system (DSS)**, which is used to improve decision making. A DSS can help develop accurate forecasts of customer demand, recommend stocks and bonds for an investment portfolio, and schedule shift workers to minimize cost while meeting customer service goals. Again, a software defect in a DSS can be devastating to an organization or its customers.

Software is also used to control many industrial processes in an effort to reduce costs, eliminate human error, improve quality, and shorten the time it takes to make products. For example, steel manufacturers use process-control software to capture data from sensors about the equipment that rolls steel into bars and about the furnace that heats the steel before it is rolled. Without process-control computers, workers could react to defects only after the fact and would have to guess at the adjustments needed to correct the process. Process-control computers enable the process to be monitored for variations from operating standards and to eliminate product defects *before* they can be made. Any defect in this software can lead to decreased product quality, increased waste and costs, or even unsafe operating conditions for employees. (Some consequences of these defects are discussed later in this chapter.)

Software is also used to control the operation of many industrial and consumer products—automobiles, medical diagnostic and treatment equipment, televisions, radios, stereos, refrigerators, and washers. A software defect can have relatively minor consequences, such as clothes not drying long enough, or it can lead to much more serious damage, such as a patient being overexposed to powerful X-rays.

As a result of the increasing use of computers and software in business, many companies are now in the software business whether they like it or not. The quality of software, its usability, and its timely development are critical to almost everything businesses do. The speed with which an organization develops software (whether in-house or contracted out) can put it ahead of or behind its competitors. Software problems may have been frustrating in the past, but mismanaged software can now be fatal to a business, causing it to miss product delivery dates, increase product development costs, and deliver products that have poor quality.

Business executives must struggle with ethical questions of how much effort and money they should invest to ensure high-quality software. A manager who takes a short-term, profit-oriented view may feel that any additional time and money spent on quality assurance delays a new product's release, sales revenues, and profits. However, a different manager may consider it unethical not to fix all known problems before putting a product on the market and charging customers for it.

Other key questions for executives are whether their products could cause damage and what their legal exposure would be if they did. Fortunately, software defects are rarely lethal and few personal injuries are related to software failures. However, the use of software does introduce a new dimension to product liability that concerns many executives, as explained in the following Legal Overview.

LEGAL OVERVIEW

Software Product Liability

Software product litigation is certainly not new. One lawsuit in the early 1990s involved a financial institution that became insolvent because defects in a purchased software application caused errors in its integrated general ledger system, in customers' passbooks, and in loan statements. Dissatisfied depositors responded by withdrawing more than $5 million.[12] In a 1992 case involving an automobile manufacturer, a truck stalled because of a software defect in the fuel injector. In the ensuing accident, a young child was killed, and a state supreme court later justified an award of $7.5 million in punitive damages against the manufacturer.[13]

continued

The liability of manufacturers, sellers, lessors, and others for injuries caused by defective products is commonly referred to as **product liability**. There is no federal product liability law; instead, it is found mainly in common law (made by state judges) and in Article 2 of the Uniform Commercial Code, which deals with the sale of goods.

If a software defect causes injury to purchasers, lessees, or users of the product, the injured parties may be able to sue as a result. Injuries range from physical mishaps and death to loss of revenue or an increase in expenses due to a business disruption caused by a software failure. Software product liability claims are frequently based on strict liability, negligence, breach of warranty, or misrepresentation.

Strict liability means that the defendant is held responsible for injuring another person, regardless of negligence or intent. The plaintiff must prove only that the software product is defective or unreasonably dangerous and that the defect caused the injury. There is no requirement to prove that the manufacturer was careless or negligent, or to prove who caused the defect. All parties in the chain of distribution—the manufacturer, subcontractors, and distributors—are strictly liable for injuries caused by the product and may be sued.

Defendants against a strict liability action may use several legal defenses, including the doctrine of supervening event, the government contractor defense, and an expired statute of limitations. Under the doctrine of supervening event, the original seller is not liable if the software was materially altered after it left the seller's possession and the alteration caused the injury. To establish the government contractor defense, a contractor must prove that the precise software specifications were provided by the government, the software conformed to the specifications, and the contractor warned the government of any known defects in the software. Finally, there are also statutes of limitations for claims of liability, which means that an injured party must file suit within a certain time after the injury occurs.

When sued for **negligence**, a software supplier is not held responsible for every product defect that causes customer or third-party loss. Instead, responsibility is limited to harmful defects that could have been detected and corrected through "reasonable" software development practices. Even when a contract may be written expressly to protect against supplier negligence, courts may disregard such terms as unreasonable. Negligence is an area of great risk for software manufacturers or organizations with software-intensive products.

The defendant in a negligence case may either answer the charge with a legal justification for the alleged misconduct or demonstrate that the plaintiffs' own actions contributed to their injuries (**contributory negligence**). If proved, the defense of contributory negligence can reduce or totally eliminate the amount of damages the plaintiffs receive. For example, if a person uses a pair of pruning shears to trim his fingernails and ends up cutting off a fingertip, the defendant could claim contributory negligence.

A **warranty** assures buyers or lessees that a product meets certain standards of quality. A warranty of quality may be either expressly stated or implied by law. Express warranties can

continued

be oral, written, or inferred from the seller's conduct. For example, sales contracts contain an implied warranty of merchantability, which requires that the following standards be met:

- The goods must be fit for the ordinary purpose for which they are used.
- The goods must be adequately contained, packaged, and labeled.
- The goods must be of an even kind, quality, and quantity within each unit.
- The goods must conform to any promise or affirmation of fact made on the container or label.
- The quality of the goods must pass without objection in the trade.
- The goods must meet a fair average or middle range of quality.[14]

If the product fails to meet its warranty, the buyer or lessee can sue for **breach of warranty**. Of course, most dissatisfied customers will first seek a replacement, a substitute product, or a refund before filing a lawsuit.

Software suppliers frequently write warranties to attempt to limit their liability in the event of nonperformance. Although the software is warranted to run on a given machine configuration, no assurance is given as to what that software will do. Even if the contract specifically excludes the commitment of merchantability and fitness for a specific use, the court may find such a disclaimer clause unreasonable and refuse to enforce it or refuse to enforce the entire contract. In determining whether warranty disclaimers are unreasonable, the court attempts to evaluate if the contract was made between two "equals" or between an expert and a novice. The relative education, experience, and bargaining power of the parties and whether the sales contract was offered on a take-it-or-leave-it basis are considered in making this determination.

The plaintiff must have a valid contract that the supplier did not fulfill in order to win a breach of warranty claim. Because the software supplier writes the warranty, this claim can be extremely difficult to prove. For example, in 1993, M.A. Mortenson Company had a new version of bid-preparation software installed for use by its estimators. During the course of preparing one new bid, the software allegedly malfunctioned several times, each time displaying the same cryptic error message. Nevertheless, the estimator submitted the bid and Mortenson won the contract. Afterward, Mortenson discovered that the bid was $1.95 million lower than intended and filed a breach-of-warranty suit against Timberline Software, makers of the bid software. Timberline acknowledged the existence of the bug, but the courts all ruled in favor of Timberline, ruling that the license agreement that came with the software explicitly barred recovery of the losses claimed by Mortenson.[15] Even if breach of warranty can be proven, the damages are generally limited to the amount of money paid for the product.

Intentional misrepresentation occurs when a seller or lessor either misrepresents the quality of a product or conceals a defect in it. For example, a cleaning product might be advertised as safe to use in confined areas, but users pass out from the fumes. A buyer or lessee who is subsequently injured can sue the seller for intentional misrepresentation or fraud. Advertising, salespersons' comments, invoices, and shipping labels are all forms of representation. Most software manufacturers use limited warranties and disclaimers to avoid any claim of misrepresentation.

Software Development Process

Developing information system software is not a simple process. It requires completing many activities that are themselves complex, with many dependencies among the various activities. System analysts, programmers, architects, database specialists, project managers, documentation specialists, trainers, and testers are all involved in large software projects. Each of these groups has a role to play, and has specific responsibilities and tasks. In addition, each group makes decisions that can affect the software's quality and the ability to use it effectively.

Many software developers have adopted a standard, proven work process (or **software development methodology**) that enables systems analysts, programmers, project managers, and others to make controlled and orderly progress in developing high-quality software. A methodology defines activities in the software development process and the individual and group responsibilities for accomplishing these activities. It also recommends specific techniques for accomplishing the various activities, such as using a flowchart to document the logic of a computer program. A methodology also offers guidelines for managing the quality of software during the various stages of development. If an organization has developed such a methodology, it is applied to any software development that the company undertakes.

As with most things in life, it is far safer and cheaper to avoid software problems at the beginning than to attempt to fix the damages after the fact. Studies at IBM, TRW, and GTE have shown that the cost to identify and remove a defect in an early stage of software development can be 100 times less than removing a defect in an operating piece of software that has been distributed to hundreds or thousands of customers.[16] This increase in cost is due to two reasons. First, if a defect is uncovered in a later stage of development, some rework of the deliverables produced in preceding stages is necessary. Second, the later the error is detected, the greater the number of people affected by the error. Consider the cost to communicate the details of a defect, distribute and apply software fixes, and possibly retrain end users for a software product that has been sold to hundreds or thousands of customers. Thus, most software developers try to identify and remove errors early in the development process as a cost-saving measure and as the most efficient way to improve software quality.

Products containing inherent defects that harm the user are the subjects of product liability suits. The use of an effective methodology can protect software manufacturers from legal liability for defective software in two ways. First, an effective methodology reduces the number of software errors that might occur. Second, if an organization follows widely accepted development methods, negligence on its part is harder to prove. However, even a *successful* defense against a product liability case can cost hundreds of thousands of dollars in legal fees. Thus, failure to develop software reasonably and consistently can be quite serious in terms of liability exposure.

Software quality assurance (QA) refers to methods within the development cycle that guarantee reliable operation of the product. Ideally, these methods are applied at each stage throughout the development cycle. However, some software manufacturing organizations that haven't established a formal, standard approach to QA consider testing to be their only QA method. Instead of checking for errors throughout the development process, they rely primarily on testing just before the product ships to ensure some degree of quality.

Several types of tests are used in software development, as discussed in the following sections.

Dynamic Software Testing

Software is developed in units called subroutines or programs. These units, in turn, are combined to form large systems. When a programmer completes a unit of software, one QA measure is to test the code by actually entering test data and comparing the results to the expected results. This is called **dynamic testing**. There are two forms of dynamic testing:

- **Black-box testing** involves viewing the software unit as a device that has expected input and output behaviors but whose internal workings are unknown (a black box). If the unit demonstrates the expected behaviors for all the input data in the test suite, it passes the test. Black-box testing takes place without the tester having any knowledge of the structure or nature of the actual code. For this reason, it is often done by someone other than the person who wrote the code.

- **White-box testing** treats the software unit as a device that has expected input and output behaviors but whose internal workings, unlike the unit in black-box testing, are known. White-box testing involves testing all possible logic paths through the software unit with thorough knowledge of its logic. The test data must be carefully constructed to make each program statement execute at least once. For example, if you wrote a program to calculate an employee's gross pay, you would develop data to test cases in which the employee worked less than 40 hours and other cases in which the employee worked more than 40 hours (to check the calculation of overtime pay).

Other Types of Software Testing

Other forms of testing include the following:

- **Static testing**—Special software programs called static analyzers are run against the new code. Rather than reviewing input and output, the static analyzer looks for suspicious patterns in programs that might indicate a defect.
- **Integration testing**—After successful unit testing, the software units are combined into an integrated subsystem that undergoes rigorous testing to ensure that all the linkages among the various subsystems work successfully.
- **System testing**—After successful integration testing, the various subsystems are combined to test the entire system as a complete entity.
- **User acceptance testing**—This independent testing is performed by trained end users to ensure that the system operates as they expect.

Capability Maturity Model Integration for Software

Capability Maturity Model Integration (CMMI) is a process improvement approach defined by the Software Engineering Institute at Carnegie Mellon University in Pittsburgh. It defines the essential elements of effective processes. The model is general enough to be used to evaluate and improve almost any process, and it is frequently used to assess software development practices. CMMI defines five levels of software development maturity (see Table 7-1) and identifies the issues that are most critical to software quality and process improvement. Identifying an organization's current maturity level enables it to specify actions to improve its future performance. The model also enables an organization to track, evaluate, and demonstrate its progress over the years.

After an organization decides to adopt CMMI, it must conduct an assessment of its software development practices (often using outside resources to ensure objectivity) and determine where they fit in the capability model. The assessment identifies areas for improvement and the action plans needed to upgrade the development process. Over the course of a few years, the organization and software engineers can raise their performances to the next level by executing the action plan.

As the maturity level increases, the organization improves its ability to deliver good software on time and on budget. For example, the Lockheed Martin Missile and Defense Systems converted to CMMI between 1996 and 2002, with outstanding improvements. Lockheed increased software productivity by 30 percent, reduced unit software costs by 20 percent, and cut the costs of finding and fixing software defects by 15 percent.[17]

CMMI can also be used as a benchmark for comparing organizations. In the awarding of software contracts, particularly with the government, organizations that bid on the contract may be required to have adopted CMMI and perform at a certain level.

Table 7-1 defines the five maturity levels of CMMI and shows how 350 organizations were assessed between April 2002 and August 2004.[18]

TABLE 7-1 CMMI maturity levels

Maturity level	Definition	Percentage of organizations at this level (as of August 2005)
Initial	Process unpredictable, poorly controlled, and reactive	5%
Managed	Process characterized for projects and is often reactive	36%
Defined	Process characterized for the organization and is proactive	29%
Quantitatively managed	Process measured and controlled	5%
Optimizing	Focus is on continuous process improvement	25%

Source: Capability Maturity Model Integration (CMMI) Overview, Carnegie Mellon University, www.sei.cmu.edu/cmmi/adoption/pdf/cmmi-overview05.pdf, October 29, 2005.

KEY ISSUES IN SOFTWARE DEVELOPMENT

Although defects in any system can cause serious problems, the consequences of software defects in certain systems can be deadly. In these systems, the stakes involved in creating quality software are raised to the highest possible level. The ethical decisions involving a trade-off—if one might even be considered—between quality and other factors such as cost, ease of use, or time to market require extremely serious examination. The next sections discuss safety-critical systems and the special precautions companies must take in developing them.

Development of Safety-Critical Systems

A **safety-critical system** is one whose failure may cause injury or death. The safe operation of many safety-critical systems relies on the flawless performance of software; these systems control an automobile's antilock brakes, nuclear power plant reactors, airplane navigation, roller coasters, elevators, and numerous medical devices, to name just a few. The process of building software for such systems requires highly trained professionals, formal and rigorous methods, and state-of-the-art tools. Failure to take strong measures to identify and remove software errors from safety-critical systems "is at best unprofessional and at worst can lead to disastrous consequences."[19] However, even with these precautions, the software associated with safety-critical systems is still vulnerable to errors that can lead to injuries or death.

In June 1994, for example, a Chinook helicopter took off from Northern Ireland with 25 British intelligence officials to a security conference in Inverness. Just 18 minutes into its flight, the helicopter crashed on the peninsula of Kintyre in Argyll, Scotland, killing everyone aboard. A handwritten memo by a senior Ministry of Defense procurement officer revealed problems with the Chinook, "many of which were traced eventually back to software design and systems integration problems which were experienced from February to July 1994." In particular, the engine management software, which controlled the acceleration and deceleration of the engines, was suspect.[20]

One of the most widely cited software-related accidents in safety-critical systems involved a computerized radiation therapy machine called the Therac-25. This medical linear accelerator was designed to deliver either protons or electrons in high-energy beams that would destroy tumors with minimal impact on the surrounding healthy tissue. Between June 1985 and January 1987, six known accidents involving massive overdoses were caused by software errors in the Therac-25, leading to serious injuries and even death.[21] The Therac bug was subtle. If a fast-typing operator mistakenly selected X-ray mode and then used a particular editing key to change to electron mode, the display would appear to show a proper setting, when in reality the Therac had been configured to focus electrons at full power to a tiny spot on the body.[22]

The Therac-25 case illustrates that accidents seldom have a single root cause and that if only the symptoms of a problem are fixed, future accidents may still occur. With the benefit of hindsight, it is clear that poor decisions were made about the development and use of the Therac-25 machine. The vendor decided to eliminate hardware backup safety mechanisms that were present in earlier versions of the machine. The vendor also elected to reuse parts of earlier Therac-20 software in a different Therac-25 machine. The software was poorly documented and system testing was incomplete. The vendor failed to take a lead role in investigating the initial accident, insisting that the cause must have been human error. Finally, hospitals did not push hard enough for answers and continued to use the Therac-25 machine, even when they knew that accidents had occurred.

When developing safety-critical systems, a key assumption must be that safety will *not* automatically result from following your organization's standard development methodology. Safety-critical software must go through a much more rigorous and time-consuming development process than other kinds of software. All tasks—including requirements definition, systems analysis, design, coding, fault analysis, testing, implementation, and change control—require additional steps, more thorough documentation, and more checking and rechecking. As a result, safety-critical software takes much longer to complete and is much more expensive.

The key to doing this additional work is to appoint a project safety engineer who has explicit responsibility for the system's safety. The safety engineer uses a logging and monitoring system to track hazards from a project's start to its finish. The hazard log is used at each stage of the software development process to assess how it has accounted for detected hazards. Safety reviews are held throughout the development process, and a robust configuration management system tracks all safety-related documentation to keep it consistent with the associated technical documentation. Informal documentation is not acceptable for safety-critical system development; formal documentation is required, including verification reviews and signatures.

The increased time and expense of completing safety-critical software can draw developers into ethical dilemmas. For example, the use of hardware mechanisms to back up or verify critical software functions can help ensure safe operation and make the consequences of software defects less critical. However, such hardware may make the final product more expensive to manufacture or harder for the user to operate, making the product less attractive than a competitor's. Companies must weigh these issues carefully to develop the safest possible product that also appeals to customers. Another key issue is deciding when the QA staff has performed enough testing. How much testing is enough when you are building a product that can cause loss of human life? At some point, software developers must determine that they have completed sufficient QA activities and then sign off to indicate their approval. Determining how much testing is sufficient demands careful decision making.

When designing, building, and operating a safety-critical system, a great deal of effort must go into considering what can go wrong, the likelihood and consequences of such occurrences, and how risks can be averted or mitigated. One approach to answering these questions is to conduct a formal risk analysis. **Risk** is the probability of an undesirable event occurring times the magnitude of the event's consequences if it does happen. These consequences include damage to property, loss of money, injury to people, and death. For example, if an undesirable event has a 1 percent probability of occurring and its consequences would cost $1,000,000, then the risk can be determined as 0.01 x $1,000,000, or $10,000. This risk would be considered greater than that of an event with a 10 percent probability of occurring, at a cost of $100 (0.10 x $100 = $10). Risk analysis is important for safety-critical systems, but is useful for other kinds of software development as well.

Another key element of safety-critical systems is **redundancy**, the provision of multiple interchangeable components to perform a single function in order to cope with failures and errors. A simple redundant system would be an automobile with a spare tire or a parachute with a backup chute attached. A more complex system used in IT is a redundant array of independent disks (RAID), which is commonly used in high-volume data storage for file servers. RAID systems use many small-capacity disk drives to store large amounts of data and to provide increased reliability and redundancy. Should one of the drives fail, it can be removed and a new one inserted in its place. Data on the failed disk can be rebuilt automatically without the server ever having to be shut down, because the data has been stored elsewhere.

N-version programming is a form of redundancy that involves the execution of a series of program instructions simultaneously by two different systems. The systems use different algorithms to execute instructions that accomplish the same result. The results from the two systems are then compared; if a difference is found, another algorithm is executed to determine which system yielded the correct result. In some cases, instructions for the two systems are written by programmers from two different companies and run on different hardware devices. The rationale for N-version programming is that both systems are highly unlikely to fail at the same time under the same conditions. Thus, one of the two systems should yield a correct result. IBM employs N-version programming to reduce disk sector failures in data storage devices. Two pieces of code in the same application save a piece of data and then compare the data to ensure that no errors occurred.[23]

During times of widespread disaster, improper planning for redundant systems can lead to major problems. For example, Hurricane Katrina knocked out 2.5 million telephone lines, four TV stations, and 36 radio stations.[24] There were inadequate backup communication systems to replace the failed systems.

After an organization determines all pertinent risks to a system, it must decide what level of risk is acceptable. This decision is extremely difficult and controversial because it involves forming personal judgments about the value of human life, assessing potential liability in case of an accident, evaluating the surrounding natural environment, and estimating the system's costs and benefits. System modifications must be made if the level of risk in the design is judged to be too great. Modifications can include adding redundant components or using safety shutdown systems, containment vessels, protective walls, or escape systems. Another approach is to mitigate the consequences of failure by devising emergency procedures and evacuation plans. In all cases, organizations must ask how safe is safe enough, if human life is at stake.

Manufacturers of safety-critical systems must sometimes decide whether to recall a product when data indicates a problem. For example, automobile manufacturers have been known to weigh the cost of potential lawsuits against that of a recall. Drivers and passengers in affected automobiles (and in many cases, the courts) have not found this approach very ethical. Manufacturers of medical equipment and airplanes have had to make similar decisions, which can be complicated when data cannot pinpoint the cause of the problem. For example, there was great controversy in 2000 over the use of Firestone tires on Ford Explorers; numerous tire blowouts and Explorer rollovers caused multiple injuries and deaths. However, it was difficult to determine if the rollovers were caused by poor automobile design, faulty tires, or improperly inflated tires. Consumers' confidence in both products was nevertheless shaken.

Reliability is the probability of a component or system performing without failure over its product life. For example, if a component has a reliability of 99.9 percent, it has one chance in one thousand of failing over its lifetime. Although this chance of failure may seem very low, remember that most systems are made up of many components. As you add more components to the system, it becomes more complex and the chance of a failure increases. For example, imagine that you are building a complex system made up of seven components, each with 99.9 percent reliability. If none of the components has redundancy built in, the product has a 93.8 percent probability of operating successfully with no component malfunctions over its lifetime. If you build the same type of system using 10 components, each with 99.9 percent reliability, the overall probability of operating without an individual component failure falls to less than 60 percent! Thus, building redundancy into systems that are both complex and safety-critical is imperative.

One of the most important and difficult areas of safety-critical system design is the human interface. Human behavior is not nearly as predictable as the reliability of hardware and software components in a complex system. The system designer must consider what human operators might do to make a system work less safely or effectively. The challenge is to design a system that not only works as it should, but leaves the operator little room for erroneous judgment. For instance, a self-medicating pain relief system must allow a patient to press a button to receive more pain reliever, but the system also must regulate itself to prevent an overdose. Additional risk can be introduced if a designer does not anticipate the information an operator needs and how the operator will react under the daily pressures of actual operation, especially in a crisis. Some people keep their wits about them and perform admirably in an emergency, but others may panic and make a bad situation worse.

Poor interface design between systems and humans can greatly increase risk, sometimes with tragic consequences. For example, in July 1988, the guided missile cruiser *U.S.S. Vincennes* mistook an Iranian Air commercial flight for an enemy F-14 jet fighter and shot down the airliner over international waters in the Persian Gulf, killing almost 300 people. Some investigators blamed the tragedy on the confusing interface of the $500 million Aegis radar and weapons control system. The Aegis radar on the *Vincennes* locked onto an Airbus 300, but it was misidentified as a much smaller F-14 by its human operators. The Aegis operators also misinterpreted the system signals and thought that the target was descending, even though the airbus was actually climbing. A third human error was made in determining the target altitude—it was wrong by 4000 feet. As a result of this combination of human errors, the *Vincennes* crew thought the ship was under attack and shot down the plane.[25]

Quality Management Standards

The International Organization for Standardization (ISO), founded in 1947, is a worldwide federation of national standards bodies from some 100 countries. The ISO issued its 9000 series of business management standards in 1988. These standards require organizations to develop formal quality management systems that focus on identifying and meeting the needs, desires, and expectations of their customers.

The **ISO 9000** standard serves many industries and organizations as a guide to quality products, services, and management. Approximately 350,000 organizations have ISO 9000 certification in more than 150 countries. Although companies can use the standard as a management guide for their own purposes in achieving effective control, the bottom line for many is having a qualified external agency say that they have achieved ISO 9000 certification. Many companies and government agencies specify that a company must be ISO 9000-certified to win a contract from them.

To obtain this coveted certificate, an organization must submit to an examination by an external assessor and fulfill the following requirements:

1. Have written procedures for everything it does.
2. Follow those procedures.
3. Prove to an auditor that it has fulfilled the first two requirements. This proof can require observation of actual work practices and interviews with customers, suppliers, and employees.

The various ISO 9000 series of standards address the following activities:

ISO 9001	Design, development, production, installation, servicing
ISO 9002	Production, installation, servicing
ISO 9003	Inspection and testing
ISO 9000-3	The development, supply, and maintenance of software
ISO 9004	Quality management and quality systems elements

Failure mode and effects analysis (FMEA) is an important technique used to develop any ISO 9000-compliant quality system. FMEA is used to evaluate reliability and determine the effect of system and equipment failures. Failures are classified according to their impact on a project's success, personnel safety, equipment safety, customer satisfaction, and customer safety. The goal of FMEA is to identify potential design and process failures early in a project when they are relatively easy and inexpensive to correct. A failure mode describes how a product or process could fail to perform the desired functions described by the customer. An effect is an adverse consequence that the customer might experience. Unfortunately, most systems are so complex that there is seldom a one-to-one relationship between cause and effect. Instead, a single cause may have multiple effects and a combination of causes may lead to one effect or multiple effects.

Another example of a quality management standard comes from the aviation industry. Over the past 25 years, the role of software in aircraft manufacturing has expanded from controlling simple measurement devices to almost all aircraft functions, including navigation, flight control, and cockpit management. Given this expanded use of software, the aviation industry and its regulatory community needed an effective means of evaluating safety-critical software. To meet this need, the Radio Technical Commission for Aeronautics (RTCA) developed DO-178B/EUROCCAE ED-128 as the evaluation standard for the international aviation community. The RTCA is a private, nonprofit organization that develops recommendations for use by the Federal Aviation Administration (FAA) and the private sector. Aviation software developers use the RTCA standard to achieve a high level of confidence in safety-critical software.

Table 7-2 provides a useful checklist for an organization that wants to upgrade the quality of the software it produces. The preferred response to each question is *yes*.

TABLE 7-2 Manager's checklist for improving software quality

Questions	Yes	No
Has senior management made a commitment to quality software?	____	____
Have you used CMMI to evaluate your organization's software development process?	____	____
Have you adopted a standard software development methodology?	____	____
Does the methodology place a heavy emphasis on quality management and address how to define, measure, and refine the quality of the software development process and its products?	____	____
Are software project managers and team members trained in the use of this methodology?	____	____
Are software project managers and team members held accountable for following this methodology?	____	____
Is a strong effort made to identify and remove errors as early as possible in the software development process?	____	____
In the testing of software, are both static and dynamic testing used?	____	____
Are white-box testing and black-box testing used?	____	____
Has an honest assessment been made to determine if the software being developed is safety-critical?	____	____
If the software is safety-critical, are additional tools and methods employed, and do they include the following: project safety engineer, hazard logs, safety reviews, formal configuration management systems, rigorous documentation, risk analysis processes, and the FMEA technique?	____	____

Summary

1. Why do companies require high-quality software in business systems, industrial process control systems, and consumer products?

    High-quality software systems are needed because they are easy to learn and easy to use. They perform quickly and efficiently to meet their users' needs, operate safely and dependably, and have a high degree of availability that keeps unexpected downtime to a minimum.

    Such high-quality software has long been required to support the fields of air traffic control, nuclear power, automobile safety, health care, military and defense, and space exploration. Now that computers and software have become an integral part of our lives, more and more users are demanding high quality in their software. They cannot afford system crashes, lost work, or lower productivity, nor can they tolerate security holes through which intruders can spread viruses, steal data, and shut down Web sites.

2. What ethical issues do software manufacturers face in making trade-offs between project schedules, project costs, and software quality?

    Software manufacturers are under extreme pressure to reduce the time to market of their products. They are driven by the need to beat the competition in delivering new functionality to users, to begin generating revenue to recover the cost of development, and to show a profit for shareholders. The resources and time budgeted to ensure quality are often cut under the intense pressure to ship the new product. When forced to choose between adding more user features or doing more testing, most software development managers decide in favor of more features. After all, they reason, defects can always be patched in the next release, which will give customers an automatic incentive to upgrade. The additional features will make the release more useful and therefore easier to sell to customers. Many customers challenge whether the decision to cut quality is ethical.

3. What are the four most common types of software product liability claims, and what actions must plaintiffs and defendants take to be successful?

    Software product liability claims are frequently based on strict liability, negligence, breach of warranty, or misrepresentation. Strict liability means that the defendant is responsible for injuring another person, regardless of negligence or intent. The plaintiff must prove only that the software product is defective or unreasonably dangerous and that the defect caused the injury. When sued for negligence, a software supplier is not held responsible for every product defect that causes customer or third-party loss. Instead, responsibility is limited to harmful defects that could have been detected and corrected through "reasonable" software development practices.

    If the product fails to meet its warranty, the buyer or lessee can sue for breach of warranty. The plaintiff must have a valid contract that the supplier did not fulfill in order to win a breach of warranty claim. Intentional misrepresentation occurs when a seller or lessor either misrepresents the quality of a product or conceals a defect in it. Most software manufacturers use limited warranties and disclaimers to avoid any claim of misrepresentation.

4. What are the essential components of a software development methodology, and what are its benefits?

A software development methodology defines the activities in the software development process, defines individual and group responsibilities for accomplishing objectives, recommends specific techniques for accomplishing the objectives, and offers guidelines for managing the quality of the products during the various stages of the development cycle.

Using an effective development methodology enables a manufacturer to produce high-quality software, forecast project completion milestones, and reduce the overall cost to develop and support software. It also protects software manufacturers from legal liability for defective software in two ways: it reduces the number of software errors that could cause damage, and makes it more difficult to prove negligence.

5. How can Capability Maturity Model Integration improve an organization's software development process?

CMMI defines five levels of software development maturity and identifies the issues that are most critical to software quality and process improvement. Its use can improve an organization's ability to predict and control quality, schedule, costs, cycle time, and productivity when acquiring, building, or enhancing software systems. CMMI also helps software engineers to analyze, predict, and control selected properties of software systems.

6. What is a safety-critical system, and what actions are required during its development?

A safety-critical system is one whose failure may cause injury or death. In the development of safety-critical systems, a key assumption is that safety will *not* automatically result from following an organization's standard software development methodology. Safety-critical software must go through a much more rigorous and time-consuming development and testing process than other kinds of software; the appointment of a project safety engineer and the use of a hazard log and risk analysis are common.

Self-Assessment Questions

1. The impact of a software defect can be quite subtle or very serious. True or False?

2. _____ is the degree to which the attributes of a software product enable it to meet the needs of its users.

3. Which of the following is *not* a major cause of poor software quality?

   a. Developers do not know how to design quality into software or do not take the time to do it.

   b. Programmers make mistakes in turning design specifications into lines of code.

   c. Software manufacturers are under extreme pressure to reduce the time to market of their products.

   d. Many organizations avoid buying the first release of a major software product.

4. A standard, proven work process for the development of high-quality software is called a(n) _____ .

5. The cost to identify and remove a defect in an early stage of software development can be 100 times less than removing a defect in an operating piece of software that has been distributed to many customers. True or False?

6. _____ is a form of testing a unit of software by entering test data and comparing the actual results to the expected results.

    a. Dynamic testing

    b. Static testing

    c. Integration testing

    d. System testing

7. _____ is an approach that defines the essential elements of an effective process and outlines a system for continuously improving software development.

    a. ISO 9000

    b. FMEA

    c. CMMI

    d. DO-178B

8. A system whose failure may cause injury or even death is called _____ .

9. _____ is the probability of an undesirable event occurring times the magnitude of the event's consequence if it does happen.

    a. Risk

    b. Redundancy

    c. Reliability

    d. Availability

10. A reliability evaluation technique that can determine the effect of system and equipment failures is _____ .

Review Questions

1. What is a software defect?

2. How does redundancy reduce the risk associated with a system component?

3. What must a defendant prove to win a strict liability claim? What legal defenses are commonly used by defendants?

4. What must a defendant prove to win a negligence claim? What legal defenses are commonly used by defendants?

5. What is breach of warranty? What must the plaintiff prove to win a breach of warranty claim? How can a software manufacturer defend against potential breach of warranty claims?

6. What is the difference between quality management and quality assurance?

7. What is a software development methodology? What is its purpose?

8. Why is it critical to identify and remove defects early in the software development process?

9. Identify the various types of testing used in software development.

10. What is a safety-critical system? What additional precautions are needed when developing such a system?

11. What is the purpose of the ISO 9000 quality standard? How is it similar to and different from CMMI?

12. What is FMEA? What is its goal?

Discussion Questions

1. Identify your top three criteria for a quality system. Briefly discuss your rationale for selecting these criteria.

2. Explain why the cost to identify and remove a defect in the early stages of software development might be 100 times less than removing a defect in software that has been distributed to hundreds of customers.

3. You are considering using N-version programming with two software development firms and hardware devices for the navigation system of a guided missile. Briefly describe what this means and outline several advantages and disadvantages of this approach.

4. Discuss the implications to a project team of classifying a piece of software as safety-critical.

5. You have been asked to draft a boilerplate warranty for a software contractor that will absolutely protect the firm from being sued successfully for negligence or breach of contract. Is this possible? Why or why not?

6. Discuss why an organization may elect to use a separate, independent team for quality testing rather than the group of people that originally developed the software.

7. You are considering contracting for the development of new software that is essential to the success of your midsized manufacturing firm. One candidate firm boasts that its software development practices are at level 4 of CMMI. Another firm claims that all its software development practices are ISO-9000 compliant. How much weight should you give to these certifications when deciding which firm to use? Do you think that a firm could lie or exaggerate its level of compliance with these standards?

What Would You Do?

1. Read the fictitious Killer Robot case at the Web site for the Online Ethics Center for Engineering & Science at *www.onlineethics.com* (look under Computer Science and Internet cases). The case begins with the manslaughter indictment of a programmer for writing faulty code that resulted in the death of a robot operator. Slowly, over the course of many articles, you are introduced to several factors within the corporation that contributed to the accident. After reading the case, answer the following questions:

   a. Responsibility for an accident is rarely defined clearly and rarely is traceable to one or two people or causes. In this fictitious case, it is clear that a large number of people share responsibility for the accident. Identify all the people you think were at least partially responsible for the death of Bart Mathews, and why you think so.

   b. Imagine that you are the leader of a task force assigned to correct the problems uncovered by this accident. Develop a list of the top 10 critical actions to take to avoid future problems. What process would you use to identify the most critical actions?

c. If you were in Ms. Yardley's position, what would you have done when Ray Johnson told you to fake the test results? How would you justify your decision?

2. You are the project manager for developing the latest release of your software firm's flagship product. The product release date is just two weeks away and enthusiasm for the product is extremely high among your customers. Stock market analysts are forecasting sales of more than $25 million per month. If so, earnings per share will increase by nearly 50 percent. There is just one problem: two key features promised to customers in this release have several bugs that would severely limit the features' usefulness. You estimate that at least six weeks are needed to find and fix the problems. In addition, even more time is required to find and fix 50 additional, less severe bugs uncovered by the QA team. What do you recommend to management?

3. You have been assigned to manage software that controls the shutdown of chemical reactors, but your manager insists it is not safety-critical software. The software senses temperatures and pressures within a 50,000-gallon stainless steel vat and dumps in chemical retardants to slow down the reaction if it gets out of control. In the worst possible scenario, failure to stop a runaway reaction would result in a large explosion that would send fragments of the vat flying and spray caustic soda in all directions. Your manager points out that the stainless steel vat is surrounded by two sets of protective concrete walls and that the reactor's human operators can intervene in case of a software failure. He feels that these measures would protect the plant employees and the surrounding neighborhood if the shutdown software failed. Besides, he argues, the plant is already more than a year behind its scheduled start-up date. He cannot afford the additional time required to develop the software if it is classified as safety-critical. How would you work with your manager and other appropriate resources to decide whether the software is safety-critical?

4. You are a senior software development consultant with a major consulting firm. Two years ago, you conducted the initial assessment of the ABCXYZ Corporation's software development process. Using CMMI, you determined their level of maturity to be level 1. Since your assessment, the organization has spent a lot of time and effort following your recommendations to raise its level of process maturity to the next level. The organization appointed a senior member of the IT staff to be a process management guru and paid him $150,000 per year to lead the improvement effort. This senior member adopted a methodology for standard software development and required all project managers to go through a one-week training course at a total cost of more than $2 million.

Unfortunately, these efforts did not significantly improve process maturity because senior management failed to hold project managers accountable for actually using the standard development methodology in their projects. Too many project managers convinced senior management that the new methodology was not necessary for their project and would just slow things down. However, you are concerned that when senior management learns that no real progress has been made, they will refuse to accept partial blame for the failure and instead drop all attempts at further improvement. You want senior management to ensure that the new methodology is used on all projects—no more exceptions. What would you do?

5. You are the CEO for a small, struggling software firm that produces educational software for high school students. Your latest software is designed to help students improve their SAT and ACT scores for getting into college. To prove the value of your software, a group of 50 students who had taken the ACT test were tested again after using your software for just

two weeks. Unfortunately, there was no dramatic increase in their scores. A statistician you hired to ensure objectivity in measuring the results claimed that the variation in test scores was statistically insignificant.

A small core group of educators and systems analysts will need at least six months to start again from scratch and design a viable product. Programming and testing could take another six months. Another option would be to go ahead and release the current version of the product and then, when the new product is ready, announce it as a new release. This would generate the cash flow necessary to keep your company afloat and save the jobs of 10 or more of your 15 employees.

Given this information about your company's product, what would you do?

Cases

1. New Wireless Technologies Reduce Medical Errors

In December 1995, seven-year-old Ben Kolb checked into Martin Memorial Hospital in Stuart, Florida, to undergo a simple ear surgery. During the procedure, the anesthesiologist injected the boy with an epinephrine solution 1000 times more concentrated than what had been specified. The boy went into cardiac arrest and died the following day.

As a result of cases like Ben Kolb's, medical institutions—which have traditionally dragged their heels when considering new IT systems—are turning to new wireless technologies for solutions. Forecasts originally predicted that 60 to 70 percent of medical institutions would adapt wireless technology by 2007. In fact, this percentage was already reached by mid-2005.

Hospitals use wireless technologies to prevent medical errors in a variety of ways. For example, doctors use handheld devices to access and update patients' medical records at their bedsides. Nurses scan bar codes on medications and on patients' wrist bands. By accessing the physician's order entry system, nurses ensure that they are administering the correct dosage and medication. Cell phones and wireless badges also improve communication between hospital staff.

Hospitals report that these technologies have had a significant effect on patient safety. For example, Beloit Memorial Hospital in Wisconsin reported a 67 percent reduction in errors four months after implementing a wireless medication administration system. Boston-based Brigham and Women's Hospital announced a 55 percent reduction in medical errors.

Few other studies have been carried out to date, yet medical practitioners believe that wireless systems will reduce errors for several reasons. Doctors prescribe more than 15,000 different drugs, and hundreds of them are similar in spelling or sound. Bad handwriting or bad hearing translates into incorrect dosages and medications. The system also helps hospital staff identify medications that may elicit allergic reactions or may interact unfavorably with other drugs the patient is taking. The system also increases the likelihood that medications will be administered on time.

However, a few critical studies have warned of the potential harmful effects of instituting these systems. One study reported that Brigham and Women's Hospital actually experienced a drastic increase in the number of potential medical errors, which were prevented by alert nurses who caught the errors prior to administrating the medication. The problem was traced to a bug in the physician ordering software that affected orders of potassium infusions.

An article in the *Journal of the American Medical Association* (*JAMA*) in March 2005 pointed out other flaws in computerized physician order entry systems (CPOE). These systems rely on the integration of existing IT and paper-based systems, and problems with this integration can produce medical errors. Furthermore, the CPOE interface may not suit the work flow process at the hospital. The physician, for example, might be forced to view 20 screens to review all the medications a single patient is taking.

"Good computerized physician order entry systems are, indeed, very helpful and hold great promise; but, as currently configured, there are at least two dozen ways in which CPOE systems significantly, frequently, and commonly facilitate errors—and some of those errors can be deadly," said Ross Koppel, author of the *JAMA* article.

Despite Koppel's assessments, the medical community is racing forward with wireless systems, with good reason. A report issued by the Institute of Medicine (IOM) of the National Academies finds that at least 44,000 Americans are killed each year in hospitals due to medical error, and this figure climbs each year. Medical errors kill more Americans each year than highway accidents, AIDS, or breast cancer.

Because wireless systems hold out the hope of combating this growing problem, even critics like Koppel do not suggest their abandonment. Instead, critics warn against blind faith in technology and ask healthcare professionals to use these systems carefully to minimize both human and machine-produced medical errors.

Questions:

1. How should hospital staff and IT staff work together to ensure the success of safety-critical wireless systems designed to reduce medical error?

2. Wireless systems can record who administered the medicine, the dosage administered, the time of day, and other details. Is such clear accountability good or bad for the health industry? Do you think nurses or other healthcare practitioners may be hesitant to use the systems because of potential legal consequences?

3. Can you identify additional safety measures that should be built into these wireless systems?

2. Prius Plagued by Programming Error

In May 2005, the National Highway Traffic Safety Administration (NHTSA) revealed that it had opened an investigation following complaints of engines stalling in Toyota's Prius hybrid. Motorists reported that while they were driving or stuck in traffic, warning lights flashed on and the gas engine suddenly switched off. In all cases, the motorists were able to maneuver to the side of the road safely because the electric motor, the steering, and the brake system continued to function.

As the second most fuel-efficient car in the United States, the Prius delivers better mileage per gallon by shifting from its gas engine to its electric motor. The electronic control unit is responsible for the smooth transition between the gas engine and the motor. Toyota eventually determined that an error in the software used by this electronic control unit caused it to malfunction and the gas engine to stall.

Today, the average car is equipped with 30 to 40 microprocessors, so software quality is a critical issue for automakers. With 35 million lines of code per car, the potential for error is enormous. IBM automotive software specialist Stavros Stefanis reports that as many as one-third of all warranty claims result from software glitches and electronic defects. "It's a big headache for the automakers," said Stefanis.

This headache is compounded by the fact that software controls safety-related systems such as engine performance, steering, antilock brakes, and air bags. A programming error could have an enormous financial impact on an automaker in major lawsuits and recalls.

So, when NHTSA opened its investigation, Toyota, a company that has established a reputation for reliability, proved extremely cooperative. In fact, the company had already issued a service bulletin to owners of 2004 and 2005 Prius hybrids in October 2004. Toyota conducted an internal investigation, and in October 2005 recalled 75,000 of the more than 88,000 2004 and 2005 Prius hybrids sold in the United States. The company recalled an additional 85,000 units sold abroad.

Toyota spokeswoman Allison Takahashi reported that NHTSA determined passenger safety was not at issue: "We are voluntarily initiating a customer-service campaign to assure that this unusual occurrence does not cause inconvenience." Following the announcement of the campaign, NHTSA promptly closed its investigation.

The investigation, however, did not slow sales of the most popular hybrid on the market. 2005 sales were up 200 percent over the previous year when the recall was announced. In fact, Toyota had to concentrate a good deal of its software development efforts on projects that would increase production to meet the demand for new hybrids in the United States.

Although the recall's effect on sales was probably negligible, the cost of the recall may be significant. Toyota will pay for advertising and other costs involved in contacting Prius owners about the recall, and then will pay for the repairs. The recall may also serve to warn automakers about the costs of human errors that go undetected during software production.

Questions:

1. Do you agree with NHTSA's assessment that problems with the Prius were not a safety-critical issue? In such cases, who should decide whether a software bug creates a safety-critical issue—the manufacturer, consumers, government agencies, or some other group?

2. How would the issue be handled differently if it were a safety-critical matter? Would the issue be handled differently if the costs involved were not so great?

3. As the amount of hardware and software embedded in the average car continues to grow, what steps can automakers take to minimize warranty claims and ensure customer safety?

3. Patriot Missile Failure

The Patriot is an Army surface-to-air missile system that defends against aircraft and cruise missiles, and more recently, against short-range ballistic missiles. The system was designed in the 1960s and enhanced in the 1980s. The short-range antimissile capability was incorporated into the Patriot PAC-2 version of the missile.

Following the Iraqi invasion of Kuwait in August 1990, the United States deployed the Patriot PAC-2 missile to Saudi Arabia during Operation Desert Shield. At the start of Desert Shield, the U.S. arsenal included only three PAC-2 missiles. PAC-2 production was accelerated so that by January 1991, 480 missiles were available. Patriot battalions were deployed to Saudi Arabia and then to Israel to defend key assets, military personnel, and citizens against Iraqi Scud missiles. Iraq launched 81 modified-Scud ballistic missiles into Israel and Saudi Arabia during the conflict.

The Iraqis modified the Scud missile to increase its range and boost its speed by as much as 25 percent. They reduced the weight of the warhead, enlarged the fuel tanks, and modified its flight so that all of the fuel was burned during the early phase of flight, rather than continuously. As a result of these modifications, the missiles became structurally unstable and often broke into pieces in the upper atmosphere. This instability made the warhead extremely difficult to intercept; also, radar mistakenly could lock onto pieces of the fragmented missile rather than the warhead.

The Patriot missile is launched and guided to the target through three phases. First, the missile guidance system turns the Patriot launcher to face the incoming missile. Second, the computer control system guides the missile toward the incoming missile. Third, the Patriot missile's internal radar receiver guides it to intercept the incoming missile.

During Desert Shield, the Patriot's radar constantly swept the sky for any object that had the flight characteristics of a Scud missile. The range gate is an electronic detection device within the radar system that uses the last observation of an object to forecast an area in the air space where the radar system should next see the object, if it truly was a Scud. This forecast is a function of the object's observed velocity and the time of the last radar detection. If the range gate determined that the detected target was a Scud, and if the Scud was in the Patriot's firing range, the Patriot battery fired its missiles.

On February 11, 1991, the Patriot Project Office received data from Patriots deployed in Israel, identifying a 20 percent shift in the system's radar range gate after the system had been running for eight consecutive hours. This shift was significant; it meant that the target was no longer in the center of the range gate and the probability of successfully tracking the target was greatly reduced. The Army knew that the Patriot system could not track a Scud with a range gate shift of 50 percent or more.

In the Patriot radar system, time is kept from time of system start-up and measured by the system's internal clock in tenths of a second. The longer the system ran, the less accurate the elapsed time calculation became. Consequently, after the Patriot computer control system ran continuously for extended periods, the range gate made an inaccurate estimate of the area in the air space where the radar system should next see the object. This error could cause the radar to lose the target and fool the system into thinking there was no incoming Scud.

Army officials assumed that other Patriot users were not running their systems for eight hours or more at a time, and that the Israeli experience was an anomaly. However, the Patriot Project Office analyzed the Israeli data and confirmed some loss in targeting accuracy. As a result, they made a software change to compensate for the inaccurate time calculation. This change was included in a modified software version that was released on February 16, 1991.

The Patriot Project Office sent a message to Patriot users on February 21, 1991, informing them that very long run times could cause a shift in the range gate, resulting in difficulty tracking the target. The message also advised users that a software change was on the way that would improve the system's targeting. However, the message did not specify what constituted "very long run times." Patriot Project Office officials assumed that users would not continuously run the batteries for so long that the Patriot would fail to track targets. Therefore, they did not think that more detailed guidance was required.

On February 25, 1991, a Patriot missile defense system operating at Dhahran, Saudi Arabia, failed to track and intercept an incoming Scud. The enemy missile subsequently hit an Army barracks and killed 28 Americans. The ensuing investigation revealed that at the time of the incident, the Patriot battery had been operating continuously for more than 100 hours. The long run time resulted in an inaccurate time calculation, which in turn caused the range gate to shift so much that the system could not track, identify, or engage the incoming Scud.

By cruel fate, modified software that fixed the inaccurate time calculation arrived in Dhahran the very next day. Army officials attributed the delay to the difficulties of arranging air and ground transportation during wartime.

The Army did not have the luxury of collecting definitive performance data during Operation Desert Storm. After all, they were operating in a war zone, not a test range. As a result, there was insufficient and conflicting data on the effectiveness of the Patriot missile. At one extreme was an early report that claimed the Patriot destroyed about 96 percent of the Scuds engaged in Saudi Arabia and Israel. (This presumably did not include Scuds the Patriot failed to engage due to the software error.) At the other extreme, in only about 9 percent of the engagements did observers actually see a Scud destroyed or disabled after a Patriot detonated nearby. Of course, some "kills" could have been effected out of their range of vision.

Questions:

1. With the benefit of hindsight, what steps could have been taken during development of the Patriot software to avoid the problems that led to the loss of life? Do you think these steps would have improved the Patriot's effectiveness enough to make it obvious that the missile was a strong deterrent against the Scud? Why or why not?

2. What ethical decisions do you think the U.S. military made in deploying the Patriot missile to Israel and Saudi Arabia and in reporting the effectiveness of the Patriot system?

3. What key lessons can be taken from this example of safety-critical software development and applied to the development of business information system software?

End Notes

[1] Crawford, Michael, "Trend Micro Bug Down To Over-Quick Testing," *TechWorld*, www.techworld.com/news/index.cfm?RSS&NewsID=3559, April 26, 2005.

[2] Kallender, Paul, "Trend Micro Lowers Forecast, Blames Software Bug: The Flaw Affected Thousands Of Its Customers," *Computerworld*, www.computerworld.com/softwaretopics/os/windows/story/0,10801,103183,00.html, July 14, 2005.

[3] Leyden, John, "PC-cillin Killed My PC," *The Register*, www.theregister.co.uk/2005/04/25/pc-cillin_duff_update, April 25, 2005.

[4] Keizer, Greg, "Apple Patches Bevy Of Tiger, Panther Bugs (Again)," *TechWeb News*, www.informationweek.com, September 25, 2005.

[5] Keizer, Greg, "RealNetworks Fixes Linux RealPlayer Flaw," *TechWeb News*, www.informationweek.com, October 3, 2005.

[6] Keizer, Greg, "Kaspersky Says It's Fixed AV Scanner Flaw," *TechWeb News*, www.informationweek.com, October 4, 2005.

[7] Keizer, Greg, "Microsoft Plans 9 Patches Next Week," *TechWeb News*, www.informationweek.com, October 7, 2005.

8 McMillan, Robert, "Symantec AntiVirus Scan Carries Critical Bug," *CIO*, www.cio.com, October 7, 2005.

9 Keizer, Greg, "Google Plugs Cross-Scripting Security Hole," *TechWeb News*, www.informationweek.com, October 10, 2005.

10 Sanders, Tom, "Windows Chokes on Latest Microsoft Patch," *StickyMinds*, www.stickyminds.com, October 17, 2005.

11 Humphrey, Watts S., "Why Quality Pays," *Computerworld*, www.computerworld.com, May 20, 2002.

12 Bierha, Bruce A., "Faulty Software or Puffery," *Computerworld*, www.computerworld.com, March 28, 1994.

13 Kaner, Cern, "Bad Software – Who Is Liable?," keynote address of Quality Assurance Institute Regional Conference, Seattle, www.badsoftware.com, June 1998.

14 Cheeseman, Henry R., *Contemporary Business Law*, 3rd Edition, page 362, Prentice-Hall, Upper Saddle River, NJ, 2000.

15 "Software Company Not Liable for $2 Million Bug in Bid," *Engineering Times*, www.nspe.org, July 2000.

16 Boehm, Barry W., *Software Engineering Economics*, Englewood Cliffs, NJ, Prentice-Hall, 1981.

17 Web site for Carnegie Mellon Software Engineering Institute, www.sei.cmu.edu/cmm, October 29, 2005.

18 Capability Maturity Model Integration (CMMI) overview, Carnegie Mellon University, www.sei.cmu.edu/cmmi/adoption/pdf/cmmi-overview05.pdf, October 29, 2005.

19 Bowen, Jonathan P., "The Ethics of Safety-Critical Systems," www.cs.rdg.ac.uk, October 29, 2005.

20 Collins, Tony, "Minister Denies Chinook Claim," *Computer Weekly*, www.computerweekly.com, July 1, 1999.

21 Leveson, Nancy G. and Turner, Clark S., "An Investigation of the Therac-25 Accidents," *IEEE Computer*, Vol. 26, No. 7, pages 18-41, July 1993.

22 "Getting Serious with Year 2000, Sometimes Bugs Can Be Deadly," *InfoWeek*, www.infoweek.com, April 13, 1998.

23 Mearian, Lucas, "Bulletproof Storage," *Computerworld*, www.computerworld.com, April 11, 2005.

24 Gross, Grant, "FCC Head: Hurricane Shows Need for Redundant Telecom," *Computerworld*, www.computerworld.com, September 22, 2005.

25 Church, George J., "High-Tech Horror," *Time*, www.time.com, July 18, 1988.

Sources for Case 1

"Panacea or Pandora's Box: Penn Study Shows that Computerized Physician-Order Entry Systems Often Facilitate Medication Errors," University of Pennsylvania Health System, Department of Public Affairs Web site, www.uphs.upenn.edu/news/News_Releases/mar05/CPOE.htm, March 08, 2005.

Berger, Robert G. and Kichak, J. P., "Computerized Physician Order Entry: Helpful or Harmful," *Journal of the American Medical Informatics Association*, www.pubmedcentral.nih.gov/articlerender.fcgi?artid=353014, March-April 2004.

Havenstein, Heather, "Wireless Leaders & Laggards: Health Care," *Computerworld*, www.computerworld.com/printthis/2005/0,4814,101711,00.html, May 16, 2005.

Sources for Case 2

Associated Press, "Toyota Recalls 160,000 Prius Hybrids," *Washington Post*, www.washingtonpost.com/wp-dyn/content/article/2005/10/14/AR2005101400127.html, October 14, 2005.

Duvall, Mel, "Software Bugs Threaten Toyota Hybrids," *Baseline*, www.baselinemag.com/article2/0,1397,1843934,00.asp?kc=BANKT0209KTX1K0100464, August 4, 2005.

Nauman, Matt, "Toyota to Fix Software Problem on 75,000 Prius Hybrids," *San Jose Mercury News*, www.siliconvalley.com/mld/siliconvalley/news/local/12893923.htm, October 13, 2005.

Sources for Case 3

Carlone, Ralph V., "Patriot Missile Defense—Software Problem Led to Systems Failure at Dhahran, Saudi Arabia," GAO Report B-247094, www.fas.org, February 24, 1992.

"Data Does Not Exist to Say Conclusively How Well Patriot Performed," GAO Report B-250335, www.fas.org, September 22, 1992.

"Taiwan Interested in Latest Patriot Missiles," *Asian Political News*, www.findarticles.com, November 30, 1998.

"U.S. To Sell 14 Upgraded Missile Systems to S. Korea," *Asian Political News*, www.findarticles.com, November 15, 1999.

"Frontline: The Gulf War: Weapons: MIM -104 Patriot," PBS Web site, www.pbs.org, January 20,1996.

CHAPTER **8**

EMPLOYER/EMPLOYEE ISSUES

VIGNETTE

Microsoft and Tata Promote Chinese Outsourcing

In June 2005, Tata Consultancy Services, Microsoft Corporation, and three government-owned Chinese software development companies announced plans to form a joint venture to provide IT outsourcing services—IT consulting services on a contract basis—both within China and abroad.

With China's annual IT budget currently at $23 billion and growing 15 to 18 percent per year, many global IT companies have set their sights on this emerging market. For Microsoft, the venture is one of a string of recent deals indicating that the software giant is finally getting off the sidelines in Asia and joining the game. Although the software development projects will not be restricted exclusively to Microsoft technologies, the venture will help the company gain ground at a time when Beijing is looking toward Linux-based solutions.

However, India-based Tata, a global IT consulting and business outsourcing company, will be the major stakeholder in the venture. The company established its Chinese branch in 2002 and currently employs 250 engineers in Hangzhou. The venture gives Tata the opportunity to expand further into China's domestic market and to increase its potential pool of qualified employees.

In recent years, Tata and other Indian outsourcing companies have grown rapidly. Their strategic approach has been to take advantage of lower wages in India and acquire U.S. clients by offering software development and maintenance at reduced rates. As a result of their rapid growth, however, the demand for qualified IT professionals is catching up with the supply, and wages have risen rapidly. With 50,000 new technology graduates a year, China will serve as a source of cheap IT labor and allow the Indian companies to keep their costs down.

The joint venture is also advantageous to the Chinese, whose companies have faced several obstacles to breaking into the global IT outsourcing market, including poor English-language skills and a reputation for ignoring intellectual property rights. By partnering with a well-entrenched firm, they can gain both experience and credibility.

The deal is clearly good for all companies involved. The big loser is the U.S. IT labor force, which—thanks in part to the investment of an American IT giant—will lose jobs to the Chinese.[1,2]

LEARNING OBJECTIVES

As you read this chapter, consider the following questions:

1. What are contingent workers, and how are they frequently employed in the information technology industry?
2. What key ethical issues are associated with the use of contingent workers, including H-1B visa holders and offshore outsourcing companies?
3. What is whistle-blowing, and what ethical issues are associated with it?
4. What is an effective whistle-blowing process?

USE OF NONTRADITIONAL WORKERS

Every two years, the Bureau of Labor Statistics (BLS) develops 10-year projections of economic growth, employment by industry and occupation, and the composition of the labor force. These projections are widely used in career guidance, planning education and training programs, and studying long-range employment trends. During the period from 2002 to 2012, total employment is projected to increase by 21.3 million jobs, or 15 percent, slightly less than the previous decade (20.7 million jobs, 17 percent). According to the BLS, employment growth will be concentrated in the service-providing sector of the economy; nine of the 10 occupations with the fastest wage and employment growth will be in the health and information technology fields. Table 8-1 shows the projected demand for these occupations.[3]

Meanwhile, the number of declared computer science majors and master's candidates has dropped 33 percent and 25 percent, respectively, since 2002, in spite of the forecast for an increased need for workers in this field.[4] As a result, IT firms and organizations that use IT products and services are concerned about a shortfall in the number of U.S. workers to fill these positions.

TABLE 8-1 Industries with fastest employment growth (2002–2012, numbers in thousands of jobs)

	EMPLOYMENT		CHANGE		ANNUAL GROWTH
Industry	2002	2012	Number	Percent	Rate (%)
Software publishers	256.0	429.7	173.7	67.9	5.3
Management, scientific, and technical consulting services	731.8	1,137.4	405.6	55.4	4.5
Community care facilities for the elderly and residential care facilities	695.3	1,077.6	382.3	55.0	4.5
Computer systems design and related services	1,162.7	1,797.7	635.0	54.6	4.5
Employment services	3,248.8	5,012.3	1,763.5	54.3	4.4
Individual, family, community, and vocational rehabilitation services	1,269.3	1,866.6	597.3	47.1	3.9
Ambulatory healthcare services except offices of health practitioners	1,443.6	2,113.4	669.8	46.4	3.9
Water, sewage, and other systems	48.5	71.0	22.5	46.4	3.9

TABLE 8-1 Industries with fastest employment growth (2002–2012, numbers in thousands of jobs) (continued)

	EMPLOYMENT		CHANGE		ANNUAL GROWTH
Industry	2002	2012	Number	Percent	Rate (%)
Internet services, data processing, and other information services	528.8	773.1	244.3	46.2	3.9
Child day-care services	734.2	1,050.3	316.1	43.1	3.6

Source: United States Department of Labor, Bureau of Labor Statistics, "BLS Releases 2002-2012 Employment Projections," www.bls.gov/news.release/ecopro.nr0.htm, February 11, 2004.

Facing a likely long-term shortage of trained and experienced workers, employers will increasingly turn to nontraditional sources to find IT workers with skills that meet their needs. These sources include contingent workers, H-1B workers, and outsourced offshore workers. Employers will have to make ethical decisions about whether to recruit new and more skilled workers from these sources or to spend the time and money to develop their own staff to meet the needs of their business. The workers affected by these decisions will demand to be treated fairly and equitably.

Contingent Workers

The **contingent workforce** includes independent contractors, workers brought in through employment agencies, on-call or day laborers, and on-site workers whose services are provided by contract firms. The exact number of contingent workers is unknown, but it probably represents 4 to 7 percent of the U.S. workforce, a total of 6 to 10 million people.[5]

A firm most often uses contingent workers when it has pronounced fluctuations in its staffing needs for technical experts on important projects, key resources in new product development, consultants in organizational restructuring, and workers in the design and installation of new information systems. Typically, these workers join a team of full-time employees and other contingent workers for the life of the project and then move on to their next assignment. Whether they work, when they work, and how much they work depends on the company's need for them. They have neither an explicit nor implicit contract for continuing employment. In a way, they are the information-age equivalent of migrant farm workers, planning a move to another employer when their current project winds down.

An organization typically obtains contingent workers in two ways: through temporary help or employee leasing. Firms that provide temporary help recruit, train, and test their employees in a wide range of job categories and skill levels and then assign them to clients. Temporary employees fill in during vacations and illnesses, meet temporary skill shortages, handle seasonal or other special workloads, and help staff special projects. However, they are not considered official employees of the company, nor are they eligible for company benefits such as vacation, sick pay, and medical insurance.

Temporary working arrangements might appeal to people who want maximum flexibility in their work schedule and a variety of work experiences. Because temporary workers do not receive additional compensation through company benefits, they are often paid a higher hourly wage than full-time employees doing equivalent work.

In **employee leasing**, a business outsources all or part of its workforce to a professional employer organization. The goal is to outsource human resource activities and associated costs, such as payroll, training, and the administration of employee benefits, to the employee leasing company. Employee leasing companies operate with a minimal administrative, sales, and marketing staff to keep overall costs down and pass the savings on to their clients. In a **coemployment relationship**, two employers have actual or potential legal rights and duties with respect to the same employee or group of employees. Employee leasing companies are subject to special regulations regarding workers' compensation and unemployment insurance. Because the workers are technically employees of the leasing firm, they can be eligible for some company benefits through the firm.

Advantages of Using Contingent Workers

When a firm employs contingent workers, it usually does not pay for benefits such as retirement, medical costs, and vacation time. A company can continually adjust the number of contingent workers to stay consistent with its business needs. Thus, the company can lay off contingent workers when they are no longer busy or needed. The company cannot do the same with full-time employees without creating a great deal of ill will and lowering employee morale. Moreover, because many contingent workers already are specialists in a particular task, the firm does not customarily incur training costs. Therefore, the use of contingent workers enables the firm to meet its staffing needs more efficiently, lower its labor costs, and respond more quickly to changing market conditions.

Disadvantages of Using Contingent Workers

On the downside, contingent workers may lack a strong relationship with the firm, which can result in low commitment to the company and its projects along with a high turnover rate. Although temporary workers don't need technical training if they are specially qualified for a temporary job, they do gain valuable practical experience working within a particular company's structure and culture, which is lost when the workers depart at the project's completion.

Deciding When to Use Contingent Workers

When management decides to use contingent workers for a project, it should recognize the trade-off it is making between completing a single project quickly and cheaply versus developing people in its own organization. If the project requires unique skills that are probably not necessary for future projects, there is little reason to invest the additional time and costs required to develop those skills in full-time employees.

If the staffing required for a particular project is truly temporary and the workers will not be needed for future projects, the use of contingent workers is a good approach. In such a situation, using contingent workers avoids the need to hire new employees and then fire them when staffing needs go down.

Management should think twice about using contingent workers when they are likely to learn corporate processes and strategies that are key to the company's success. It is next to impossible to control contingent workers from passing on such information to subsequent employers. This can be damaging if the next employer is a key competitor.

Although contingent workers can often be the most flexible and cheapest way to get a job done, their use can raise ethical and legal issues about the relationships among the staffing firm, its employees, and its customers, including the potential liability of customers for withholding payroll taxes, payment of employee retirement benefits and health insurance premiums, and administration of worker's compensation to the staffing firm's employees. Depending on how closely workers are supervised and how the job is structured, contingent workers can be viewed as permanent employees by the Internal Revenue Service, the Labor Department, or a state's worker compensation and unemployment agencies.

For example, Microsoft agreed to pay a $97 million settlement in 2001 to some 10,000 so-called permatemps, temporary workers employed for an extended length of time as software testers, graphic designers, editors, technical writers, receptionists, and office support staffers. Some had worked at Microsoft for several years. The *Vizcaino v. Microsoft* class action was filed in federal court in 1992 by eight former workers who claimed that they and thousands more permatemps had been illegally shut out of a stock purchase plan that allowed employees to buy Microsoft stock at a 15 percent discount.[6] Microsoft shares had skyrocketed in value throughout the 1990s and split eight times. The sharp appreciation in the stock price meant that, had they been eligible, some temporary workers in the lawsuit could have earned more money from stock gains than they received in salary while at Microsoft.

The *Vizcaino v. Microsoft* lawsuit dramatically illustrated the cost of misclassifying employees and violating laws that cover workers, compensation, taxes, unemployment insurance, and overtime. The key lesson was that, even if workers sign an agreement that they are contractors and not employees, the deciding factor is not the agreement but the degree of control the company exercises over them. The following questions can help determine whether someone is an employee:

- Does the person have the right to control the manner and means of accomplishing the desired result?
- How much work experience does the person have?
- Does the worker provide his own tools and equipment?
- Is the person engaged in a distinct occupation or an independently established business?
- Is the method of payment by the hour or by the job?
- What degree of skill is required to complete the job?
- Does the person hire employees to help?

In general, a worker hired for a highly specific skilled position is more likely to be classified as an independent contractor if the worker sues. The Microsoft ruling means that employers must exercise care in their treatment of contingent workers. If companies surrender control of contingent workers to agencies, then the agencies must hire and fire the workers, promote and discipline them, do performance reviews, decide wages, and tell them what to do on a daily basis. Read the manager's checklist in Table 8-2 for questions that pertain to the use of contingent workers. The preferred answer to each question is *yes*.

TABLE 8-2 Manager's checklist for the use of contingent employees

Questions	Yes	No
Have you reviewed the definition of an employee in your company's pension plan and policies to ensure it is not so broad that it encompasses contingent workers, thus entitling them to benefits?	____	____
Are you careful not to use contingent workers on an extended basis? Do you make sure that the assignments are finite, with break periods in between?	____	____
Do you use contracts designating the worker as a contingent worker?	____	____
Are you aware that the actual circumstances of the working relationship determine whether a worker is considered an employee in various contexts, and that a company's definition of a contingent worker may not be accepted as accurate by a government agency or a court?	____	____
Do you avoid telling contingent workers where, when, and how to do their jobs?	____	____
Do you make sure that contingent workers use their own equipment and resources, such as computers and e-mail accounts?	____	____
Do you avoid training your contingent workers?	____	____
When leasing employees from an agency, do you let the agency do its job? Do you avoid asking to see résumés and getting involved with compensation, performance feedback, counseling, or day-to-day supervision?	____	____
If you lease employees, do you use a leasing company that offers its own benefits plan, deducts payroll taxes, and provides required insurance?	____	____

H-1B Workers

An H-1B is a temporary working visa granted by the U.S. Citizenship and Immigration Services (USCIS) for people who work in specialty occupations—jobs that require a four-year bachelor's degree or higher in a specific field, or the equivalent experience. Many companies turn to **H-1B workers** to meet critical business needs or to obtain essential technical skills and knowledge that are not readily found in the United States. H-1B workers may also fill temporary shortages of needed skills. Employers often need H-1B professionals to provide special expertise in overseas markets or on projects that enable U.S. businesses to compete globally. A key requirement is that employers must pay H-1B workers the prevailing wage for U.S. workers to do equivalent jobs.

People can work for a U.S. employer as H-1B employees for a maximum continuous period of six years. After the H-1B expires, the foreign worker must remain outside the United States for one year before another H-1B petition can be approved. H-1B temporary professionals make up less than 0.1 percent of the U.S. workforce of more than 140 million people, but nearly 40 percent are employed as computer programmers.[7] The top five source countries for H-1B workers are India, China, Canada, the United Kingdom, and the Philippines.

Each year the U.S. Congress sets a federal cap on the number of H-1B visas to be granted (see Table 8-3), although frequently the number of visas issued varies greatly from this cap. The cap applies only to certain IT professionals, such as programmers and engineers at private technology companies. A large number of foreign workers are exempt from the cap, including scientists hired to teach at American universities, government research labs, and nonprofit organizations. In 2005, Congress approved an additional 20,000 visas beyond the annual cap, specifically for foreign nationals who have earned graduate degrees at U.S. institutions. (U.S. companies can also hire up to 10,500 Australian citizens under a new E-3 visa program.)[8]

TABLE 8-3 Number of H-1B visas granted by USCIS

Fiscal year (October 1–September 30)	H-1B visas cap
1998	65,000
1999	115,000
2000	115,000
2001	195,000
2002	195,000
2003	195,000
2004	65,000
2005	65,000

As concern increases about employment in the IT sector, displaced workers challenge whether the United States needs to continue importing thousands of H-1B workers each year. Most business managers, however, say such criticisms conceal the real issue, which is to find qualified people, wherever they are, for increasingly challenging work. Some human resource managers and educators are concerned that the continued use of H-1B may be a symptom of a larger, more fundamental problem—the United States is not developing sufficient IT employees with the right skills to meet its corporate needs.

One critical issue when considering H-1B is that even highly skilled and experienced H-1B workers can require help using English as a second language. Communication in most business settings is often fast paced and full of idiomatic expressions; workers who are not fluent in English may find it difficult and uncomfortable to participate. Even a simple lunch with coworkers can be distressing. As a result, H-1B workers might prefer to remain isolated and work alone. Even worse, they may create their own cliques and stop trying to acclimate, which can hurt a project team's morale and lead to division. Managers and coworkers should strive to help improve H-1B workers' English skills and cultural understanding and be sensitive to their heritage and needs. H-1B workers should feel at ease and be able to interact socially to feel like true members of their team.

Heads of U.S. companies continue to complain that they have trouble finding enough qualified employees and have urged the USCIS to loosen the reins on visas for qualified workers.[9] They warn that reducing the number of visas will encourage them to move work to foreign countries where they can find the workforce they need.

H-1B Application Process

Most ethical companies make hiring decisions based on how well an applicant fulfills the job qualifications. Such companies consider the need for obtaining an H-1B visa *after* deciding to hire the best available candidate. To receive an H-1B visa, the person must have a job offer from an employer who is also willing to offer sponsorship. There are two application stages: the Labor Condition Attestation (LCA) and the H-1B visa application. The company files an LCA with the Department of Labor (DOL), stating the job title, geographic area in which the worker is needed, and salary to be paid. The DOL's Wage and Hour Division administers LCAs to ensure that the foreign worker's wages will not undercut those of an American worker. After the LCA is approved, the employer may then apply to the USCIS for the H-1B visa candidate, identifying who will fill the position and stating the person's skills and qualifications for the job. A candidate cannot be hired until the USCIS has processed the application, which can take several days or several months.

Companies whose H-1B contingent makes up more than 15 percent of their workforce face further hurdles before they can hire more. They must prove that they first tried to find U.S. workers before they can hire H-1Bs—for example, they can show copies of employment ads they placed in newspapers or periodicals. They must also confirm that they are not hiring an H-1B worker after having laid off a similar U.S. worker. Employers must attest to such protections by affirmatively filing with the DOL and maintaining a public file. Failure to comply with DOL regulations can result in an audit and fines in excess of $1000 per violation, payment of back wages, and ineligibility to participate in immigration programs.

To prevent delays, the American Competitiveness in the Twenty-First Century Act contains a provision that allows current H-1B holders to start working for employers as soon as their petitions are filed. Therefore, a company that wants to hire a critical person who already has H-1B status can do so in a matter of weeks.

Using H-1B Workers Instead of U.S. Workers

To remain world leaders, U.S. firms must be able to attract the best and brightest workers from all over the globe. Most H-1B workers are brought to the United States to fill a legitimate gap that cannot be filled with the existing pool of workers. Some managers reason that as long as skilled foreign workers can be found to fill critical positions, why spend thousands of dollars and take months to develop their current U.S. workers? Although such logic may appear sound for short-term hiring decisions, its implementation can come across as callous and coldhearted, making it difficult for laid-off workers and recent IT graduates to accept. It also does nothing to develop the strong core of permanent IT workers that the United States needs as the economy expands and capital investment increases in IT products and services. Heavy reliance on the use of H-1B workers lessens the incentive for U.S. companies to educate and develop their own workforces.

Potential Exploitation of H-1B Workers

Salary abuse occurs, even though companies applying for H-1B visas must offer a wage that is not five percent less than the average salary for the occupation. Because wages in the IT field vary greatly, unethical companies can get around the average salary requirement. Determining an appropriate wage is an imprecise science at best. For example, an H-1B worker may be classified as an entry-level IT employee and yet fill a position of an experienced worker who would make $10,000 to $30,000 more per year.

Until Congress approved the Visa Reform Act of 2004, there were few investigations into H-1B salary abuses. The act increased the H-1B application fee by $2000, earmarked $500 of each payment for antifraud efforts, and defined a modified wage-rate system that allows for greater variances in pay to visa holders.[10] Investigations are typically triggered by complaints from H-1B holders, but the government can conduct random audits or launch an investigation based on information from third-party sources. If found guilty of underpayment, an employer must reimburse the underpaid employee for back wages. As a result of investigations, however, only $2 million was paid to workers in fiscal 2003 to make up for underpayments.[11] A recent study, "The Bottom of the Pay Scale: Wages for H-1B Computer Programmers FY 2004," found that H-1B workers in computer jobs were paid an average of $13,000 less than U.S. workers in the same jobs.[12]

H-1B visa workers must also be concerned with what happens at the end of the six-year visa term. If a worker is not granted a green card, the firm loses the worker without having developed a permanent employee. The stopgap nature of the visa program leaves both applicants and sponsoring companies unfulfilled and provides foreign IT workers no certainty of what their futures hold. Many foreign workers find themselves unemployed and are forced to uproot their families and return home; their lives are turned upside down. In some countries, the returning worker is even stigmatized as being unable to make it in the United States.

Offshore Outsourcing

Outsourcing—sometimes called managed services or facilities management—is yet another approach to meeting staffing needs. With **outsourcing**, companies receive services from an outside organization that has expertise in providing a specific function. A company contracts with the outside firm to perform this specialized function on an ongoing basis. Examples include contracting with an organization to operate a computer center, support a telecommunications network, or staff a computer help desk. Coemployment legal problems with outsourcing are minimal, because the company that contracts for services generally does not supervise or control the contractor's employees. The primary rationale for outsourcing is to lower costs, but companies also use it to obtain strategic flexibility and to focus on their core competencies. The idea of outsourcing was adopted by IT executives in the 1970s as they began to supplement their IT staff with contractors and consultants. This trend eventually led to outsourcing entire IT business units to such organizations as Accenture, Electronic Data Systems, and IBM, which took over the operation of data centers and other IT functions.

Offshore outsourcing is a variation of outsourcing in which work is done by an organization whose employees are in a foreign country. IT professionals can do much of their work anywhere—on a company's premises or thousands of miles away in a foreign country. In addition, companies can save up to 70 percent on some projects by reducing labor costs through offshore outsourcing. As a result, and because large staffs of experienced IT resources are readily available in certain foreign countries, the use of offshore outsourcing is increasing in the IT industry. American Express, Aetna U.S. Healthcare, Compaq, General Electric, IBM, Microsoft, Motorola, Shell, Sprint, and 3M are examples of big companies that employ offshore outsourcing for functions such as help desk support, network management, and information systems development. A 2004 survey estimated that 104,000 U.S. software and service jobs were moved overseas in 2003, but that's only 1 percent of the estimated 10.5 million IT jobs in the United States.[13]

As more businesses become comfortable with moving their key processes offshore, U.S. IT service providers are forced to lower prices. Many U.S. software firms set up development centers in low-cost foreign countries where they can obtain well-trained resources. Intuit, maker of the Quicken tax preparation software, currently has facilities in Canada and Great Britain and is setting up another in Bangalore, India.[14] Accenture, IBM, and Microsoft maintain large development centers in India. Cognizant Technology Solutions is headquartered in Teaneck, New Jersey, but operates primarily from technology centers in India.

Because of the high cost of U.S.-based application developers and the ease with which customers and suppliers can communicate, it is now quite common to use offshore outsourcing for major programming projects. Contract programming is flourishing in Brazil, Bulgaria, Canada, China, Ireland, Israel, Malaysia, Malta, Mexico, the Philippines, Poland, Russia, and Singapore, among other countries. But India, with its rich talent pool, English-speaking citizenry, and low labor costs, is widely acknowledged as the best source of programming skills outside Europe and North America. India exports software to more than 100 countries and its companies now employ more than 400,000 software engineers. ABN Amro, a Dutch financial services giant, selected Indian companies such as Tata Consultancy Services, Infosys Technologies, and Patni Computer Systems as part of a group of vendors to handle its infrastructure, application development, and maintenance requirements. The total value of the work could exceed $2 billion if ABN Amro moves more than 2000 IT jobs to India as planned.[15]

Table 8-4 shows the leading countries that provide offshore IT services for U.S. firms. Table 8-5 lists several IT offshore outsourcing firms. Of course, any work done at a relatively high cost in the United States is subject to outsourcing, not just IT work. For example, "Deloitte Research estimates that by the end of this decade, approximately 850,000 financial services positions (15 percent of the U.S. total) will be absorbed offshore."[16]

TABLE 8-4 Leading countries for providing offshore IT services

Country	Comments
India	Low cost, highly skilled labor pools
Canada	Close to United States, no language barrier, highly skilled labor pool
China	Low cost, large pool of skilled labor, lack of English-language proficiency
Poland	Low overall cost of business operations
Czech Republic	An emerging contender
Russia	Unpredictable political and business climate

Source: McDougal, Paul, "India, Canada, and China Are Top Outsourcing Destinations: Study," *InformationWeek*, www.informationweek.com, September 21, 2005.

TABLE 8-5 Partial list of offshore IT outsourcing firms

Firm	Number of employees	Headquarters	Recent annual revenues	Key clients
Tata Consultancy Group	49,000	India	> $2 billion	
Infosys Technologies	49,000	Bangalore, India	$2.1 billion	
Wipro	42,000	Bangalore, India	$1.9 billion	
Satyam Computer Services, Ltd		Hyderabad, India	$800 million	Texas Instruments
HCL Infosystems	30,000	New Delhi, India		
EPAM systems	1,000	Princeton, NJ, and Budapest, Hungary	NA	Empire Blue Cross Blue Shield, Mandalay Resort Group, Visa International
Luxoft	1,400	Moscow, Russia	$25 million	Boeing, Dell, Caterpillar, IBM
Aplana	200	Moscow, Russia	NA	General Electric, Procter & Gamble
Hubport Interactive	NA	Davao City, Philippines	NA	Rancho Palos Verdes Golf and Country Club
Outsourcing Solutions, Inc.	NA	Cebu City, Philippines	NA	Kauai Publishing Company, Mother Earth News, and Sound Choice
Interxion	170	Amsterdam, The Netherlands	NA	Coca Cola, IBM, Siemens
Delphi Technologies	NA	Dublin, Ireland	NA	IBM, Siebel Systems

Pros and Cons of Offshore Outsourcing

Wages that an American worker would consider low represent an excellent salary in other parts of the world, and some companies feel they would be foolish not to exploit such an opportunity. Why pay a U.S. IT worker a six-figure salary, they reason, when they can use offshore outsourcing to hire four Indian-based workers for the same cost?[17] However, this attitude might represent a short-term point of view: offshore demand is driving up salaries in India by roughly 14 percent per year, and these increases are expected for the next several years. Indian offshore suppliers must therefore charge more for their services to cover the increase. If current salary trends continue, Indian labor rates will equal U.S. costs by 2020. At that point, cost savings would no longer be an incentive for U.S. offshore outsourcing.

Another benefit of offshore outsourcing is its potential to dramatically speed up development efforts. For example, the state of New Mexico contracted the development of a tax system to Syntel, one of the first U.S. firms to successfully launch a global delivery model that enables workers to make progress on a project around the clock. With technical teams working from networked facilities in different time zones, Syntel executes a virtual "24-hour workday" that saves its customers money, speeds projects to completion, and provides continuous support for key software applications.[18]

In determining how much money and time they will save, however, firms should consider that each project will require additional time to select an offshore vendor and incur additional costs for travel and communications. In addition, organizations find that it can take years of ongoing effort and a large up-front investment to develop a good working relationship with an offshore outsourcing firm. Finding a reputable vendor can be especially difficult for a small or midsized firm that lacks experience in identifying and vetting contractors.

Many of the same ethical issues that arise in considering H-1B and contingent workers apply to offshore outsourcing. For example, managers must decide whether to use offshore outsourcing firms or spend the time and money to develop their own staff and meet their business needs. Workers affected by these decisions will demand fair and equitable treatment. Like other contingent workers, offshore outsourcing employees gain valuable and practical experience from working within a particular company's structure and culture; that experience is lost when the employee is reassigned after a project's completion. Finally, offshore outsourcing does not advance the development of permanent IT workers in the United States, which increases its dependency on foreign workers to build the IT infrastructure of the future. Many of the jobs that go overseas are entry-level positions that help develop employees for future, more responsible positions.

The use of offshore outsourcing does not lack for supporters. A 2005 study by the Information Technology Association of America (ITAA) reached the conclusion that outsourcing IT jobs offshore will create 337,000 new jobs throughout the U.S. economy by 2010. Its reasoning is that, although IT jobs are lost through outsourcing, more jobs are ultimately created because U.S. companies can lower their IT-related costs, freeing up money to spend on other things, which results in lower prices, lower interest rates, and higher spending throughout the economy. Advocates say this added activity spurs job creation, increases the competitiveness of U.S. companies, expands markets for U.S. goods overseas, and enables U.S. firms to produce more for less.[19]

Many people do not share the ITAA's optimistic outlook on the impact of outsourcing. "The industry's relentless downsizing, healthcare cost shifting, job exporting, and visa importing strategies are causing tech workers to be less optimistic about their futures in one of America's most important industries," said Marcus Courtner, president of WashTech CWA, a Seattle-based alliance of technology workers.[20]

The difficulty of communicating directly with people over long distances can make offshore outsourcing perilous, especially when key team members speak English as a second language. To improve the chances that an offshore outsourcing project will succeed, a company must carefully evaluate the outsourcing firm to ensure that its employees meet five basic prerequisites:

- They have expertise in the technologies involved in the project.
- They provide a project manager who speaks the employer company's native language.

- A large staff is available.
- They have a good telecommunications setup.
- Good on-site managers are available from the outsourcing partner.

Successful projects require day-to-day interaction between software development and business teams, so it is essential for the hiring company to take a hands-on approach to project management. Companies cannot afford to outsource responsibility and accountability.

Although it has its advocates, offshore outsourcing doesn't always pay off. For example, a software developer in Cambridge, Massachusetts, went to India in search of cheap labor but instead found lots of problems. Customs officials charged huge tariffs when the company tried to ship the necessary development software and manuals, and the programmers weren't nearly as experienced as they claimed to be. The code produced in India was inadequate. The company had to send a representative to India for months to work with the programmers there and correct the problems.

Offshore outsourcing also tends to upset domestic staff, especially those who have to be laid off in favor of low-wage workers outside the United States. Surviving members of a department who suddenly have to work with foreign nationals can become bitter and non-productive, and morale can become extremely low.

Cultural differences can also cause misunderstandings among project members. Indian programmers, for instance, are known for keeping quiet even when they notice problems. And subtle differences in gestures can cause confusion. In the United States, shaking your head from side to side means *no*. In the south of India, it means *yes*.

The potential compromise of customer data is yet another outsourcing issue. For example, there is evidence that the personal details of Australian citizens (including names, addresses, telephone numbers and other ID numbers, and employment information) were taken from customer databases in India and offered for sale on the black market.[21] Most countries have laws designed to protect the privacy of their citizens. Firms that outsource must take precautions to protect private data, regardless of where it is stored or processed or who handles it. One useful tool for protecting data is the Statement on Auditing Standards (SAS) No. 70, *Service Organizations*, an internationally recognized standard developed by the American Institute of Certified Public Accountants (AICPA). A successful SAS 70 audit report demonstrates that an outsourcing firm has effective internal controls in accordance with the Sarbanes-Oxley Act of 2002.[22]

The following list provides several tips for companies that are considering offshore outsourcing:

- Set clear, firm business specifications for the work to be done.
- Assess the probability of political upheavals or factors that might interfere with information flow and ensure that the risks are acceptable.
- Assess the basic stability and economic soundness of the outsourcing vendor and what might occur if the vendor encounters a severe financial downturn.
- Establish reliable satellite or broadband communications between your site and the outsourcer's location.
- Implement a formal version-control process, coordinated through a quality assurance person.

- Develop and use a dictionary of terms to encourage a common understanding of technical jargon.
- Require vendors to supply project managers at the client site to overcome cultural barriers and facilitate communication with offshore programmers.
- Require a network manager at the vendor site to coordinate the logistics of using several communications providers around the world.
- Obtain advance agreement on the structure and content of documentation to avoid manuals that explain how the system was built, not how to maintain it.
- Carefully review a current copy of the outsourcing firm's SAS 70 audit report to ascertain its level of control over information technology and related processes.

WHISTLE-BLOWING

Like contingent workers, whistle-blowing is a significant topic in any discussion of ethics in IT. Both issues raise ethical questions as well as social and economic implications. How these issues are addressed can have a long-lasting impact not only on the people and employers involved, but the entire IT industry.

Whistle-blowing is an effort to attract public attention to a negligent, illegal, unethical, abusive, or dangerous act by a company that threatens the public interest. In some cases, whistle-blowers are employees who act as informants on their company, revealing information to enrich themselves or to gain revenge for some perceived wrong. In most cases, however, whistle-blowers act ethically in an attempt to correct what they think is a major wrongdoing, often at great personal risk.

The whistle-blower usually has special information about what is happening based on personal expertise or a position of employment within the offending organization. Sometimes, the whistle-blower is not an employee but a person with special knowledge gained from reliable sources. For example, Bev Harris is a literary publicist and writer whose book *Black Box Voting: Ballot Tampering in the 21st Century* is highly critical of modern election equipment. Diebold Election Systems, which provides systems that tally votes and register voters in many counties around the country, has been the target of her wrath for several years. A whistle-blower suit filed by Harris alleged that Diebold:

- Engaged in unfair business practices
- Made false claims about its product
- Used uncertified software in elections in California
- Sold systems that are capable of being easily hacked

Harris cites a batch of internal Diebold memos and Diebold source code she discovered on the Internet as her sources of information. "She went from anonymous activist to media darling soon after posting the documents online."[23] The attorney general of California took up Harris' whistle-blower claim against Diebold, eventually settling with the company for $2.6 million. As one might imagine, Bev Harris is not popular with Diebold and the election officials around the country who selected and installed Diebold equipment. She has been called a fruitcake and muckraker.[24]

Whistle-blowers risk their own careers and might even affect the lives of their friends and family. In extreme situations, whistle-blowers must choose between protecting society or remaining silent.

Protection for Whistle-Blowers

Whistle-blower protection laws allow employees to alert the proper authorities to employer actions that are unethical, illegal, or unsafe, or that violate specific public policies. Unfortunately, no comprehensive federal law protects all whistle-blowers. Instead, numerous laws each protect a certain class of specific whistle-blowing acts in various industries. To make things even more complicated, each law has different filing provisions, administrative and judicial remedies, and statutes of limitations (which set time limits for legal action). Thus, the first step in reviewing a whistle-blower's claim of retaliation is for an experienced attorney to analyze the various laws and determine if and how the employee is protected, and then determine what procedures to follow in filing a claim.

From the perspective of the whistle-blower, a short statute of limitations is a major weakness of many whistle-blower protection laws. Failure to comply with the statute of limitations is a favorite defense of firms accused of wrongdoing in whistle-blower cases.

The False Claims Act is a federal law that provides strong protection for whistle-blowers. See the Legal Overview for more information about this act.

LEGAL OVERVIEW

False Claims Act

The **False Claims Act**, also known as the "Lincoln Law," was enacted during the U.S. Civil War to combat fraud by companies that sold supplies to the Union Army. War profiteers shipped boxes of sawdust instead of guns, for instance, and swindled the Union Army into purchasing the same cavalry horses several times. When it was enacted, the act's goal was to entice whistle-blowers to come forward by offering them a share of the money recovered.

The **qui tam** ("who sues on behalf of the king as well as for himself") provision of the False Claims Act allows a private citizen to file a suit in the name of the U.S. government, charging fraud by government contractors and other entities who receive or use government funds. In qui tam actions, the government has the right to intervene and join the legal proceedings. If the government declines, the private plaintiff may proceed alone. Some states have passed similar laws concerning fraud in state government contracts.

Several types of cases can be filed as qui tam actions, including mischarging for services, product and service substitution, false certification of entitlement for benefits, and false negotiation to justify an inflated contract. Mischarging is the most common form of qui tam case. For example, an IT contractor might overcharge hundreds of hours of programming time as part of a government contract, or a physician might overcharge the government for medical services that a nurse actually performed.

continued

Violators of the False Claims Act are liable for three times the dollar amount that the government is defrauded. They can also receive civil penalties of $5000 to $10,000 for each instance of a false claim. A qui tam plaintiff can receive between 15 and 30 percent of the total recovery from the defendant, depending on how helpful the person was to the success of the case. Successful recoveries for a qui tam case have been as high as $150 million. Over the years, whistle-blower lawsuits brought by private citizens have helped the government recover more than $3.5 billion from companies that defrauded taxpayers. For example, a former Oracle vice president filed a whistle-blower lawsuit charging that the company fraudulently billed the federal government for training over a six-year period. Oracle agreed to pay $8 million to settle the charges and the whistle-blower received $1.6 million of the total settlement amount.[25]

The False Claims Act provides strong whistle-blower protection. Any person who is discharged, demoted, harassed, or otherwise discriminated against because of lawful acts of whistle-blowing is entitled to all relief necessary "to make the employee whole." Such relief may include job reinstatement, double back pay, and compensation for any special damages, including litigation costs and reasonable attorney's fees.

The provisions of the False Claims Act are quite complicated, so it is unwise to pursue a claim without legal counsel. However, because the potential for significant financial recovery is good, attorneys are generally willing to assist.

Besides the False Claims Act, several other laws provide protection for whistle-blowers. Some are aimed at specific industries, as described in the following sections.

Environmental Protection and Whistle-Blowing

Special provisions have been added to several federal environmental laws to protect whistle-blowers and encourage reporting of any wrongdoing that would damage the environment. These laws include the Clean Air Act, the Toxic Substances Control Act, the Clean Water Act, the Safe Drinking Water Act, and the Comprehensive Environmental Response, Compensation, and Liability Act, which established the Superfund. All of these laws were passed between 1955 and 1980.

Whistle-Blowing Protection for Nuclear Workers

The Energy Reorganization Act safeguards workers in the nuclear power and nuclear weapons industries. These workers are provided whistle-blower protection because of the potential for serious harm to people who work with or live near facilities that produce or handle nuclear technology. Almost any employee in these industries, whether employed in the private sector or by some branch of government, has the power to report any safety problem, environmental violation, or other illegal activity without fear of reprisal. In addition, any whistle-blower who believes he has been the victim of retaliation is protected under the law.

Whistle-Blowing Protection for Private-Sector Workers

Under state law, an employee can traditionally be terminated for any reason or no reason, in the absence of an employment contract. However, many states have created laws that prevent workers from being fired because of an employee's participation in "protected" activities. One such activity is the filing of a qui tam lawsuit under the provisions of the False Claims Act, in which a worker charges fraud by government contractors or others who receive or use government funds. States that recognize the public benefit of such cases offer protection to whistle-blowers; for example, they can file claims against their employers for retaliatory termination and are entitled to jury trials. If successful, they can receive punitive damage awards.

Dealing with a Whistle-Blowing Situation

Each potential whistle-blowing case involves different circumstances, issues, and personalities. Two people in the same company and the same situation may have different values and concerns that cause them to react differently, and both reactions might be ethical. Therefore, it is impossible to outline a definitive step-by-step procedure of how to behave in a whistle-blowing situation. This section provides a general sequence of events and highlights key issues that the potential whistle-blower should consider.

1. *Assess the seriousness of the situation*—Before considering whistle-blowing, a person should have specific knowledge that his company or a coworker is acting unethically and that the action represents a *serious* threat to the public interest. The employee should carefully and informally seek trusted resources outside the company and ask for their assessment. Do they also see the situation as serious? Their point of view may help the employee see the situation from a different perspective and alleviate concerns. On the other hand, the outside resources may reinforce the employee's initial suspicions, forcing a series of difficult ethical decisions.

2. *Begin documentation*—An employee who identifies an illegal or unethical practice should reserve judgment and begin to compile adequate documentation to establish wrongdoing. The documentation should record all events, facts, and insights about the situation. This record helps to construct a chronology of events if legal testimony is required in the future. An employee should identify and copy all supporting memos, correspondence, manuals, and other documents *before* taking the next step. Otherwise, records may disappear and become inaccessible. The employee should maintain documentation and keep it up to date throughout the process.

3. *Attempt to address the situation internally*—An employee should next attempt to address the problem internally by providing a written summary to the appropriate managers, including a statement that they either responded or clearly chose not to respond. Ideally, the employee can expose the problem and deal with it from inside the organization, creating as few bad feelings as possible. The focus should be on disclosing the facts and how the situation affects others. The employee's goal should be to fix the problem, not to place blame. Given the potential negative impact of whistle-blowing on the employee's future, this step should not be dismissed or taken lightly. Fortunately, many problems are solved at this point, and further, more drastic actions by the employee are unnecessary. On the other hand, managers who are engaged in unethical or illegal behavior might not welcome an employee's questions or concerns.

4. *Consider escalation within the company*—The employee's initial attempt to deal with the situation internally may be unsuccessful. At this point, the employee may rationalize that he has done all that is required by raising the issue. Some people who elect to take no further action continue to wrestle with their consciences; they can develop ulcers, drug or alcohol problems, or lose peace of mind. Others may feel so strongly about the situation that they must take further action. Thus, a determined and conscientious employee may feel forced to choose between escalating the problem and going over the manager's head or going outside the organization to deal with the problem. The employee may feel compelled to sound the alarm on the company because there appears to be no chance to solve the problem internally.

Going over an immediate manager's head can put one's career in jeopardy. Supervisors may retaliate against a challenge to their management, although some organizations may have an effective corporate ethics officer who can be trusted to give the employee a fair and objective hearing. Or, a senior manager with a reputation for fairness and some responsibility for the area of concern might step in. However, in all but the most enlightened of work environments, the challenger is likely to be fired, demoted, or reassigned to a less desirable position or job location. Such actions send a loud signal throughout an organization that loyalty is highly valued and that challengers will be dealt with harshly. Whether reprisal is ethical depends in large part on the legitimacy of the employee's issue. If the employee attempted to treat an issue much more seriously than was warranted, then the employee does deserve some sort of reprisal for exercising poor judgment and accusing management of unethical behavior.

If senior managers refuse to deal with a legitimate problem, the employee's options are to drop the matter or go outside the organization to try to remedy the situation. Even if a senior manager agrees with the employee's position and overrules the immediate boss, the employee may want to request a transfer to avoid working for the same person.

5. *Assess the implications of becoming a whistle-blower*—If the employee feels he has made a strong attempt to resolve the problem internally but achieved no results, he must stop and fully assess whether he is prepared to go forward and blow the whistle. Depending on the situation, significant legal fees may be needed to air or bring charges against an agency or company that has unlimited time and legal resources and a lot more money. An employee who chooses to proceed might be accused of having a grievance with the employer or of trying to profit from the accusations. The employee may be fired and may lose the confidence of coworkers, friends, and even family.

A potential whistle-blower must attempt to answer many ethical questions:

- Given the potentially high price, does the whistle-blower really want to proceed?
- Has the employee exhausted all means of dealing with the problem? Is whistle-blowing all that is left?
- Is the employee violating an obligation to be loyal to the employer and work for its best interests?

- Will the public exposure of corruption and mismanagement in the organization really correct the underlying cause of these problems and protect others from harm?

From the moment an employee becomes known as a whistle-blower, a public battle may ensue. Whistle-blowers can expect attacks on their personal integrity and character and negative publicity in the media. Because friends and family will hear these accusations, they should be notified beforehand and consulted for advice before the whistle-blower goes public. This notification helps prevent friends and family from being surprised at future actions by the whistle-blower or the employer.

The whistle-blower should also consider consulting support groups, elected officials, and professional organizations. For example, the National Whistle-blower Center provides referrals for legal counseling and education about the rights of whistle-blowers.

6. *Use experienced resources to develop an action plan*—A whistle-blower should consult with competent legal counsel who have experience in related cases. They will determine which statutes and laws apply, depending on the agency, employer, state involved, and the nature of the case. Legal counsel should also know the statute of limitations for reporting the offense as well as the whistle-blower's protection under the law. Before blowing the whistle, the employee should get an honest assessment of the soundness of his legal position and an estimate of the costs of a lawsuit.

7. *Execute the action plan*—A whistle-blower who chooses to pursue the matter legally should do so based on the research and decisions of legal counsel. Sometimes, if the whistle-blower wants to remain unknown, the safest course of action is to leak information anonymously to the press. The problem with this approach is that anonymous claims are often not taken seriously. In most cases, working directly with appropriate regulatory agencies and legal authorities is more likely to get results, including the imposition of fines, halting of operations, or other actions that draw the offending organization's immediate attention.

8. *Live with the consequences*—Whistle-blowers must be on guard against retaliation, such as being discredited by coworkers, threatened, or "set up." For example, management may attempt to have the whistle-blower transferred, demoted, or fired for breaking some minor rule, such as arriving late to work or leaving early. To justify their actions, management may argue that such behavior has been ongoing. The whistle-blower might need a good strategy and a good attorney to counteract such actions and take recourse under the law.

Summary

1. What are contingent workers, and how are they frequently employed in the IT industry?

    The contingent workforce includes independent contractors, workers brought in through employment agencies, on-call or day laborers, and on-site workers whose services are provided by contract firms. Employers hire contingent workers to obtain essential technical skills or knowledge that is not easy to find in the United States or to meet temporary shortages of needed skills.

2. What key ethical issues are associated with the use of contingent workers, including H-1B visa holders and offshore outsourcing companies?

    Some people contend that employers exploit contingent workers, especially H-1B foreign workers, to obtain skilled labor at less than competitive salaries. Others believe that the use of H-1B workers is required to keep the United States competitive.

    The use of contingent workers enables a firm to meet its staffing needs more efficiently, lower its labor costs, and respond more quickly to changing market conditions. Facing a likely long-term shortage of trained and experienced workers, employers will increasingly turn to nontraditional sources to find IT workers with skills that meet their needs. Employers will need to make ethical decisions whether to recruit new and more skilled workers from these sources or to spend the time and money to develop their current staff to meet the needs of their business. The workers affected by these decisions will demand to be treated fairly and equitably.

    The use of contingent workers can raise ethical and legal issues about the relationship among the staffing firm, its employees, and its customers, including the potential liability of the customer for the withholding of payroll taxes, payment of employee retirement benefits and health insurance premiums, and administration of worker's compensation of the staffing firm's employees.

3. What is whistle-blowing, and what ethical issues are associated with it?

    Whistle-blowing is an employee's effort to attract public attention to a negligent, illegal, unethical, abusive, or dangerous act by his or her company. A potential whistle-blower must consider many ethical implications, including whether the high price of whistle-blowing is worth it; whether all other means of dealing with the problem have been exhausted; whether whistle-blowing violates the obligation of loyalty that employees owe to companies; and whether public exposure of the problem will actually correct its underlying cause and protect others from harm.

4. What is an effective whistle-blowing process?

    An effective process includes the following steps: (1) assess whether whistle-blowing is warranted, (2) begin documentation, (3) attempt to address the situation internally, (4) consider escalating the situation within the company, (5) assess the implications of becoming a whistle-blower, (6) use experienced resources to develop an action plan, (7) execute the plan, and (8) live with the consequences.

Self-Assessment Questions

1. The U.S. Bureau of Labor Statistics predicts an overabundance of U.S. IT workers over the next decade. True or False?

2. An employment situation in which two employers have actual or potential legal rights and duties with respect to the same employee is called _____ .

3. Which of the following is *not* an advantage for organizations that employ contingent workers?

   a. The firm does not have to offer employee benefits.

   b. Training costs are kept to a minimum.

   c. It provides a way to meet fluctuating staffing needs.

   d. The contingent worker's experience may be useful to the next firm that hires the worker.

4. If contractors sign an agreement stating they are not employees, the firm is protected from future claims for company benefits. True or False?

5. The _____ sets the cap on the number of H-1B visas to be granted each year, and the _____ is responsible for granting these visas.

6. Which of the following statements about the salary of H-1B employees is false?

   a. Companies must offer a wage that is not five percent less than the average salary for the occupation.

   b. The determination of average salary is a precise science.

   c. The number of annual investigations of H-1B salary abuse has been small (less than 150 per year).

   d. Salary investigations can be triggered by complaints from H-1B holders, and the government can conduct random audits.

7. Outsourcing is an organization's practice of using workers in a foreign country. True or False?

8. Which of the following is *not* cited as a business advantage of offshore outsourcing?

   a. lower labor costs

   b. potential to speed up development efforts

   c. the opportunity to learn different languages, cultures, and ways of operating

   d. the ability to tap into a large, well-educated labor pool

9. Which of the following statements about whistle-blowing is true?

   a. Whistle-blowing is fairly common and draws attention to all sorts of issues, from the mundane to the serious.

   b. There is almost universal government protection for whistle-blowers.

   c. A short statute of limitations is one major weakness in many whistle-blower protection laws.

   d. The False Claims Act is intended to discourage whistle-blowers from coming forward with frivolous claims.

10. It is almost impossible to outline an exact step-by-step procedure for how one should deal with a whistle-blowing situation. True or False?

Review Questions

1. What does the U.S. Bureau of Labor Statistics forecast for future employment needs in the IT industry?

2. What is a contingent workforce, and in what kinds of roles is it employed most effectively?

3. What is a coemployment relationship? What key factors determine whether someone is considered an employee for the purpose of receiving benefits such as profit sharing and worker's compensation?

4. What is an H-1B visa? What are the requirements for a worker to qualify for one?

5. What is the application process for an H-1B visa? Why is it divided into two phases?

6. What is offshore outsourcing? What are some advantages and disadvantages of taking this approach in the area of application development?

7. What is whistle-blowing? Under what conditions is it justified?

8. What legal protection is available to whistle-blowers? What is a frequent weakness in these statutes?

9. List and briefly describe the stages of the whistle-blowing process.

Discussion Questions

1. Briefly discuss the advantages and disadvantages of using H-1B workers. What ethical issues surround the use of H-1B workers?

2. Briefly discuss the advantages and disadvantages of being an H-1B worker.

3. What advice would you offer a manager within your company who is considering outsourcing to meet a critical, ongoing business need?

4. What factors would you weigh in deciding whether to employ offshore outsourcing on a project?

5. Outline a process you would follow to select an offshore outsourcing firm for a major project.

6. Which steps of the whistle-blowing process are most important? Why?

What Would You Do?

1. Your firm has just added six H-1B workers to your 50-person department. You have been asked to help get one of the workers "on board." Your manager wants you to introduce him to other workers, provide basic company background and information, show him how work gets done, and show him the surrounding community (including where to live, where to shop, restaurants, and recreational activities). Your goal is to help the new worker be productive as soon as possible. How would you feel about taking this responsibility? How would you help make the new employee productive? Would this be a priority for you?

2. Dr. Jeffrey Wigand is a whistle-blower who was fired from his position of vice president for research and product development at Brown & Williamson Tobacco Company in 1993. He was interviewed for a segment of the CBS show *60 Minutes* in August 1995, but the network made a highly controversial decision not to air the interview as initially scheduled. CBS management was worried about the possibility of a multibillion dollar lawsuit for tortious interference; that is, interfering with Wigand's confidentiality agreement with Brown &

Williamson. The interview finally aired on February 4, 1996, after the *Wall Street Journal* published a confidential November 1995 deposition that Wigand gave in a Mississippi case against the tobacco industry, which repeated many of the charges he made to CBS. In the interview, Wigand said that Brown & Williamson had scrapped plans to make a safer cigarette and continued to use a flavoring in pipe tobacco that was known to cause cancer in laboratory animals. Wigand also charged that tobacco industry executives testified untruthfully before Congress about tobacco product safety. Wigand suffered greatly for his actions; he lost his job, his home, his family, and his friends.

Visit Wigand's Web site at *www.jeffreywigand.com* and answer the following questions. (You may also want to watch *The Insider*, a 1999 movie based on Wigand's experience.)

   a.   What motivated Wigand to take an executive position for a tobacco company and then five years later denounce the industry's efforts to minimize the health and safety issues of tobacco use?

   b.   What whistle-blower actions did Dr. Wigand take?

   c.   If you were in Dr. Wigand's position, what would you have done?

3.   Microsoft is a major user of temporary workers. To minimize legal issues, Microsoft sought to ensure that temporary workers were not mistaken about their place within the company. The temporary agencies that Microsoft used provided workers with handbooks that laid out the ground rules in explicit detail. Temporary workers were barred from using company-owned athletic fields—for insurance reasons, the agency explained. At some Microsoft facilities, temps were told they could not drive their cars to work because it would create parking problems for regular workers. Instead, they were told to take the bus. Temps were also told not to buy goods at the company store or participate in social clubs such as chess, tai chi, or rock climbing that were open only to regular employees. They were not permitted to attend parties given for regular employees, a private screening of the latest *Star Wars* film, or company meetings at the Kingdome stadium in Seattle. In addition, their e-mail addresses were to contain an *a-* to indicate their nonpermanent status in the company.

Imagine that you are a senior manager in Human Resources at Microsoft and that you have been asked to respond to temporary workers' complaints about working conditions. How would you handle this?

4.   Catalytic Software, a U.S.-based IT outsourcing firm with offices in Redmond, Washington, and Hyderabad, India, wants to tap India's large supply of engineers as contract software developers for IT projects. However, instead of just outsourcing projects to local Indian software development companies, as is the common practice of U.S. companies, Catalytic is developing a self-contained company community near Hyderabad. Spread over 500 acres, the community of New Oroville will be a self-sustaining residential and office community that is expected to house about 4000 software developers and their families, as well as 300 support personnel for sanitation, police, and fire.

The goal of this high-tech city is to knock down barriers that large-scale technology businesses encounter in India. By building a company community, Catalytic ensures that it will have enough qualified employees to staff round-the-clock shifts. The company expects this facility to attract and keep top professionals from all over the world. Building a company town also solves the problem of transportation of a 4000-person staff. Because of the terrible roads, the commute from Hyderabad, 25 kilometers from New Oroville, would take roughly an hour.

Catalytic will provide private homes with private gardens, all within a short walk of work, school, recreation, shopping, and public facilities. Each house will include cable TV, telephones, and a fiber-optic data pipeline that connects to the Internet so that employees can work efficiently even at home. (Employees will be awarded bonuses for working overtime.) New Oroville will have four indoor recreational complexes and six large retail complexes. There will be ample green space, including five parks for outdoor exercise and recreation.

You have just completed a job interview with Catalytic Software for a position as project manager and have been offered a 25 percent raise to join the company. Your position will be based in Redmond and will involve managing U.S.-based projects for customers. The position requires that you spend the first year with Catalytic in the New Oroville facility to learn their methods, culture, and people. (You can take your entire family or accept a pair of three-week company-paid trips back to Redmond.) Why do you think the temporary assignment in New Oroville is a requirement? What else do you need to know in considering this position? Would you accept it? Why or why not?

5. The Recording Industry Association of America (RIAA) tracks down companies whose employees illegally create digital jukeboxes by downloading tunes onto company file servers and sharing them with coworkers. Similarly, the Business Software Alliance (BSA) is a watchdog group that represents most major software manufacturers in combating the illegal copying of software. Both associations get most of their tips from disgruntled ex-employees who call their anonymous hot lines. The associations then negotiate a monetary settlement with the offending firm in lieu of prosecuting them for violation of copyright.

   If you were an employee of a company that ran afoul of the RIAA or BSA, would you report your company or would you attempt to work out the issue internally? What if there were a $10,000 reward? What if the company had recently let you go to reduce its expenses?

Cases

1. Economic Revival Raises H-1B Controversy

In April 2005, Microsoft chairman Bill Gates called on the federal government to scrap its cap on H-1B visas during a visit to Capitol Hill. Gates claimed that restrictions on these visas prevent Microsoft from staffing its U.S.-based operations. The United States, Gates claims, does not have enough smart, qualified professionals to fulfill the need of its growing IT industry.

Congress, too, has recognized this need. During the 1990s and early 2000s, the number of H-1B visas that Congress approved annually reached as high as 195,000. Following the bursting of the dot-com bubble, unemployed American IT professionals began to clamor against importing IT talent. In 2004, Congress reduced the number of H-1B visas to 65,000. Gates argued that all restrictions on H-1B visas should be lifted.

Yet many in the IT industry disagree with Gates. Information Builders CEO Gerald Cohen quickly shot down Gates' remarks. "He's full of it," Cohen said in an interview to *Computerworld*. "He's going there because it's just cheaper. He can find all the engineers he wants in this country."

Cohen and others argue that IT companies bring in H-1B workers to replace higher-paid American IT workers. Federal law stipulates that H-1B workers must be offered a wage that is not five percent less than the average salary for the occupation. Yet, some companies purportedly get around the "average salary requirement" by classifying a position that would normally be filled by an experienced worker as entry level. Furthermore, the government categorizes IT jobs using a system that does not correspond well to job classifications in the private sector. The Department of Labor also receives complaints that IT companies fail to pay H-1B workers between contracts, reducing their annual wage.

A *Computerworld* survey found that between 2001 and 2003, the decline in H-1B wages mirrored the wage decrease suffered by U.S. workers. Then, in 2003, a Foote Partners study discovered that H-1B wages continued to decline at a rate of 2 percent, while U.S. wages rose 6.2 percent.

In 2005, with the economy in recovery, Congress decided to issue an additional 20,000 H-1B visas. The Senate Judiciary Committee planned to increase the H-1B visas by 30,000 in 2006. In the hopes of discouraging companies who might bring in foreign engineers simply to replace highly paid American equivalents, the legislation raised the application fee for an H-1B visa from $185 to $3185.

Gates argued that by preventing smart people from getting into the country and being hired by IT companies, the United States risks losing its strategic position in technological innovation. Cohen and others argue the opposite. They say that unrestricted H-1B visas and other attempts to "offshore"—or move IT jobs outside of the United States—will destroy native IT talent. They also argue that enrollment in computer science programs has decreased, and will continue to do so, as high-tech wages drop and unemployment rises.

"Why do you have declining computer science majors?" Cohen asked. "Because every parent is saying, 'Why major in computer science when all the jobs are going offshore?' It feeds itself. And I guarantee you, if it doesn't stop, in a couple years you're not going to have much of an IT industry here."

Yet, even if Congress takes an aggressive stand in limiting H-1B visas, no legislation will be able to combat the growing trend toward offshoring. Even Cohen admits that his own company has been forced to offshore some IT tasks to remain competitive. By the time Cohen steps down as CEO, he predicts that his company will have migrated 50 percent of its IT operations overseas.

Questions:

1. In your opinion, will raising the application fee deter employers from hiring H-1B workers for the sole purpose of reducing labor costs?

2. What effects do Bill Gates and Gerard Cohen believe H-1B visa restrictions will have on the development of a skilled IT labor force in the United States? What effect do you think these restrictions might have?

3. In light of the growing trend toward offshore outsourcing, what purpose, if any, do H-1B visa restrictions serve?

2. American Engineer Blows the Whistle on Airbus' Superjumbo A380

The Airbus A380, the largest passenger airliner ever built, flew its maiden voyage on April 27, 2005. This superjumbo double-decker surpasses Boeing's 747 in size, and orders are already pouring in. Although the project is behind schedule and over budget, the company is well on its way to breaking even. *Popular Science* named the Airbus A380 as the 2005 Grand Award Winner in the "Best of What's New."

Indeed, the aircraft is a titanic project—complete with one lethal flaw, according to Joseph Mangan, an aerospace engineer from Kansas City, Kansas.

In February 2004, Mangan moved to Vienna to work as an aerospace manager for TTTech Computertechnik, an Austrian subcontractor that supplies the computer chips that control cabin pressure in the A380. Mangan says that by March he confronted his employers about problems with the documentation submitted to the Federal Aviation Authority and the European Aviation Safety Agency (EASA), the two organizations that were certifying the chips. By the summer, Mangan insists that he had discovered serious defects in the software and repeatedly requested that TTTech correct the software before continuing with the certification process. Finally, in September 2004, Mangan revealed the design flaw to Airbus and EASA during an official audit. Mangan was fired a few days later.

The flaw in the software, Mangan believes, could lead to rapid loss of cabin pressure, leaving pilots, crew, and passengers with little time to don their air masks. Experts believe that such depressurization may have contributed to the 2005 crash of a Boeing 737 in Greece that killed 121 people and the 2002 crash of China Airlines Boeing 747, which killed 225 people.

A typical passenger jet has two outflow valves that control cabin pressure. To achieve redundancy and ensure safety, manufacturers install three separate motors, each with a different chip, to operate each outflow valve. For example, the Boeing 777 uses chips manufactured by Advanced Micro Devices, Motorola, and Intel. In addition, jets typically allow pilots to manually override this system.

To reduce the superjumbo A380's weight, Airbus decided to install four outflow valves, each operated by only one motor, and each motor driven by a TTTech controller chip. Mangan says he discovered that these chips were executing unpredictable commands when fed certain data. If one chip fails, he contends, all four will fail. So, while Airbus claims to have achieved redundancy by installing four outflow valves, Mangan vehemently rejects this assertion. His nightmare is that the failure of all four valves will cause the 555-seat jetliner to crash.

Knowing what he knows, Mangan said he felt a moral obligation to warn the public. In the event of a crash, Mangan could also be held legally accountable. After determining that a metal strip from a Continental DC-10 sliced the tires of a Concord jet, causing a crash in 2000, French prosecutors went after the American mechanic who installed the strip.

TTTech insists that Mangan was fired for poor job performance and is exacting revenge by trying to destroy the company. The company has filed both civil and criminal charges against Mangan, claiming that the information he released to the media was proprietary and that he has damaged the company's reputation.

Austria, unlike the United States, offers little protection for whistle-blowers. TTTech was able to obtain a gag order from an Austrian judge. Mangan is currently facing jail time for failing to pay a $185,000 fine for violating the gag order.

In the meantime, an EASA investigation concluded that TTTech's chip was unacceptable and would have to be fixed before Airbus received certification for the A380. TTTech's CEO Stefan Poledna says the company never received any indication from EASA that their chip was noncompliant. TTTech did identify and fix a glitch, Poledna admits, but as part of the routine software development and review process.

Airbus claims that it has examined the issue internally and found Mangan's allegations to be baseless. Yet, Airbus has also dismissed previous concerns about the use of glass laminates for the plane's fuselage. This double denial has raised fears that Airbus' titanic project, with its ambitious use of new technologies, may be headed toward an iceberg of its own.

Questions:

1. Describe the importance of redundancy to such safety-critical systems as the Airbus cabin.

2. If TTTech were located in Kansas City rather than Vienna, what protection would Joseph Mangan receive as a whistle-blower?

3. In your opinion, how well did Joseph Mangan deal with his situation as a whistle-blower?

3. Software to Assist in Managing Contingent Workers

Many organizations are turning to recruitment and applicant tracking systems (RATS), software packages that can manage and control money spent on contingent workers. These systems can track which workers and suppliers provide quality service, how much each department spends on contingent workers, and how many contingent workers are on the job. This information improves hiring decisions by enabling managers to better gauge consulting costs, service costs, and quality.

Most RATS packages centralize the hiring process, keeping track of which suppliers provide employees and determining where the hiring company can get the most leverage in cost reduction and services. Many packages offer the added benefit of automating the approval of job requisitions. Some even have built-in checks and balances that prevent managers from circumventing the company's hiring rules. A common violation is that a department manager hires someone for a project because he wants to get the person a job or the person is a friend of a friend. Most of the packages also offer back-end integration to financial and human resources systems to streamline payment processes.

Some RATS allow hiring managers to rate contingent workers, which provides feedback to suppliers and lets other hiring managers in the same company determine whether a contingent employee should be rehired in another department. Managers can then point and click on the jobs they need to fill and the agencies they prefer to use. If the system keeps a history of all workers, a manager can even request a specific worker who performed well on a previous assignment or find someone the company used three months ago. Order processing for a contingent worker can take less than a minute. Future planned enhancements for RATS include benefits administration, training, and validation of skill sets that potential employees purport to have. These enhancements will further streamline the hiring process.

Even though many companies can achieve significant savings and other benefits by using RATS, it is often difficult to get human resources, hiring managers, and other company employees to use it because it requires a significant culture change. Companies frequently resist changing the vendor-management concept, and hiring managers ask why they can't use a certain consulting service anymore. Suppliers also resist the change to a more automated system, fearing a loss of control and potential income.

Questions:

1. What key advantages are associated with the use of a RATS?

2. Does a RATS increase or decrease the chances that a company will deal more ethically with contingent workers? Why do you think so?

3. What additional features do you think the RATS application needs to increase its effectiveness? Can you think of any added features that would help managers better address the concerns of full-time workers about the use of contingent workers?

End Notes

[1] Kirby, Carrie, "Big Joint Venture in China: Microsoft, Others to Provide Global Outsourcing Service," *San Francisco Chronicle*, www.sfgate.com/cgi-bin/article.cgi?f=/c/a/2005/07/01/BUG4GDHF9R1.DTL, July 1, 2005.

[2] McDougall, Paul, "Microsoft and TCS to Open Outsourcing Center in China," *InformationWeek*, www.informationweek.com/showArticle.jhtml?articleID=164904223, June 30, 2005.

[3] United States Department of Labor, Bureau of Labor Statistics, "BLS Releases 2002-2012 Employment Projections," www.bls.gov/news.release/ecopro.nr0.htm, February 11, 2004.

[4] Computer Research Association, Taulbee Survey, 2004.

[5] Taleo Research, "The Contingent Workforce Management Report," November 2005.

[6] Ninth Circuit Court of Appeals, Re: Vizcaino v. Microsoft, www.techlawjournal.com/courts/vizcaino/19990512.htm, May 12, 1999.

[7] Gardner, David W., "U.S. Expects 20,000 H1B Visas to Go Quickly," *TechWeb.com*, www.techweb.com, May 11, 2005.

[8] Thibodeau, Patrick, "IT Groups Push Congress to Raise H-1B Visa Limits," *Computerworld*, www.computerworld.com, October 3, 2005.

[9] Gardner, David W., "U.S. Expects 20,000 H1B Visas to Go Quickly," *TechWeb.com*, www.techweb.com, May 11, 2005.

[10] Thibodeau, Patrick, "H-1B Investigations Are Expected to Increase," *Computerworld*, www.computerworld.com, March 21, 2005.

[11] Thibodeau, Patrick, "H-1B Investigations Are Expected to Increase," *Computerworld*, www.computerworld.com, March 21, 2005.

[12] Gross, Grant, "U.S. Senate Approves H-1B Visa Increase," *Computerworld*, www.computerworld.com, November 4, 2005.

[13] Gross, Grant, "Survey: Tech Salaries Down Last Year," *PC World*, www.pcworld.com, December 23, 2004.

[14] "Intuit Opens Indian Development Center, Releases Quicken 2006," *InformationWeek*, www.informationweek.com, August 5, 2005.

[15] McDougal, Paul, "ABN Amro Set to Unveil Largest IT Offshore Outsourcing Deal," *InformationWeek*, www.informationweek.com, August 31, 2005.

[16] Cipolla, Vin, "Workforce Insights," Veritude Web site, www.veritude.com/ResourceCenter/ResourceView.aspx?id=1023, November 24, 2005.

[17] Rai, Saritha, "Outsourcers in India Fight for Skilled Labor," *International Herald Tribune*, www.iht.com, November 2, 2005.

[18] Radosevich, Lynda, "Offshore Development Shipping Out," *IT Week*, www.zdnet.co.uk/itweek, September 1, 1996.

[19] Carlson, Caron, "ITAA Study Pushes for Outsourcing," *eWeek*, www.eweek.com, October 31, 2005.

[20] Koprowski, Gene L., "Unions Step Up Organizing of IT Workers, Outsourcing Fight," *eWeek*, September 5, 2005.

[21] Bajkowski, Julian, "Report: Black Market Growing for Offshore Data," *Computerworld*, www.computerworld.com, August 16, 2005.

[22] "Statement on Accounting Standards (SAS) No. 70 - About SAS 70," www.sas70.com/index2.htm.

[23] Nicolosi, Michelle, "Voter Machine Maker Settles Over Her Whistle-Blower Suit," *Seattle Post-Intelligencer*, www.seattlepi.nwsource.com, January 25, 2005.

[24] Nicolosi, Michelle, "Voter Machine Maker Settles Over Her Whistle-Blower Suit," *Seattle Post-Intelligencer*, www.seattlepi.nwsource.com, January 25, 2005.

[25] Evers, Joris, "Brief: Oracle Pays $8 M to Settle Suit over Training Charges," *Computerworld*, www.computerworld.com, May 16, 2005.

Sources for Case 1

Mark, Roy, "Gates Rakes Congress on H1B Visa Cap," *IT Management*, itmanagement.earthweb.com/career/article.php/3500986, April 27, 2005.

Tennant, Don, "Q&A: Information Builders CEO Blasts Gates' H-1B Stand," *Computerworld*, www.computerworld.com/careertopics/careers/story/0,10801,101493p3,00.html, May 2, 2005.

Thibodeau, Patrick, "The H-1B Equation," *Computerworld*, www.computerworld.com/careertopics/careers/story/0,10801,100059,00.html, February 28, 2005.

Sources for Case 2

Evans-Pritchard, Ambrose, "Airbus Whistleblower Faces Prison," *The Telegraph*, www.telegraph.co.uk/money/main.jhtml?xml=/money/2005/10/15/ccairb15.xml&menuId=242&sSheet=/money/2005/10/15/ixcoms.html, October 15, 2005.

Mangan, Joseph, "Silence Kills," www.eaawatch.net.

Pae, Peter, "A Skeptic Under Pressure," *The Los Angeles Times*, www.latimes.com/business/la-fi-whistleblower27sep27,0,7486292.story, September 27, 2005.

Sources for Case 3

Schwartz, Karen D., "Managing Human Capital," *Computerworld*, www.computerworld.com, April 16, 2001.

"Automating Internal Project Management and Staffing," *Computerworld*, www.computerworld.com, April 16, 2001.

Jones, Katherine, and Alschufer, David, Aberdeen Group, "Hourly Hiring Management Systems: Improving the Bottom Line for Hourly Worker-Centric Enterprises," Aberdeen White Paper, www.aberdeen.com, June 2002.

Goodridge, Elisabeth, "Workforce Management Draws VC Dollars," *InformationWeek*, www.informationweek.com, October 29, 2001.

CHAPTER 9

THE IMPACT OF INFORMATION TECHNOLOGY ON THE QUALITY OF LIFE

QUOTE

The [healthcare] industry is about 10 years behind in the application of technology and at least 10 to 15 years behind in leadership capability from the technology and perhaps the business perspective.[1]

—Cliff Dodd, CIO of Kaiser Permanente, a U.S. healthcare delivery system

VIGNETTE

Technological Advances Create Digital Divide in Healthcare

Healthcare costs are skyrocketing, with premiums rising annually at a double-digit rate. A host of factors are to blame, including rising prescription and hospital costs, expanding coverage mandated by state and federal governments, excessive administrative expenses, and—last but not least—expensive technological advances. Yet, some of these advances are designed to decrease healthcare costs, at least in theory.

The Centers for Medicare and Medicaid Services are in the process of testing a technological advance called the telemedicine system, which allows chronically ill patients to monitor their health daily and upload data through the Internet into a database that monitors their condition. With early intervention, patients can avoid health crises as well as expensive hospital visits. Patients purchase home monitoring systems such as glucose tests and a device called Health Buddy, which collects the readings and transmits them via a phone line to the Health Hero database. The Health Hero Network then reviews the data and automatically reports potential problems to health practitioners.

The goal of the program is to cut Medicare expenses by five percent per patient, but the question is how it will affect the medical expenses of the participants. Will the patients have to pay for all or part of the monitoring devices and the Health Buddy? Will the system increase or decrease annual patient costs?

The irony of technological advances in medicine is that while they can prolong and improve quality of life, they also create new standards and expectations and ultimately could either push the healthcare system toward a financial crisis or increase the disparity between rich and poor. Patients who cannot pay for these new technologies fall through the gap of the healthcare system's very own "digital divide." The success of programs such as Health Buddy depends not only on whether they decrease Medicare expenses, but whether they lower patient costs.[2,3]

LEARNING OBJECTIVES

As you read this chapter, consider the following questions:

1. What impact has IT had on the standard of living and worker productivity?
2. What is being done to reduce the negative influence of the digital divide?
3. What impact has IT had on reducing the costs of healthcare?

THE IMPACT OF IT ON THE STANDARD OF LIVING AND PRODUCTIVITY

The most widely used measurement of the material standard of living is gross domestic product (GDP) per capita, adjusted for inflation. The GDP actually represents the total annual output of a nation's economy. According to this measurement, the standard of living varies greatly among groups within a country and from nation to nation. Overall, industrialized nations tend to have a higher standard of living than developing countries.[4]

In the United States and in most Western nations, the standard of living has improved for a long time. However, its rate of change varies as a result of business cycles that affect prices, wages, employment levels, and the production of goods and services.[5] Disasters such as hurricanes, tsunamis, and war can also reduce the standard of living. The worst economic downturn in U.S. history occurred during the Great Depression of the 1930s, when real per-capita GDP declined by approximately one-third and the unemployment rate reached 25 percent.[6]

Productivity is defined as the amount of output produced per unit of input, and is measured in many different ways. For example, productivity in a factory might be measured by the number of labor hours it takes to produce an item, while productivity in the service sector might be measured by the annual revenue an employee generates, divided by the employee's annual salary. Most countries have been able to produce more goods and services over time not by requiring a proportional increase in input, but by making production more efficient. These gains in productivity led to increases in the GDP-based standard of living because the average hour of labor produced more goods and services. In the United States, labor productivity growth has averaged roughly 2 percent per year for the past century, meaning that living standards have doubled about every 36 years.

Table 9-1 shows the annual change in labor productivity since 1839. After averaging 1.9 percent per year from 1889 to 1937, productivity during the 1950s soared to 3.6 percent per year as modern management techniques and automated technology made workers far more productive. Worker productivity remained at a high level of 2.6 percent during the 1960s. For the next 22 years, the average growth rate dipped below 1.5 percent. From 1995 to 2005, the productivity rate increased to 2.1 percent, slightly above the historical average.[7]

TABLE 9-1 Comparison of labor productivity rates (compounded aggregate growth rate)

Years	United States	France	Germany
1839–1937	1.9%		
1950–1960	3.6%	5.2%	6.3%
1960–1973	2.6%	5.1%	5.0%
1973–1980	1.0%	3.4%	3.3%
1980–1990	1.4%	2.9%	1.9%

TABLE 9-1 Comparison of labor productivity rates (compounded aggregate growth rate) (continued)

Years	United States	France	Germany
1990–1995	1.1%	1.5%	2.1%
1995–2000	2.0%	1.1%	1.5%
2000–2005	2.1%		

Source: McKinsey Global Institute, "Whatever Happened to the New Economy?," November 2002.

Innovation is a key factor in productivity improvement, and IT has played an important role in enabling such innovation. Progressive management teams use IT, new technology, and capital investment to implement innovations in products, processes, and services.[8]

In the early days of IT (the 1960s), productivity improvements were easy to measure. For example, midsized companies had a dozen or more accountants whose full-time jobs were to perform payroll-related accounting. When businesses learned to apply automated payroll systems, fewer accounting employees were needed. The productivity gains from such IT investments were obvious.

Today, organizations are trying to further improve business processes that have already gone through several rounds of improvement and that have employed IT for years. Organizations are also using newer, better IT capabilities to help workers who already have an assortment of personal productivity applications on their desktop computers, laptops, and personal digital assistants (PDAs). Instead of eliminating workers, companies are saving workers small amounts of time each day. Whether these saved minutes really result in improved worker productivity is a matter for debate. Many analysts argue that workers merely use the extra time to do some small task they didn't have time to do before, such as respond to e-mail they would have ignored. These minor gains make it harder today to quantify the benefits of IT investments on worker productivity.

The relationship between investment in information technology and U.S. productivity growth is also more complex than you might think at first. Consider the following facts:

- The rate of productivity from 1995 to 2005 is only slightly higher than the long-term U.S. rate and not nearly as high as it was for two decades following World War II. So, although the recent increase in productivity is welcome, it is not statistically significant.
- Average labor productivity in the United States remained relatively high despite a reduced level of investment in IT from 1999 to 2004. If there was a simple, direct relationship, the productivity rate should have decreased.
- The U.S. productivity rate remained higher than that in France and Germany (as well as in many other industrialized countries) from 1999 to 2004, even though those two countries did not reduce their level of IT investment.[9] Again, this argues against a simple, direct relationship.

One explanation for the previous points is the possible lag time between the application of innovative IT solutions and the capture of significant productivity gains. IT can enhance productivity in fundamental ways by allowing firms to make radical changes in work processes, but such major changes can take years to complete because firms must make substantial complementary investments in retraining, reorganizing, changing reward systems, and the like. Furthermore, the effort to make such a conversion can divert resources from normal production activities, which can actually reduce productivity, at least temporarily. For example, researchers examined a sample of 527 large U.S. firms from 1987 to 1994 and found that the benefits of applying IT grow over time and can take at least five to seven years to fully realize.[10]

Another explanation for the complex relationship between IT investment and U.S. productivity growth is that many other factors influence worker productivity rates besides IT. Table 9-2 summarizes fundamental ways in which companies can increase productivity, and the following list summarizes additional factors that can affect national productivity rates.

TABLE 9-2 Fundamental drivers for productivity performance

Reduce the amount of input required to produce a given output by:	Increase the amount of output produced by a given amount of input by:
Consolidating operations to better leverage their scale	Selling higher-value goods
Improving performance by becoming more efficient	Selling more goods to increase capacity and use of existing resources

- Labor productivity growth rates differ according to where a country is in its business cycle—expansion or contraction.[11] Times of expansion enable firms to gain full advantage of economies of scale and full production. Times of contraction present fewer investment opportunities.
- Outsourcing and offshore outsourcing can skew productivity if the contracting firm has different productivity rates from the outsourcing firm.
- U.S. regulations make it easier for companies to hire and fire workers and to start and end business activities than in many other industrialized nations. This flexibility makes it easier for U.S. firms to relocate workers to more productive positions.[12]
- More competitive markets for goods and services can provide greater incentives for technological innovation and adoption as firms strive to keep ahead of competitors.[13]
- In today's service-based economy, it is difficult to measure the real output of such services as accounting, customer service, and consulting.
- The greatest benefits from IT investments don't always yield tangible results, such as cost savings and reduced head count, but intangible benefits such as improved quality, reliability, and service.

As you can see, it is difficult to quantify how much the use of IT has contributed to worker productivity. Ultimately, however, the issue is academic. There is no way to compare organizations that don't use IT with those that do, because there is no such thing as a noncomputerized airline, financial institution, manufacturer, or retailer. Businesspeople analyze the expected return on investment to choose which IT option to implement, but they recognize that IT is required to remain in business. At this point, trying to measure its precise impact on worker productivity is like trying to measure the impact of telephones or electricity.

The Digital Divide

When people talk about standards of living, they are often referring to a level of material comfort measured by the goods, services, and luxuries that are available to a person, group, or nation—factors beyond the GDP-based measurement of standards of living. Some of these indicators are:

- Average number of calories consumed per person per day
- Availability of clean drinking water
- Average life expectancy
- Literacy rate
- Availability of basic freedoms
- Number of people per doctor
- Infant mortality rate
- Crime rate
- Rate of home ownership
- Availability of educational opportunities

Another indicator of the standard of living is the availability of technology. The **digital divide** is a term used to describe the gulf between people who do and don't have access to modern information and communications technology, such as computers and the Internet. For example, of the roughly 1 billion Internet users worldwide, only 20 million (2 percent) are estimated to live in less developed nations.

The digital divide exists not only between more and less developed countries but between economic classes in the same country, the educated and uneducated, and people who live in cities versus those in rural areas. Consider the following examples.

- A 2000 study found that American whites and Asian-Americans were almost twice as likely to have Internet access as African-Americans and Hispanics. Households with annual incomes in excess of $75,000 were more than 10 times as likely to have Internet access than those with annual incomes of $15,000 or less.[14]
- A 2005 European Union (EU) survey found clear evidence of a digital divide within EU member countries based on age, income, and education. The survey found that 85 percent of students aged 16 to 24 used the Internet, while only 13 percent of people between 55 and 74 went online. Only 25 percent of people who had not completed high school used the Internet, compared with 77 percent of college or university graduates. Average Internet use across the European Union stood at 47 percent, ranging from users in Sweden (82 percent) to those in Greece (20 percent).[15]
- There is no economical way to provide telecommunication service to the 700 million rural people in India who have an annual GDP of less than $500 per capita. In many of India's rural communities, one must travel more than 5 miles to the nearest telephone.[16]
- Many parts of Africa lack basic facilities such as drinking water, roads, and electricity. Easy access to modern computers and information technology is simply not realistic.[17]

Many people think that the digital divide must be bridged for a number of reasons. Clearly, health, crime, and other emergencies could be handled better if a person in trouble had easy access to a communications network. Access to IT and communications technology can greatly enhance learning and provide a wealth of educational and economic opportunities. Much of the vital information people need to manage their career, retirement, health, and safety is increasingly provided by the Internet. High-tech access can give a country's industries a competitive advantage over its less fortunate neighbors.

A number of current U.S. programs, including E-Rate, E-Tech, and optimization technologies, are designed to help eliminate the digital divide. These programs are discussed in the following sections.

E-Rate Program

In the United States, the Education Rate (E-Rate) Program was authorized as part of the Telecommunications Act of 1996. One of its goals was to help schools and libraries obtain access to state-of-the-art services and technologies at discounted rates. The Federal Communications Commission (FCC) ruled that the program would be supported with up to $2.25 billion per year from a fee charged to telephone customers. The FCC also established the private, nonprofit Universal Service Administrative Company (USAC) to administer the program. Unfortunately, the program has not gone well. Following a year-long investigation, a House subcommittee in 2005 approved a bipartisan staff report detailing abuse, fraud, and waste in E-Rate. In one infamous example, USAC disbursed $101 million from 1998 to 2001 to equip Puerto Rico's 1540 schools with high-speed Internet access, but a review found that few computers were actually connected to the Internet and $23 million worth of equipment was found in unopened boxes in a warehouse.[18]

E-Tech Program

The No Child Left Behind Act (NCLB) became law in 2002. Its stated goal is to end the achievement gap between rich and poor students and white and minority students while improving everyone's academic performance. In recognition that the ability to use computers and access the Internet is a requirement to succeed in the U.S. educational system and the global workforce, the NCLB requires each state to have an Enhancing Education through Technology (Ed-Tech) program. The Ed-Tech program has the following goals:

- Improve student academic achievement through the use of technology in schools.
- Assist children in crossing the digital divide by ensuring that every student is technologically literate by the end of eighth grade.
- Encourage the effective integration of technology with teacher training and curriculum development to establish successful, research-based instructional methods.

To help achieve these goals, the U.S. Department of Education awarded $480 million in Ed-Tech grants to individual states in 2005, and outlined a seven-step approach to achieve a successful Ed-Tech program:

- Strengthen leadership.
- Consider innovative budgeting.

- Improve teacher training.
- Support e-learning and virtual schools.
- Encourage broadband access.
- Move toward digital content.
- Integrate data systems.[19]

States have begun innovative and successful programs to help achieve the Ed-Tech program goals, but there have also been some unsuccessful initiatives. Overall, more than 25 percent of all K–12 public schools now provide some form of individual online instruction to supplement regular classes or provide for special needs.[20]

Optimization Technologies

New information technologies can be used with little capital cost to reduce the digital divide. These **optimization technologies** can make computing and communication better, cheaper, faster, and more available to larger segments of the world's population.[21] For example, scientists in the Media Lab at the Massachusetts Institute of Technology (MIT) have developed a prototype of an easy-to-use laptop computer that could be mass-produced for less than $100 each. Its features include a hand crank for generating power in places where electricity is not readily available. As these devices become available and more widely distributed, they can only be truly effective if they reach impoverished people instead of being diverted by criminals and corrupt governments.[22]

Media Lab Asia is a large academic research program dedicated to bringing the benefits of new information and communications technologies to millions of people in Asia, Africa, and Latin America. It began as an offshoot of MIT's Media Lab and received much of its initial funding from the government of India; more than a dozen leading technical and business universities are participants. The first regional laboratories were built in Mumbai-Pune, Tamilnadu, Kanpur-Lucknow, New Delhi, and Kharagpur. Each lab conducts several core projects and is networked by high-bandwidth connections to the other labs to enable collaboration.[23]

When IT and its benefits are available to everyone—regardless of their economic status, geographic location, language, or social status—they can enhance the sharing of ideas, culture, and knowledge. How much will these benefits raise the standard of living in underdeveloped countries? Could the end of the digital divide change the way people think about themselves in relation to the rest of the world? Could such enlightenment, coupled with a better standard of living, contribute to a reduction in violence, poverty, poor health, and even terrorism?

THE IMPACT OF IT ON HEALTHCARE COSTS

Healthcare costs are soaring out of control. The United States spent $1.7 trillion on healthcare in 2003, which is more than four times what the Pentagon spent on national defense.[24] This spending represented 15.3 percent of the U.S. GDP, a greater percentage than in other industrialized nations, even though nearly 45 million Americans are uninsured and other countries provide health insurance to all their citizens (see Table 9-3).[25] U.S. citizens spent an additional $230 billion of their own money on healthcare.[26] Current estimates are that national healthcare spending will more than double to over $4 trillion per year by 2016, with one in every five dollars spent in the United States on healthcare.[27]

TABLE 9-3 Healthcare spending in industrialized countries as percentage of GDP

Country	Healthcare spending as percentage of GDP
United States	15.3%
Switzerland	10.9%
Germany	10.7%
Canada	9.7%
France	9.5%

Source: "Health Insurance Cost," National Coalition on Health Care Web site, www.nchc.org/facts/cost.shtml.

As healthcare costs go up, so do the insurance premiums. Employers today pay 78 percent more for healthcare insurance than they did in 2000, and employees pay 64 percent more. And the rates continue to increase by double digits on an annual basis.[28] To reduce the impact of healthcare premiums on corporate profits, many U.S. companies are shifting the costs to employees and some are eliminating healthcare coverage altogether for retirees. For example, in December 2005, General Motors reached an agreement with the United Auto Workers to cut healthcare coverage for retirees, saving about $3 billion a year. Many experts point to two primary causes for cost increases: the use of more expensive technology and the shielding of patients from the true cost of medical care.

The development and use of new medical technology, such as new diagnostic procedures and treatments, has increased spending and "accounts for one-half to two-thirds of the increase in healthcare spending in excess of general inflation."[29] Although many new diagnostic procedures and treatments are at least moderately more effective than their older counterparts, they are also more costly. Even when new procedures and treatments cost less (for example, magnetic resonance imaging), they stimulate much higher rates of use because they are more effective or cause less discomfort to patients.[30]

Patients sometimes overuse medical resources because they appear to be free or almost free, thanks to the share of medical bills that are paid by third parties, such as insurance companies and governments. A patient who doesn't have to pay for a medical test or procedure is probably less likely to consider its cost-to-benefit ratio. Attempts by insurance companies to rein in those costs have led to a blizzard of paperwork but have proven ineffective.

To really gain control over soaring healthcare costs, patient awareness and technology costs must be improved and brought under control. In the meantime, however, the improved use of IT in the healthcare industry can lead to significant cost reductions in a number of ways.

Electronic Health Records

For an industry that depends on highly sophisticated technology for diagnostics and treatment, healthcare has been slow to implement IT. The healthcare industry invests about $3000 for each worker in IT, compared with $7000 per worker for industry in general.[31]

One tremendous opportunity for improving healthcare through IT is in the capture and recording of patient data. Before seeing a physician, most patients are given a clipboard and pen with a standard form to complete. Some people must wonder: "This is the same form I filled out last time; what did they do with the data from my last visit?" It is nearly impossible to pull together the paper trail created by a patient's interactions with various healthcare entities into a clear, meaningful, consolidated view of the person's health history. This lack of integration can result in diagnostic and medication errors and the ordering of duplicate tests that dramatically raise the cost of healthcare.

This lack of integration can even compromise patient safety.[32] For example, suppose that a person with an allergy to aspirin has a heart attack while driving, pulls off to the side of the road, dials 911, and then loses consciousness. Paramedics who arrive on the scene are likely to give the victim aspirin to open the blood vessels and arteries, which could trigger an allergic reaction. The trauma of the reaction, coupled with the heart attack, could kill the victim. If the United States had a comprehensive healthcare information network, such medical errors could be drastically reduced. Healthcare experts agree that "going digital" will reduce errors that kill up to 98,000 patients a year.[33]

For example, an **electronic health record (EHR)** is a summary of health information generated by each patient encounter in any healthcare delivery setting. The EHR includes patient demographics, medical history, immunizations, laboratory data, problems, progress notes, medications, vital signs, and radiology reports. Healthcare professionals can use an EHR to generate a complete electronic record of a clinical patient encounter.[34] Surprisingly, only 13 percent of U.S. hospitals have electronic patient record systems, even though experts agree that $78 billion to $400 billon could be saved each year by using such records in a network with open communications standards.[35,36] EHRs could incorporate data from any healthcare entity a patient uses and make the data easily accessible to other healthcare professionals.[37]

In 2004, the Medical Records Institute asked employees at 463 healthcare organizations why they favored EHRs. Table 9-4 compares the survey responses of IT staff with those of medical staff.

TABLE 9-4 Primary reasons to implement electronic health records

Motivations	IT managers and professionals	Physicians and nurses
The need to improve clinical processes or workflow efficiency	87%	85%
The need to improve quality of care	82%	87%
The need to share patient information among healthcare practitioners and professionals	89%	71%
The need to reduce medical errors and improve patient safety	81%	73%

Source: Gilhooly, Kym, "Rx for Better Health Care," *Computerworld*, January 31, 2005. (This table shows only the top four reasons of the 12 listed in the original table.)

In April 2004, President Bush issued an executive order calling for the broad adoption of interoperable EHRs and other e-health initiatives by 2014, and appointed Dr. David J. Brailer to the new position of national coordinator for health information technology.[38] Though the details of the National Health Information Network (NHIN) have yet to be defined, experts say that its key concerns are to protect patient privacy and find a consistent way to identify each patient.[39] As a result, the NHIN probably will be based on a decentralized architecture that requires neither a universal patient identifier nor a centralized repository for health data. Instead, healthcare organizations will have to adopt standards to allow interoperability among institutions and make individual health records available to the appropriate doctors, hospitals, labs, and pharmacies.[40]

One approach would be to create a sort of domain name system for patients that works much like the one that drives the Web. When a patient consents to have his medical information recorded, his name, date of birth, sex, zip code, and medical record number might be transmitted to a secure medical DNS server. A healthcare practitioner who needed to access a patient's medical records would obtain consent and then query the patient DNS system by providing basic demographic information. The DNS system could then provide a list of all medical institutions the patient has visited and his medical record number for each institution. This could enable a doctor to contact other institutions that have treated the patient and request medical records via e-mail or fax.[41] For example, Sutter Health in Sacramento, California, planned to deploy an EHR system by 2006 to connect 26 hospitals, more than 5000 physicians, and millions of patients in Northern California.[42]

Use of Mobile and Wireless Technology

The healthcare industry was actually a leader in adopting mobile and wireless technology, perhaps because of the urgency to communicate with doctors and nurses who are almost always on the move. For example, doctors were among the first large groups to start using PDAs on the job. Other common uses of wireless technology in the healthcare field include:[43]

- Providing a means to access and update EHRs at patients' bedsides to ensure accurate and current patient data
- Enabling nurses to scan bar codes on patient wristbands and on medications. An attached computer on a nearby cart is linked by a wireless network to a database that contains physician medication orders. This arrangement helps nurses administer the right drug in the proper dosage at the correct time of day.
- Using wireless devices to locate healthcare employees wherever they are

Telemedicine

Telemedicine employs modern telecommunications and information technologies to provide medical care to people who live far away from healthcare providers. This technology reduces the need for patients to travel for treatment and can enable healthcare professionals to serve more patients in a broader geographical area. There are two basic forms of telemedicine: store-and-forward and live.

Store-and-forward telemedicine involves acquiring data, sound, images, and video from a patient and then transmitting it to a medical specialist for assessment later. This type of regular monitoring does not require the presence of both patient and care provider at the same time, and having access to such information can enable healthcare professionals to recognize problems and intervene before high-risk situations become life threatening. For example, patients who have chronic diseases often don't recognize early warning signs that indicate an impending health crisis. A sudden weight gain by a patient who has suffered congestive heart failure could indicate retention of fluids, which could lead to a traumatic trip to the emergency room or even loss of life. A physician who uses telemedicine to keep tabs on such patients could make a vital difference.

For example, the Center for Medicare and Medicaid Services has begun a home-health telemedicine project of 2000 chronically ill Medicare beneficiaries in Oregon and Washington. Participants use home-monitoring devices that measure blood glucose or blood pressure levels and then feed the data into a collection device, which plugs into a phone line and automatically sends the data to a secure remote database that is monitored by healthcare professionals. If a problem is identified, the patient is immediately contacted and advised to change dosage levels, call a physician, or take other measures.[44]

Live telemedicine requires the presence of patients and healthcare providers at the same time and often involves a videoconference link that allows audio and video communications between the two sites. Live telemedicine enables Childrens Hospital in Los Angeles to provide remote triage services to other area hospitals that lack pediatric intensive-care units (ICUs). Each participating hospital is equipped with special endoscopy equipment for examining the inside of the body, vital-sign monitoring equipment, and Web-based video cameras. Patient data from an emergency room miles away is transmitted to a videoconferencing workstation at the Childrens Hospital ICU, where it can be viewed and analyzed by a pediatric specialist. Special telemedicine technology can even improve the images that doctors see on the monitors at Childrens Hospital, which can help them determine which patients need to be transported and admitted to the ICU for further treatment.[45]

Medical Information Web Sites for Lay People

Healthy people as well as those who suffer from illness need reliable information on a range of medical topics to learn more about healthcare services and to take more responsibility for their own well being. Clearly, lay people cannot become as informed as trained medical practitioners, but a tremendous amount of healthcare information is available via the Web. These sites have a critical responsibility to publish reliable and objective information. Table 9-5 gives just a small sample of Web sites that offer medical information.

TABLE 9-5 Health information Web sites

URL	Description
WebMD.com	Access to reference material and online professional publications from Thomson Healthcare Information Group, Stamford, Connecticut
cancer.org	The American Cancer Society
oncolink.upenn.edu/	The Abramson Cancer Center at the University of Pennsylvania
looksmarthealth.com	A comprehensive health Web site with information on diet and nutrition, disease treatment and therapy, herbal remedies, baby care, sex, and other topics
americanheart.org	The American Heart Association
heartdisease.about.com	A good starting place to learn about heart disease and cardiology
diabetes.org/home.jsp	The American Diabetes Association
niddk.nih.gov	The National Institute of Diabetes, Digestive, and Kidney Diseases
cdc.gov	The Centers for Disease Control and Prevention
medicinenet.com/script/main/hp.asp	Helpful source for medical information, including symptoms and signs, procedures and tests, medications, and a medical dictionary
heartburn.about.com/	Information on what causes heartburn and how to prevent it
alzheimers.org/	Alzheimer's Disease Education and Referral Center
urologychannel.com/	Provides clear, accurate information about urologic conditions (for example, erectile dysfunction, HIV, AIDS, kidney stones, STDs), including overviews, symptoms, causes, diagnostic procedures, and treatment options
osteo.org	The National Institutes of Osteoporosis and Related Bone Diseases

In addition to publicly available information on the Web, some health providers and employers offer useful online tools to members and employees that go beyond basic health information. For example, WellPoint is a leading health benefits company that serves approximately 34 million members nationwide. It formed a partnership with technology provider Subimo LLC to offer online healthcare support tools. With this technology, WellPoint members can "go online to compare, among other things, quality, safety, and cost information on hospitals nationwide; quality and risk indicators for specific health treatment options; nationwide average prices of drugs and treatment options; and coverage and costs for treatments by in-network and out-of-network healthcare providers."[46]

Subimo is responsible for capturing data and updating its databases as needed to provide accurate and current medical information. For example, a WellPoint member who needs a hip replacement can go online and find information about the surgery, other available treatment options, a list of questions to ask the physician, potential risks, nearby hospitals that perform the surgery, and quality-of-service information about the hospitals, such as the number of reported postoperative infections and other complications.[47]

Summary

1. **What impact has IT had on the standard of living and worker productivity?**

   The most widely used measurement of the material standard of living is gross domestic product (GDP) per capita, adjusted for inflation. In the United States, as in most Western nations, the standard of living has improved for a long time. However, its rate of change varies as a result of business cycles that affect prices, wages, employment levels, and the production of goods and services.

   Productivity is defined as the amount of output produced per unit of input. Most countries have been able to produce more goods and services over time not by requiring a proportional increase in input, but by making production more efficient. These gains in productivity led to increases in the GDP-based standard of living because the average hour of labor produced more goods and services. Progressive management teams use IT, new technology, and capital investment to implement innovations in products, processes, and service.

   It is harder today to quantify the benefits of IT investments on worker productivity because there can be a considerable lag between the application of innovative IT solutions and the capture of significant productivity gains. In addition, many other factors influence worker productivity rates, not just IT investments.

2. **What is being done to reduce the negative influence of the digital divide?**

   The *digital divide* is a term used to describe the gulf between people who do and don't have access to modern information and communications technology, such as computers and the Internet. The digital divide exists not only between more and less developed countries but between economic classes in the same country, the educated and uneducated, and people who live in cities versus those in rural areas.

   New information technologies can be used with little capital cost to reduce the digital divide. These optimization technologies can make computing and communication better, cheaper, faster, and more available to larger segments of the world's population. Scientists at the MIT Media Lab are among the leaders in developing optimization technologies; they have created a prototype of an easy-to-use computer that could be produced for less than $100 each. One of the lab's offshoots, Media Lab Asia, is a large academic research program dedicated to bringing the benefits of new IT and communications technologies to millions of people in Asia, Africa, and Latin America.

3. **What impact has IT had on reducing the costs of healthcare?**

   Healthcare costs are soaring out of control. As costs go up, so do healthcare insurance premiums. Many experts point to two primary causes for cost increases: the use of more expensive technology and the shielding of patients from the true cost of medical care. Improved use of IT in the healthcare industry can lead to significantly reduced costs in a number of ways:

   - Electronic health records (EHRs) of patient information can be generated from each patient visit in every healthcare setting.

   - Wireless technology can be used to access and update electronic health records at patients' bedsides, match bar-coded patient wristbands and medication packages to physician orders, and assist in voice communication.

- Telemedicine employs modern telecommunications and information technologies to provide medical care to people who live far away from healthcare providers. This technology reduces the need for patients to travel for treatment and can enable healthcare professionals to serve more patients in a broader geographical area.
- Web-based health information can help people inform themselves about medical topics.

Self-Assessment Questions

1. Which of the following statements about the standard of living is *not* true?

   a. It varies greatly from nation to nation.

   b. It varies little among groups within the same country.

   c. Industrialized nations generally have a higher standard of living than developing countries.

   d. It is frequently measured using the GDP per capita.

2. The amount of output produced per unit of input is called _____ .

   a. return on investment

   b. GDP

   c. productivity

   d. standard of living

3. Labor productivity has grown about three percent per year for the past century, which means that living standards have doubled about every 24 years. True or False?

4. The *digital divide* refers to a difference in the fundamental standard of living among developed and less-developed countries. True or False?

5. Which of the following statements about the digital divide is *not* true?

   a. It exists not only between more and less developed countries but between economic classes in the same country, the educated and uneducated, and people who live in cities versus those in rural areas.

   b. A 2000 study found that American whites and Asian-Americans were almost five times as likely to have Internet access as African-Americans and Hispanics.

   c. Average Internet use across the European Union stands at 47 percent, ranging from Sweden (at 82 percent) to Greece (20 percent).

   d. In many of India's rural communities, one must travel more than 5 miles to the nearest telephone.

6. Which of the following is a valid reason for trying to reduce the digital divide?

   a. Health, crime, and other emergencies could be better handled if people in trouble had access to a communications network.

   b. Much of the vital information that people need to manage their retirement, health, and safety is increasingly provided by the Internet.

c. Ready access to information and communications technology can provide a country with a wealth of economic opportunities and give its industries a competitive advantage.

d. all of the above

7. In the United States, the _____ was authorized as part of the Telecommunications Act of 1996, with a goal of helping schools and libraries obtain state-of-the-art services and technologies at discounted rates.

8. New information technologies can be used with little capital cost to reduce the digital divide. These technologies, which are generally referred to as _____ , make computing and communication better, cheaper, faster, and more available to larger segments of the world's population.

9. Which of the following statements about healthcare spending is *not* true?

   a. The U.S. government spent $1.7 trillion on healthcare in 2003.

   b. The United States spent four times more on healthcare in 2003 than the Pentagon spent on national defense.

   c. U.S. spending on healthcare represents 15.3 percent of its GDP.

   d. The United States spends a lesser percentage of its GDP on healthcare costs than other industrialized nations because nearly 45 million Americans are uninsured and other countries provide health insurance to all their citizens.

10. Two main reasons have been advanced as the cause of rising healthcare costs: the use of more expensive technology and the _____ of patients from the true cost of medical care.

11. To lessen the impact of healthcare premiums on corporate profits, many U.S. companies are shifting the costs to employees and some are eliminating healthcare coverage altogether for retirees. True or False?

12. Identify two fundamental barriers to the secure exchange of electronic health records.

Review Questions

1. What is the GDP and what does it measure?

2. Identify several factors besides investment in IT that can influence worker productivity.

3. Identify some inherent problems in measuring worker productivity.

4. How would you define the term *digital divide*?

5. Identify two specific examples of technologies that are designed to improve computing and communications to large segments of the world's population.

6. What benefits would come from making information technology available to everyone? Do you think that these benefits could raise the standard of living? Why or why not?

7. Define the term *telemedicine* and identify the two basic types of telemedicine.

8. List 10 key pieces of data you would expect to find in a patient's electronic health record.

9. Identify three examples of the use of wireless devices in the healthcare field.

10. Describe the Web-based medical information that is available to anyone. What additional information might be available to members of progressive healthcare organizations?

Discussion Questions

1. Is it harder today to quantify the benefits of IT investments on worker productivity? Discuss the question fully.

2. Explain how increases in worker productivity can lead to an increase in the standard of living.

3. Identify three factors that you think are significant indicators of the standard of living. Find current measures of these three factors for five different countries. Based on your indicators, which of the five countries would you say has the highest standard of living?

4. Explain why the development and use of new medical technology, such as new diagnostic procedures and treatments, has increased spending and accounts for one-half to two-thirds of the increase in healthcare spending in excess of general inflation.

5. Should the medical industry place more emphasis on using older medical technologies and containing medical costs?

6. Briefly explain how a domain name system for locating patient healthcare information might work.

7. Medical information that you obtain from Web sites must be accurate and reliable. How can you ascertain the credibility of a Web site's information?

What Would You Do?

1. Many people think that the "haves" in society must help the "have-nots." One idea is to help others to gain high-speed, reliable access to the Internet. If you were a U.S. Senator, would you support the following suggestion? Why or why not? Discuss some of the problems, technical and nontechnical, in providing this type of access.

   "Every time anyone buys a personal computer, signs up for a cell phone, or contracts for cable or satellite TV service, they pay a $20 'Internet enablement' fee. The hardware vendors and service providers match the fee. The resulting money is paid into a fund managed by a nonprofit organization and is used to provide Internet access and computers for U.S. residents who cannot afford them."

2. You have been invited to accept a six-month internship at the Media Lab in Kharagpur, which is in the Midnapore West district of the state of West Bengal in India. Your modest living expenses will be paid and you will receive $5000 U.S. at the end of your internship. The lab's goal is to bring the benefits of new information and communications technologies to millions of people in Asia, Africa, and Latin America. Kharagpur was chosen to have the first campus of the prestigious Indian Institutes of Technology (IITs). The IITs are the premier technical education institutes in India and are internationally recognized for their academic and technical excellence. Assume that you decide to accept this challenging opportunity. In what research and other activities would you participate to help bring the benefits of information and communications technology to India? Do you think that the Media Lab can make a positive impact? Why or why not?

3. You are a midlevel manager at a major metropolitan hospital and are responsible for capturing and reporting statistics regarding the cost and quality of patient care. You believe in a strict interpretation when defining various reportable incidents; as a result, your hospital's rating on a number of quality issues has declined in the six months you have held the

position. Your predecessor was more lenient and was inclined to let minor incidents go unreported or to classify some serious incidents as less serious. The quarterly quality meeting is next week, and you know that your reporting will be challenged by the chief of staff and other members of the quality review board. How should you prepare for this meeting? Should you defend your strict reporting procedures or revert to the former reporting process for the "sake of consistency in the numbers," as several people have urged?

Cases

1. Overcoming the Digital Divide in Uzbekistan

In 2003, the United Nations World Summit on the Information Society (WSIS) met to discuss the global digital divide and set a goal to have half the world connected to the Internet by 2015. In November 2005, the WSIS reconvened and reported considerable progress. Between 2000 and 2004, the number of Internet users grew from 381 million to 872 million worldwide, and developing countries' share of this figure rose from 21 percent to 39 percent.

These figures, however, come with other less reassuring trends. Internet use in developing nations is concentrated in cities rather than poorer towns and villages. Broadband connections are rare in the developing world, and only 38 percent of schools in poorer countries are connected to the Internet, compared with 94 percent in richer nations.

Uzbekistan, a developing nation in central Asia, illustrates the many obstacles to achieving WSIS's goal. The country's economy is based largely on agriculture and mining, and its authoritarian government maintains tight control of industry and education. At the time of the 2003 WSIS declaration, less than one percent of the population had Internet access through an antiquated phone system that has yet to achieve universal coverage.

Yet, Uzbekistan has been a U.S. ally in the battle against world terrorism. Beset by its own Islamic extremists, Uzbekistan represents a Muslim country with which the U.S. State Department believes it can "develop a dialogue." The countries enjoy a good trade relationship, and the United States offers humanitarian and technical assistance.

Among these U.S.-funded programs is Global Connections and Exchange (GCE), formerly known as School Connectivity for Uzbekistan. When the program began, public Internet access points were limited to about 40 clubs or cafes, most of them in the capital city of Tashkent. The average citizen could not afford the high access fees. The GCE program set out to establish Internet access in 60 schools throughout the country.

The overall goal of the program is to bridge the digital divide by helping teachers and students exploit electronic resources and opportunities. However, the Bureau of Educational and Cultural Affairs also uses the program to further its own goals. As such, the program encourages democratic curriculum reform, organizes cooperative projects with U.S. schools on such themes as diversity and tolerance, and fosters civic participation and volunteerism in local communities.

The program has had to overcome many hurdles. In Uzbekistan, teachers are not paid a living wage, so they supplement their meager earnings by tutoring students privately. Because most teachers are women and traditionally are responsible for taking care of their families, they have little time left to prepare for classes or indulge in professional development.

To address this problem, the program set up a list of Web sites that categorized online resources by grade, class subject, and theme according to national standards. This navigation tool saves teachers time in class preparation and provides needed resources to help replace frequently outdated textbooks. To attract greater interest, the program established a monthly online lesson plan contest with a cash reward. Within a year, the number of entrants expanded from 15 to 50.

Despite increased teacher involvement, the program was unable at first to achieve curriculum reform. The Uzbek educational system focuses primarily on rote memorization rather than skill building and critical thinking as foundations for learning. As a result, teachers in the program initially used the Internet to supply students with large amounts of text to read and commit to memory, without verifying the credibility of the information source. Through message boards, instant messaging, and cooperative projects with schools in the United States, teachers and students are slowly beginning to discover alternative learning methods.

The Uzbekistan government has accepted the GCE program and its objectives. Although the literacy rate is reportedly 97 percent, the educational system is in financial crisis; enrollment is dropping, and schools are unable to provide supplies or even money to pay for heating the classrooms. In the 18 months that the program has been up and running, the 60 participating schools have reported increased teacher, student, and parent involvement and decreases in absenteeism. The program provides students in poorer nations with digital skills and exposure and gives a big push forward to the schools themselves.

Despite its success, the GCE and other programs like it may be in jeopardy in Uzbekistan. The government brutally repressed protests in the city of Andijon in May 2005, resulting in more than 500 deaths, and then rejected a U.S. request to hold an inquiry into the incident, responding instead by evicting the United States from its air base. If Uzbekistan philosophically moves closer to Russia and China, efforts to overcome the digital divide may be stalled or thwarted altogether.

Questions:

1. Imagine that you have been appointed to lead the GCE program in Uzbekistan for the next three years. Identify at least six specific goals you would set for the program.

2. Explain how meeting each of these specific goals would improve the standard of living in Uzbekistan.

3. Using the Uzbekistan case as an example, describe the types of obstacles—other than economic—that schools face in overcoming the global digital divide.

2. Does IT Investment Pay Off?

Ford Motor Company recently announced that it had successfully deployed a new wireless system in its Dearborn-based F-150 truck plant. The system allows the company to increase labor productivity by moving from "push-based" scheduling, in which all parts are delivered to stations along the assembly line at a given time, to "pull-based" scheduling, in which parts arrive just in time for assembly. Material operators simply push a button when parts start to run low and the system indicates what parts need to be delivered where. This new system is one of many IT innovations intended to increase labor productivity.

Labor productivity has increased remarkably since the dawn of the computer age, as shown in Table 9-1 earlier in this chapter. MIT economist Erik Brynjolfsson has reported a strong correlation between IT capital per worker and a company's productivity. He says there is a growing consensus that IT is the most important factor in increasing productivity. Considering the impact that technology has had on the workplace, this correlation seems obvious.

Or is it? What about the cost of training Ford workers to use the new wireless system? What about the costs incurred when the hardware breaks down? What if there is a bug in the software? What if a new system makes the current one obsolete in two or three years? What if just-in-time delivery was a bad management decision and doesn't improve productivity at all?

Paul Strassmann, a former CIO of both Xerox and NASA, disagrees with Brynjolfsson and argues that IT has not improved labor productivity at all. Back in the 1990s, Strassmann developed a method to measure increased productivity based on microeconomics instead of GDP and other large-scale measurements. Strassmann argues that productivity should be taken as the ratio of the cost of goods to transaction costs, which includes administrative and other costs not directly related to the production or delivery of a good or service. Strassmann and others at Alinean LLC have found that this ratio has remained constant for the past 10 years.

In response, macroeconomists argue that Strassmann's calculations do not account for customer-added value, such as timely delivery or the creation of innovative products and services. In addition, they say technology has allowed companies to meet regulatory standards and reporting requirements established by government agencies.

Strassmann and other proponents of the microeconomic approach don't deny this oversight and don't argue against IT investment. No company can remain competitive and fail to innovate. Specifically, Strassmann and his colleagues are hoping to do away with short "build and junk" cycles, in which new information systems are scrapped before ever breaking even, and avoid ill-conceived IT investments. They want to develop microeconomic parameters that CIOs can take to their board of directors as proof they are cutting costs. Thus, while economists still debate the impact of IT investment, the discussions are at least producing tools that help businesses improve their IT decision making and perhaps even increase their productivity.

Questions:

1. Apart from the annual rate of output per worker, what are other ways of measuring labor productivity?

2. What factors determine whether a new information system will increase or decrease labor productivity?

3. Why is it so difficult to determine whether IT has increased labor productivity?

3. Telemedicine Moving into the Mainstream

One day in June 2003, a woman came to Dr. Michael Orms' office in Kotzebue, a tribal village of about 3000 people in northwest Alaska. The woman was hemorrhaging badly as a result of an ectopic pregnancy. Her condition was deteriorating rapidly.

"Due to heavy fog, there was no way we could Medevac her to a larger facility with an operating room," explained Orms, a family practitioner. "There was a high possibility that she would have died even in transport to Anchorage." There was only one option. Without an operating room, general anesthesia, or years of training as a surgeon, Orms would have to operate.

Orms called Dr. Daniel Szekely, a specialist in Anchorage, and then brought his Polycom videoconference equipment into the patient's room while the staff of the Alaska Federal Health Care Access Network (AFHCAN) set up a monitor for Szekely. Observing remotely, Szekely guided Orms step by step through a laparotomy using only local anesthesia. The woman's life was saved.

Since 1994, AFHCAN has worked with other federal agencies to improve healthcare in Alaska through telemedicine. Alaska has twice the area of the next largest U.S. state, extreme weather conditions, and a road system that does not reach many rural communities. In fact, a small plane or boat is the only way to reach more than 200 isolated villages. In those areas, telemedicine is the only way to provide many healthcare services that other Americans take for granted.

Many obstacles still delay the delivery of telemedicine to other rural areas throughout the United States. For example, while more than half the U.S. population has access to high-speed Internet connections, cable and DSL access in rural areas is considerably lower. In addition, high-speed access costs will have to drop before they are economically feasible for everyone.

In the meantime, another branch of healthcare is rushing to adopt telemedicine technologies: virtual treatment of the elderly and chronically ill is on the rise. New devices such as cell phones that test blood sugar levels and wristwatches that record heartbeats allow doctors and nurses to monitor their patients remotely on a daily basis. A concerned patient can arrange to speak to a public health nurse via phone through the computer monitor. Several studies have shown that this proactive service not only alleviates patient anxiety but reduces emergency room visits and hospital stays significantly. Translated into financial terms, these types of telemedicine services could save the government billions of dollars.

In 2006, the Veterans Administration plans to double the number of patients in its telemedicine program to 20,000. A broadband system that monitors intensive care units has been adopted by more than 150 hospitals in more than 22 states. Seven states have passed laws requiring health insurance companies to cover the costs of telemedicine.

Critics cite the limitations of telemedicine: doctors may not pick up on the subtle indicators of illness during a virtual visit, hardware can break, and software can have bugs. Yet, as telemedicine moves out of the realm of nonprofit and government-sponsored programs and into the mainstream, it clearly has a role to play in improving the quality of healthcare in this country and reducing its cost.

Questions:

1. Can you provide examples that either refute or confirm the idea that a gap exists between the healthcare services available to the wealthy and the poor in the United States?

2. Should healthcare organizations make major investments in telemedicine to provide improved services to rural areas, even if the cost of these investments cannot be justified and results in increased health costs?

3. What are the drawbacks of telemedicine? What situations might not lend themselves to telemedicine solutions?

End Notes

1 May, Thornton A., "Health Care Needs 'IT Value Docs'," *Computerworld*, February 14, 2005.

2 "Facts on the Cost of Health Care," National Coalition on Health Care Web site, www.nchc.org/facts/cost.shtml.

3 McGee, Marianne Kolbasuk, "Medicare Center Tests Telemedicine In Treating Chronic Illnesses," *InformationWeek*, www.informationweek.com/shared/printableArticleSrc.jhtml?articleID=165600662, July 6, 2005.

4 The Columbia Encyclopedia, Sixth ed., 2001–2005.

5 The Columbia Encyclopedia, Sixth ed., 2001–2005.

6 EH.net Encyclopedia, www.eh.net/encyclopedia.

7 Nasar, Sylvia, "Productivity," *The Concise Encyclopedia of Economics*, www.econlib.org/library/Enc/Productivity.html.

8 McKinsey Global Institute, "Whatever Happened to the New Economy?" November 2002.

9 "Industrial Metamorphosis," *The Economist*, October 1, 2005.

10 Brynjolfsson, E., and Hitt, L., "Computing Productivity: Firm-Level Evidence," Sloan Working paper no. 4210-01, eBusiness@MIT working paper No. 139, June 2003.

11 DiCecio, Riccardo, "Cross-Country Productivity Growth," International Economic Trends, November 2005.

12 DiCecio, Riccardo, "Cross-Country Productivity Growth," International Economic Trends, November 2005.

13 DiCecio, Riccardo, "Cross-Country Productivity Growth," International Economic Trends, November 2005.

14 Charry, Jonathan M., "Innovation, Population and the Digital Divide: Some Thoughts on the Information Age," *Computerworld*, www.computerworld.com, October 3, 2005.

15 Associated Press, "EU Survey: Age, Income, Education Are Factors in Internet Use," *InformationWeek*, www.informationweek.com, November 10, 2005.

16 Bagla, Gunjan, "Bringing IT to Rural India One Village at a Time," *CIO*, www.cio.com, May 1, 2005.

17 Blanchard, Jean-Marie, and Marine, Souheil, "Bridging the Digital Divide: An Opportunity for Growth for the 21st Century," Alcatel Web site, www.alcatel.com, August 31, 2004.

18 "Report Documents Millions Wasted by E-Rate, Offers Reform Recommendations," Web site for the Committee on Energy and Commerce, energycommerce.house.gov, October 10, 2005.

19 "National Education Technology Plan," www.nationaledtechplan.org.

20 "ESEA: Educational Technology State Grants—2005," www.ed.gov.

21 Charry, Jonathan M., "Innovation, Population and the Digital Divide: Some Thoughts on the Information Age," *Computerworld*, www.computerworld.com, October 3, 2005.

22 Editorial, "Laptop to Bridge 'Digital Divide'," *The Enquirer*, page B10, November 19, 2005.

23 Media Lab Asia—About Us, Media Lab Asia Web site, www.medialabasia.org, December 10, 2005.

[24] Gralla, Mia, "Broadband Access Could Slash Healthcare Costs by Over $1 Trillion," *InformationWeek*, www.informationweek.com, December 9, 2005.

[25] "Health Insurance Cost," National Coalition on Health Care Web site, www.nchc.org/facts/cost.shtml.

[26] "Health Insurance Cost," National Coalition on Health Care Web site, www.nchc.org/facts/cost.shtml.

[27] Pugh, Tony, "Spending on Health Care Might Double in 10 Years," *The Cincinnati Enquirer*, page A4, February 22, 2006.

[28] McGee, Marianne Kolbasuk, "No Quick Cure for Health-Care System," *InformationWeek*, www.informationweek.com, October 24, 2005.

[29] Nichols, L. M., "Can Defined Contribution Health Insurance Reduce Cost Growth?" Employee Benefit Research Institute, EBRI issue brief No. 246, June 2002.

[30] Ginsburg, Paul B., "Controlling Health Care Costs", *The New England Journal of Medicine*, Volume 351, No. 16, 1591–1593, www.nejm.com, October 14, 2004.

[31] May, Thornton A., "Health Care Needs 'IT Value Docs'," *Computerworld*, February 14, 2005.

[32] Gilhooly, Kym, "Rx for Better Health Care," *Computerworld*, January 31, 2005.

[33] Reuters, "U.S. Pushes Digital Medical Records," *The Boston Globe*, July 22, 2004.

[34] "EHR," Healthcare Information and Management Systems Society Web site, www.himss.org, December 18, 2005.

[35] Reuters, "U.S. Pushes Digital Medical Records," *The Boston Globe*, July 22, 2004.

[36] Gilhooly, Kym, "Rx for Better Health Care," *Computerworld*, January 31, 2005.

[37] Gilhooly, Kym, "Rx for Better Health Care," *Computerworld*, January 31, 2005.

[38] Gilhooly, Kym, "Rx for Better Health Care," *Computerworld*, January 31, 2005.

[39] Halamka, John D., "Health Care Needs a DNS for Patients," *Computerworld*, May 30, 2005.

[40] Gilhooly, Kym, "Rx for Better Health Care," *Computerworld*, January 31, 2005.

[41] Halamka, John D., "Health Care Needs a DNS for Patients," *Computerworld*, May 30, 2005.

[42] Gilhooly, Kym, "Rx for Better Health Care," *Computerworld*, January 31, 2005.

[43] Havenstein, Heather, "Wireless Leaders & Laggards: Health Care," *Computerworld*, May 16, 2005.

[44] McGee, Marianne Kolbasuk, "Medicare Center Tests Telemedicine In Treating Chronic Illnesses," *InformationWeek*, www.informationweek.com, July 6, 2005.

[45] McGee, Marianne Kolbasuk, "On Call: Telemedicine Put to Work in the Emergency Room," *InformationWeek*, www.informationweek.com, May 17, 2004.

[46] McGee, Marianne Kolbasuk, "WellPoint Offers New Online Tools to Help Consumers Make Smarter Health-Care Decisions," *InformationWeek*, www.informationweek.com, September 21, 2005.

[47] McGee, Marianne Kolbasuk, "WellPoint Offers New Online Tools to Help Consumers Make Smarter Health-Care Decisions," *InformationWeek*, www.informationweek.com, September 21, 2005.

Sources for Case 1

"At-a-glance: Global Digital Divides," BBC News, news.bbc.co.uk/1/shared/spl/hi/pop_ups/05/technology_global_digital_divides/html/1.stm.

"Global Connections and Exchange - Uzbekistan," International Research and Exchanges Board Web site, www.irex.org/programs/connectivity/index.asp.

Katz, Ari, "Technology & Expression in Uzbekistan," Digital Divide Network, www.digitaldivide.net/articles/view.php?ArticleID=414, July 8, 2005.

Wright, Robin and Tyson, Ann Scott, "U.S. Evicted From Air Base In Uzbekistan," *Washington Post*, www.washingtonpost.com/wp-dyn/content/article/2005/07/29/AR2005072902038.html, July 30, 2005.

Sources for Case 2

Brynjolfsson, Erik, "The IT Productivity Gap," *Optimize Magazine*, ebusiness.mit.edu/erik/Optimize/pr_roi.html, July 2003.

Dignan, Larry, "For Great IT, Focus on the Information, Not the Technology," *Baseline Magazine*, www.baselinemag.com/article2/0,1540,1872510,00.asp, October 19, 2005.

Hoffman, Thomas, "IT Productivity Debate: The Forest vs. the Trees," *Computerworld*, www.computerworld.com/printthis/2004/0,4814,88884,00.html, January 12, 2004.

Hoffman, Thomas, "Sidebar: IT Productivity: The SG&A Debate," *Computerworld*, www.computerworld.com/managementtopics/roi/story/0,10801,88823,00.html?f=x257, January 12, 2004.

Sources for Case 3

The Alaska Federal Health Care Access Network Web site, www.afhcan.org.

"Family Practitioner, Telemedicine Help Save Kotzebue Woman's Life," *The Mukluk Telegraphy*, www.anthc.org/abt/News/archive/upload/Sept_Oct03.pdf#search='Northern%20Lights%20telemedicine', September/October 2004.

Choi, Candice, "Virtual Nurse Visits Grow as Health Care Costs Rise," *New York Newsday*, www.nynewsday.com/news/local/wire/newyork/ny-bc-ny--videonurse1231dec31,0,5845612.story?coll=ny-region-apnewyork, December 31, 2005.

Wenske, Paul, "Doctor's Office Visit with Online Twist," *The Kansas City Star*, www.kansascity.com/mld/kansascity/news/local/13536204.htm, January 3, 2006.

APPENDIX

A BRIEF INTRODUCTION TO MORALITY

By Clancy Martin, Assistant Professor of Philosophy, University of Missouri—Kansas City

INTRODUCTION

This appendix offers a quick survey of various attempts by Western civilization to make sense of the ethical question "What is the good?" As you will recall from Chapter 1, *ethics* is "the discipline dealing with what is good and bad and with moral duty and obligation." How should we live our lives? How should we act? Which goals are worth pursuing and which are not? What do we owe to ourselves and to others? These are all ethical questions.

The answers to these questions are provided in what we call *moralities* or *moral codes*. The Judeo-Christian morality, for example, attempts to tell us how we should live our lives, the difference between right and wrong, how we ought to act toward others, and so on. If you ask a question like "Is it wrong to lie?", the Judeo-Christian morality has a ready answer: "Yes, it is wrong to lie; it is right to tell the truth." Speaking loosely, we could also say that, according to Judeo-Christian morality, it is *immoral* to lie and *moral* to tell the truth.

Moralities, or moral codes, differ by time and place. According to some people—8th-century BC Greeks, for example—it is not always wrong to lie, and it is not always right to tell the truth. So we are confronted with the *ethical* problem of choosing between different *moralities*. Some moralities may be better than others. It may even be true—as many thinkers have argued—that only *one* system of morality is ultimately acceptable. Thinking about ethics means thinking about the strengths and weaknesses of moralities, understanding why we might endorse one morality and reject another, and searching for better systems of morality or even "the best" morality. Especially in our own day, when globalization and accelerating advances in communication have created a cultural blending (and cultural conflicts) like never before, our ability to understand different moralities is crucial.

This appendix introduces you to the way various Western philosophers have answered the ethical question "What is the good?" Because the Western tradition is complicated enough, we have not addressed Eastern moralities and the ethical thinking of many fascinating Eastern philosophers. One of the interesting things about studying ethics is the enormous variety of moralities that humans have created and the many similarities between competing moralities. Unlike the rest of your textbook, this appendix is not specifically focused on the ethical problems created by technology. But as you read through the various moralities in the appendix, ask yourself how you would deal with the moral dilemmas you have studied and confronted in your own life.

THE KNOTTY QUESTION OF GOODNESS

Achilles kills Hector outside the gates of Troy. He binds Hector's corpse by the ankles, ties the ankles to the back of his chariot, and drags the body around the city walls. The treatment of the fallen Trojan hero by his victorious Greek enemy is so outrageous that not only Trojans, but most of Achilles' Greek allies and even the Gods, are shocked. But what is wrong with Achilles' action?

To an ancient Greek of the time, the answer would not have been obvious. When the poet **Homer (8th century BC)** tells this story in his epic *The Iliad*, his purpose is to illustrate a failure in the morality of his own day. Among Greeks of Homer's day, the prevailing moral code was: "Help to friends and harm to enemies." That code may sound naïve or ridiculously simplistic today. But for the collection of small and largely independent city-states that was ancient Greece, it was a moral code that had worked reasonably well for centuries. Yet Homer saw that different times were on the way. When the Greeks banded together, as they did to combat the Trojans, the old morality looked barbaric. There was nothing heroic about the lone Achilles dragging his vanquished enemy behind him. On the contrary, he seemed like a savage.

When a society is passing from an old moral code to a new one, or when two different cultures clash in their moral codes, the extraordinarily difficult question of which moral code is correct inevitably appears. *Ethics*, the systematic study of moral codes, is the attempt to answer that question. Almost every philosopher and most thinking people will agree that some moral codes are better than others; many philosophers and others will argue that a particular moral code is the best.

Perhaps the most famous philosopher of all time, **Socrates (470–399 BC)**, argued that there was only one true moral code, and it was simple: "No person should ever willingly do evil." Socrates thought that no harm could come to a person who always sought the good, because what truly counted in life was the caretaking of one's self or soul. But Socrates also acknowledged that identifying the good was rarely easy, and his method of constantly interrogating his friends and fellow citizens—what came to be called Socratic questioning, or the Socratic dialectic—tried to improve everyone's thinking about what one ought and ought not do.

Socrates never wrote down any of his philosophy. But his student **Plato (427–347 BC)** made Socrates the hero of almost all of his many philosophical dialogues. Plato was the first "professional" philosopher in the West: he established a school of philosophy called the Academy (where we get the word *academic*), published a great number of books both for general readers and his own students, and formed arguments on virtually every subject in philosophy (not only morality). In fact, Plato possessed such breadth that the 20th-century philosopher Lord Alfred North Whitehead wrote that "all subsequent philosophy is only a footnote to Plato."

In many of his dialogues Plato raises the question: "What is the good?" Like Homer (who was one of Plato's favorite writers), Plato lived in a time when great political, social, and cultural changes were occurring. Athens had lost the first major war in its history, trade was accelerating across the Mediterranean, and people were traveling deeper into Asia and Africa and discovering new cultures, religions, and values. Many candidates for "the good" were being offered by different thinkers: some thought that "pleasure" was the highest good, others argued that "peace" (both personal and social) and what contributed to it was the best, others argued for "flourishing" and material wealth and power, while still others endorsed "honor and fame." But Plato responded that, while all of these things might be examples of goodness, they were not good itself. What is it that makes them good? What is the nature of the property "goodness" that they all share? And because we recognize that most "goods" may also mislead us into badness—the good of pleasure is an obvious example—how shall we sort the good from the bad?

Plato's idea is that we cannot reliably say what is good and what is not until we know what goodness is. Once we have identified goodness itself, we can discriminate among particular goods and particular activities that are designed to seek the good. We will judge what is "good" and "better" by comparing it with what is "best": the truly and wholly good. And the truly and wholly good ought always and everywhere to be good. Could we say that something was truly, wholly good if it was good only in some countries and not others, during some times and not others? So, if we can identify goodness as such, Plato said, we can solve every problem posed by the clash between good and bad; that is, we can solve every problem of morality.

One way to think about Plato's insight is to see the moral importance of *standards*. We have standards for good hamburgers, for good businesses, and for good hammers, so why not have standards for good people and good actions? A standard is one way of providing a *justification* for an evaluation. Suppose Rebecca insists, "It is always wrong to kill an innocent human being." And Thomas replies, "But why?" Rebecca may justify her evaluation by appealing to a standard of rightness and wrongness. Of course, identifying that standard may prove more difficult than appealing to it, and the history of ethics, again, may be seen as the struggle to provide such a standard. The philosophers you will read about in the following sections attempted to answer Plato's knotty questions in their own ways.

RELATIVISM: WHY "COMMON SENSE" WON'T WORK

What about simply using common sense to find the good? Some 20th-century philosophers argued for what they called moral "intuitions": a kind of "consult your conscience" approach to morality. This view is initially compelling for most people; it holds that the standard for goodness demanded by Plato is accessible to all of us if we simply think through our moral decisions carefully enough. (Socrates may have been arguing for the same view.) There is a "voice" in our heads that tells us what is morally right and wrong, and if you honestly and thoroughly interrogate yourself about what you ought to do, that "voice" will praise the right action and warn you against the wrong one. Someone who says, "Do the right thing!" is invoking this common-sense notion. We all know what the right thing is, a moral intuitionist argues, if we use our common sense and are tough on ourselves. The difficulty is that we don't always want to use common sense or ask ourselves tough questions.

A Brief Introduction to Morality

Therefore, the problem of right and wrong is not so much that of moral knowledge as it is weakness of will. We *know* what we ought to do, but it is hard to make ourselves *do* it.

A crippling difficulty with this view is called the problem of relativism. *Cultural relativism* is the simple observation that different cultures employ different norms (or standards). Implicit in this view is that it is morally legitimate for different cultures to create and embrace different norms. So, for example, among the Greeks of Homer's day, lying was considered to be a virtue. Odysseus was praised specifically for his ability to lie well. In 18th-century Germany, on the other hand, lying was widely considered as morally reprehensible as theft. Some philosophers even argued that lying was just as morally foul as murder. For the relativist, lying is neither right nor wrong; rather, it can be right at a certain time and place and wrong in another. Another example is bribery. Although people in many nations condemn bribery, it is perfectly acceptable in other countries, particularly in Latin America. The relativist would say: "Bribery itself is not right or wrong. Rather, some people at some times and in some places say it is wrong, and other people say it is right, depending on the circumstances. Bribery is therefore wrong for some people, right for others."

You have probably encountered this relativism with something as simple as e-mail. The conventions that govern e-mail etiquette vary dramatically from user to user, group to group, and culture to culture. The emoticon-laced e-mail you send to a friend would be wholly inappropriate if sent to a professor. The kind of language you use in an e-mail to a college admissions officer is not what you would use to e-mail your parents or an e-mail pal in India. A practical platitude that embodies this idea is: "When in Rome, do as Romans." What is *appropriate* and what counts as a "good" e-mail (as opposed to a "bad" or offensive e-mail) depends on the conventions within its cultural context. Even e-mails have *norms*.

Moral relativists argue that all norms and values are relative to the cultures in which they are created and expressed. For the moral relativist, it makes no sense to say that there are any transcultural or transhistorical values, and that any attempt to construct them would still be informed by the particular cultural values of a person or group. All you can talk about are the values "on the ground": the values that particular cultures embrace. And common sense may be one of the best tools for discovering those values. Common sense may be the psychological embodiment of the complex structure of rules, standards, and values that are the substance of every robust culture.

But moral relativists run into trouble, because there are some moral claims they cannot consistently make. Moral relativists can say "slavery is wrong in my society" or "slavery is wrong in the 20th century," but they cannot say that slavery is always wrong. Furthermore, because they cannot appeal to transcultural standards for morality, they cannot speak of *moral progress*. Moral values (like all other values) change over time for the relativist, but they do not improve or degenerate. Yet, most of us would agree that the growing worldwide prohibition against slavery and torture, for example, is not merely a change, it is moral progress. And if we believe in moral progress, we cannot be relativists.

Egoism vs. Altruism

Throughout this book we have seen that ethics deals with the question of how we should treat one another. But some thinkers would say we have already misconstrued the question when we ask "How should we treat others?" For an *egoist*, the salient moral question is "How do I best benefit myself?", and the answer to Plato's question "What is the good?" is simply "The good is whatever is pleasing to *me*."

Egoism is usually divided into two types. *Psychological egoism* is the thesis that people always act from selfish motives, whether they should or not. *Ethical egoism* is the more controversial thesis that, whether people always act from selfish motives, they should if they want to be moral.

There is a superficial plausibility to psychological egoism, because it might appear that most of us make many of our choices for self-interested reasons. You probably decided that you wanted to go to college rather than immediately finding a job. You might respond: "No, I went to college because my parents wanted me to!" But the psychological egoist would reply: "That simply means that, for you, pleasing your parents is more important than other things that would have kept you out of college."

However, some of the problems with psychological egoism already are glaringly apparent. First, though we may make many decisions based on our own interests, it is far from obvious that *all* of our decisions are motivated by self-interest. We make many decisions, including decidedly uncomfortable ones, because we are thinking of the interests of others. It is silly to suppose that our own interests must always and implicitly conflict with those of others, as a psychological egoist believes. Why did you go to college? Because you wanted to, and your parents, teachers, and friends wanted you to. Everyone's interests happily coincided, and it is oversimplifying your complex choice to say, as a psychological egoist would, "I did it because *I* wanted to."

While considering ethical egoism, we should also look at its opposite: *altruism*. The altruist argues that the morally correct action always best serves the interest of others. Wouldn't the world be a better place, the altruist asks, if we worried about ourselves less and tried to help other people?

No one will deny that everyone benefits from altruism, but problems arise if we try to adopt altruism as a moral code. Practically speaking, it is sometimes difficult to know what best serves the interest of another, beyond helping people with the basic necessities of life. For example, a devout Southern Baptist might sincerely believe that his neighbors are condemned to hell unless they accept his religious views, and might feel an altruistic urge to convert them, despite their hesitation. Another more famous example involves a boat full of altruists lost at sea. They can only survive if one of them volunteers to be eaten, but if the only moral action is to serve the interests of others, how can any of the adrift altruists be truly moral when one of them has to die to save the rest?

Problems like these help to motivate advocates of ethical egoism. We do not reliably know the interests of others, the ethical egoist says, but we certainly know our own. And, unlike altruists, whose satisfaction is in helping others, ethical egoists try to create a happy and moral world by seeking good for themselves. The hacker who thinks she can morally break the rules because she has the smarts to do so is both a psychological egoist ("you would break the rules too, if you could") and an ethical egoist ("everyone who can break the rules to help themselves should do so"). Given the choice between self-interest and altruism, the ethical egoist takes the former.

A Brief Introduction to Morality

Of course, the only choice is not between ethical egoism and altruism. Most moral codes and most people recognize the importance of both self-interest and the interest of others. The more telling objection to ethical egoism is that it does not respect our deepest intuitions about moral goodness. If an ethical egoist can serve his own interest by performing some horrific act against another human being, and be guaranteed that the act will not interfere with his self-interest, he is morally permitted to perform that act. In fact, if he finds that he can *only* serve his interest by performing the horrific act and getting away with it, he is morally *required* to do so. An employer who could benefit from spying on her employee's e-mail would be morally required to do it if it served her long-term interest. But for most of us, such examples are sufficient to defeat ethical egoism. Moral codes are plausible only if they accommodate basic intuitions about our sense of right and wrong, and ethical egoism fails on that ground.

DEONTOLOGY, OR THE ETHICS OF LOGICAL CONSISTENCY AND DUTY

Most people find they cannot accept relativism as a moral code because of their moral intuitions that some things are *always* wrong (like slavery or the torture of innocents). For this reason, they must also abandon a "common sense" approach to morality, which relies on embedded knowledge of cultural norms. The problems with egoism and altruism are even more glaring. But don't despair—there are lots more moral theories to consider. The rest of this appendix reviews several modern attempts to articulate a consistent morality.

Immanuel Kant (1724–1804) is generally considered the most important philosopher since Aristotle. Kant's moral theory is an attempt to refine and provide a sound philosophical foundation for the strict Judeo-Christian morality of his own day. Most people, when they begin thinking about ethics in a philosophical way, find that they are some brand of Kantian. Kant's theory is called *deontology*, from the Greek word *deon*, meaning *duty*. For Kant, to do what is morally right is to do one's duty.

Understanding what one's duty requires is the difficult part, of course. Kant begins with the idea that the only thing in the world that is wholly good, without any qualification, is good will. Most good things may be turned to evil or undesirable ends, or are mixed with bad qualities. Human beings do not seem wholly good: they are a mix of good and bad. Money is a good that most of us seek, while "love of money is the root of all evil." But the will to do good—the desire or intention—must be wholly good. If we think through what we mean by "moral goodness," Kant argues, we realize that the notion of moral goodness is just another name for this will to goodness. Kant recognizes that, as the old saying goes, "the road to Hell is paved with good intentions"; he is not saying that good will must always have good consequences. (In general, Kant is suspicious of the moral worth of consequences.) But the intention to do good, before it gets tangled up in the difficulties of the world, must itself be purely good.

Morality, therefore, comes from our ability to intend that certain things happen: that is, from our ability to choose. The good choice will come from a good will. But how do we sort the good choice from the bad? Kant, following the ancient Greek philosopher Aristotle, believed that the property that makes human beings unique, and that propels us into the moral sphere, is the faculty of *reason*. Kant saw human beings as constantly torn between their passions, drives, and desires (what he called "inclinations") and the rational ability to make good choices on the basis of good and defensible reasons. For Kant, with his dim view of human nature, what we *want* to do is very rarely what we *ought* to do. But we can recognize what we ought to do by the application of reason.

Kant's derivation of the *categorical imperative*, which he argued is the fundamental principle of all morality, is notoriously complex. But the key idea is simple: reason demands consistency and rejects contradiction. Accordingly, Kant argued that the moral principle we should follow must preserve consistency in all cases and prevent any possibility of contradiction. This moral principle might be expressed as: "Act only on that maxim such that the maxim of your action can be willed to be a universal law." (Although Kant offered several different formulations of the categorical imperative, this is the most famous and most basic formulation.) Kant's prose is dense and confusing, and the categorical imperative is no exception. What does Kant mean?

Kant observed that we make choices according to rules. We tell the truth even when it is inconvenient or embarrassing because we have a rule in our heads that tells us to do so. This is an example of what Kant calls a "subjective principle of action" or a *maxim*. Other examples of maxims are "don't steal" and "keep your promises." Our heads are full of rules that we use to guide our choices. When we worry about *moral* choices, Kant tells us in the categorical imperative that we should act only on choices that "can be willed to be a universal law." That is, before acting on a maxim that informs a moral choice, one must ask: "Could this rule (this maxim) be applied to everyone, everywhere, for all time?" Kant argues that, by *universalizing* a maxim, one can see whether it generates a contradiction. If it generates a contradiction, it cannot be rational, and so it is not a legitimate expression of a good will. If it does not generate a contradiction, it looks morally permissible. When we follow the categorical imperative, Kant thinks, we are doing our (moral) duty.

Take a couple of examples. Suppose you decide to borrow money without intending to pay it back. Your maxim might be: "If I need to borrow money I should do so, even though I know I will never pay it back." Now universalize this maxim according to the categorical imperative. Suppose everyone, everywhere, always borrowed money without the intention of paying it back? Obviously no one would lend money and the very possibility of borrowing would be eliminated. It is rationally contradictory to choose to borrow money without intending to pay it back.

Or, suppose you are caught cheating and try to lie your way out of it. Your maxim is: "When caught cheating, I should lie to get out of trouble." But suppose everyone, everywhere, always lied to get out of trouble when caught cheating? To lie you must hide the truth, and in this situation, were it universalized, it would be impossible to hide the truth. Lies depend on being exceptions to the rule of truthful communication; if lies are no longer the exception but the rule, there is no more truthful communication, and a lie becomes impossible. Again, this is a rational contradiction, and we see that the lie is immoral.

Suppose, however, that you try a maxim like "Thou shalt not kill." What if everyone, everywhere, always avoided killing others? No contradiction is generated. There may be many impractical consequences of universal not-killing, but there are no logical problems with it. If you try a maxim of "Thou shalt kill," on the other hand, you see how quickly it falls apart.

It is not difficult to generate objections to this theory. If one makes maxims specific enough, it is easy to justify apparently immoral actions while following the rule of universal maxims. For example, one can easily universalize a maxim like "a woman with no money whose children are dying of pneumonia should steal penicillin if necessary to save her children's lives," yet Kant would maintain that theft is always wrong and irrational.

Kant also maintains that it is always irrational and wrong to lie, even in the attempt to save an innocent life. But to most of us that sounds absurd. Should a mother never lie, even if it means saving the life of her child? Should the Danes who lied to the Nazis about whether they were protecting Jews have told the truth? Surely not.

Perhaps the most controversial aspect of Kant's moral theory is his distinction between moral duty and happiness. Kant argues that to choose freely on the basis of what we rationally see is right—following the categorical imperative, acting from duty—is the only way we can choose *morally*. But suppose we are acting a certain way solely because it makes us happy, even though those actions happen to agree with what would otherwise be our duty. For Kant, actions motivated by inclination (with the result of happiness) are not motivated by duty, and so we should not consider them *moral* actions. For example, a suicidal person who does not shoot herself because she recognizes that it would be irrational (and thus contrary to her duty) is acting morally. However, another person who fleetingly considers shooting himself but then declines because he loves his life is not acting morally; he is merely inclining toward his happiness.

But if moral duty and happiness are opposed, it seems that only miserable people can be moral. Wouldn't it be nicer if we could have both moral worth in our actions and happy lives? This leads us to *utilitarianism*, the theory of morality that responds specifically to deontology by insisting that morality and happiness are not opposites, but the very same thing.

HAPPY CONSEQUENCES, OR UTILITARIANISM

Hedonism is the notion, first advocated by the Greek philosopher **Epicurus (342–270 BC)**, that pleasure is the greatest good for human beings. (Epicurus is the source of the word *epicurean*.) To be moral is to live the life that produces the most pleasure and avoids pain. But we should not suppose that Epicurus was arguing for a life of debauchery. Drinking too much wine, for example, though fun while it lasts, produces more pain than pleasure in the end, so Epicurus sorted pleasures into categories:

- Natural and necessary, like sleeping and moderate eating
- Natural but unnecessary, like drinking wine or playing chess
- Unnatural and unnecessary, which hurt one's body (for example, smoking cigarettes)
- Unnatural but necessary (but there are no such pleasures)

Epicurus said that we should cultivate natural and necessary pleasures, enjoy natural but unnecessary pleasures in moderation, and avoid all other sorts. The true hedonist does not seek what is immediately pleasurable, but looks for pleasures that will guarantee a long, healthy life full of them. For this reason, *friendship* is Epicurus' favorite example of a pleasure that everyone should cultivate; friendship was consistently considered one of the highest human goods among ancient Greeks.

Jeremy Bentham (1748–1832) adopted Epicurus' basic principles when he developed the theory that later became known as *utilitarianism*. In response to Plato's question "What is the good?", Bentham argued that it is easy to see what humans consider good because they are always seeking it: pleasure. But Bentham was not an egoist, and he argued that the highest good would result from a maximum of pleasure for all people concerned in any moral decision. Decisions that promote *utility* are those that create the most pleasure (the words *utility* and *pleasure* were virtually interchangeable to Bentham, though later utilitarians would ascribe many different meanings to *utility*). Whenever making a decision, the person who desires a moral result should weigh all possible outcomes, and choose the action that produces the most pleasure for everyone concerned. Bentham called this weighing of outcomes a "utilitarian calculus."

Bentham's new moral theory enjoyed enormous popularity, but brought inevitable objections. Some philosophers argued that such a theory made people look no better than swine (because they were just pursuing pleasure). Others objected that people would surely frame their moral decisions to enable them to do whatever they pleased. **John Stuart Mill (1806–1873)** responded to these objections and gave us the form of utilitarianism that, in its fundamentals, is the same moral theory that so many philosophers and economists still endorse today.

Mill argued that the good that human beings seek is not so much pleasure as happiness, and that the basic principle of utilitarianism was what he called the "Greatest Happiness Principle": that action is good which creates the greatest happiness, and the least unhappiness, for the greatest number. He also insisted that people who used this principle must adopt a disinterested view when deciding what would create the greatest happiness. He called this the perspective of "the perfectly disinterested benevolent spectator."

When making a moral decision, then, people will consider the various outcomes and make the choice that produces the most happiness for themselves and everyone else. This is not the same as asking which choice will produce the most pleasure. Accepting a job selling computer software for $55,000 a year might produce more short-term pleasure than going to graduate school, but it might not produce the most happiness. You might be broke and hungry in graduate school, but still very happy because you are progressing toward a goal and finding intellectual stimulation along the way.

The utilitarian must also ask: does this decision produce the most happiness for everyone else, and am I evaluating their happiness fairly and reasonably? Suppose that the recent graduate is again deliberating whether to go to graduate school. Her mother and her father, both attorneys, very much want her to go into the law. But she is fed up with school and will be miserable sitting in a classroom all day. She is sick of eating Ramen noodles and having roommates, and would like to drink a nice bottle of wine once in a while and buy a new car. It is true that her parents' happiness is relevant to the decision, but she must try to weigh the happiness of everyone involved. How unhappy will her parents be if she takes a few years off? How unhappy will she be back in a lecture hall? Utilitarians admit that finding the good is not always easy, but they insist that they offer a practical method for finding the good that anyone can use to solve a moral dilemma.

Utilitarianism is a kind of *consequentialism*, because we evaluate the morality of actions on the basis of their probable outcomes or consequences. For this reason utilitarianism is also what we call a *teleological* theory. Coming from the Greek word *telos*, meaning *purpose* or *end*, teleology refers to the notion that some things and processes are best understood by considering their goals. For utilitarians the goal of life is happiness, and thus they argue that the good (moral) life for humans is the happy life.

Utilitarianism is probably the most popular moral theory of the last hundred years. It is widely used by economists, because one easy way of measuring utility is by assigning dollar signs to outcomes. Today's most famous advocate of animal rights, Peter Singer, is also a well-known utilitarian. Many different versions of utilitarianism have been advanced. In *rule-utilitarianism*, we first rationally determine the general rules that will produce good outcomes, and then follow those rules. In *preference-utilitarianism*, we solve the difficult problem of what will create the most happiness for others by simply asking every person involved for their preference.

But there are many strong objections to utilitarianism. One was raised by the German philosopher **Friedrich Nietzsche (1844–1900)** in his masterpiece *Thus Spoke Zarathustra*. At the end of the book, Zarathustra asks himself if his efforts to find the good for human beings and for himself have increased his personal happiness. He responds to himself: "Happiness? Why should I strive for happiness? I strive for my work!" Nietzsche's point is that many profound and praiseworthy human goals are unquestionably moral, and yet they cannot be said to contribute to the happiness of the person who has those goals, and perhaps not even to the happiness of the greater number. It is true that Van Gogh's paintings, though they destroyed him, created a greater happiness for the rest of us. But that did not count for him as a reason to paint them—he had no idea of his own legacy. For a utilitarian, such self-sacrifice is not only confused but immoral. And yet if our moral theory has difficulty accounting for the value of Van Gogh sacrificing his happiness and everything he loved to his art, we might be in trouble.

Perhaps the most telling objection to utilitarianism is that it could be used to morally sanction a "tyranny of the majority." Suppose you could solve all of the suffering of the world and create universal happiness by flipping a switch on a black box. But, in order to power the box, you had to place one person inside it, who would suffer unspeakably painful torture. None of us would be willing to flip that switch, and yet for a utilitarian such an action would not only be permissible, it would be morally demanded.

A related objection comes from the British philosopher Bernard Williams. Suppose you are an explorer in the Amazon basin and you stumble on a tribe that is about to slaughter 20 captured warriors from another tribe. You interrupt the gruesome execution, and the tribal chief offers to release 19 prisoners in your honor, on the condition that you accept the ceremonial role of choosing one victim and killing him yourself. A utilitarian would be morally required to accept, but most of us would be morally appalled at the idea of killing a complete stranger who presented no threat to us.

Utilitarians have responded to such objections by introducing the notion of certain irreducible human *rights* into utilitarianism. The discussion now turns to these rights and their origins in *social contract theory*.

Promises and Contracts

Although there are good reasons for being suspicious of egoism of any stamp, **Thomas Hobbes (1588–1679)** was a psychological egoist, which was essential to his moral and political theory. Hobbes argued that there are two fundamental facts about human beings: (1) we are all selfish, and (2) we can only survive by banding together. You may have heard Hobbes' famous dictum that human life outside of a society—that is, in his imagined "state of nature"—is "solitary, poor, nasty, brutish, and short." We form groups for self-interested reasons because we need one another to survive and prosper. But the fact that we band together as selfish beings inevitably results in tension between people. Because resources are always scarce, there is competition, and competition creates conflict. Accordingly, if we are to survive as a group, we need rules that everyone promises to follow. These rules, which may be simple at first but become enormously complex, are an exchange of protections for freedoms. "I promise not to punch you in the nose as long as you promise not to punch me in the nose" is precisely such an exchange. You trade the freedom to throw your fists wherever you please for the protection of not being punched yourself. These rules of mutual agreement are, of course, called *laws*, and they guarantee our protections or *rights*. The system of laws and rights that make up the society is called the *social contract*.

Social contract theory builds on the Greek notion that good people are most likely encouraged by a good society. Few social contract theorists would argue that morality can be reduced to societal laws. But most would insist that it is extremely difficult to be a good person unless you are in a good society with good laws. Hobbes argued that the habit of exchanging liberties for protections would extend itself into all dimensions of a good citizen's behavior. The law is an expression of the reciprocity expressed in the Golden Rule—"do unto others as you would have them do unto you"—and so through repeated obedience to the law we would develop the habit, Hobbes thought, of treating others as we would like to be treated.

The most famous American social contract theorist was **John Rawls (1921–2002)**. Rawls argued that "justice is fairness," and for him the morally praiseworthy society distributes its goods in a way that helps the least advantaged of its members. Rawls asked us to imagine what rules we would propose for a society if, when we thought about the rules, we imagined that we had no idea what our own role in that society would be. What rules would we want for our society if we did not know whether we would be poor or rich, African-American or Native Indian, man or woman, or a teacher, plumber, or famous actor? Rawls imagined that this thought experiment—which he called standing behind "the veil of ignorance"—would guarantee fairness in the formulation of the social contract. Existing social rules and laws that did not pass this test—that no rational person would endorse if standing behind the veil of ignorance—were obviously unfair and should be changed or discarded.

Strictly speaking, social contract theory is not a moral code. But because so many of our moral decisions are made in the context of laws and rights, we should understand that the foundation of those laws and rights is a system of promises that have been made, either implicitly or explicitly, by every citizen who freely chooses to live in and benefit from a commonwealth.

A RETURN TO THE GREEKS: THE GOOD LIFE OF VIRTUE

In the 20th century, many philosophers grew increasingly suspicious of the possibility of founding a workable moral system upon rules or principles. The problem with moral rules or principles is that they self-consciously ignore the particulars of the situations in which people actually make moral decisions. For the dominant moralities of the 20th century, deontology and utilitarianism, what is moral for one person is moral for another, regardless of the many differences that undoubtedly exist between their lives, personalities, and stations. This serious weakness in prevailing moral systems caused philosophers to turn once more to the ancient Greeks for help.

Aristotle (384–307 BC) argued that it does not make sense to speak of good actions unless one recognizes that good actions are performed by good people. But good people deliberate over their actions in particular situations, each of which may differ importantly from other situations in which a person has to make a moral choice. But what is a good person?

Aristotle would have responded to this question with his famous "function argument," which posits that the goodness of anything is expressed in its proper function. A good hammer is good because it pounds nails well. A good ship is good if it sails securely across the sea. A bad ship, on the other hand, will take on water and drift aimlessly across the waves. Moreover, we can recognize the function of a thing by identifying what makes it different from other things. The difference between a door and a curtain lies fundamentally in the way they do their jobs. Human function, the particular ability that makes mankind different from all other species, is the ability to reason. The good life is the life of the mind: to be a good person is to actively think.

But to pursue the life of the mind, we need many things. We need health; we need the protection and services of a good society; we need friends for conversation. We need leisure time and enough money to satisfy our physical needs (but not so much as to distract or worry us); we need education, books, art, music, culture, and pleasant distractions to relax the mind.

This does sound like the good *life*. But how does the thinking person *act*? Presumably, Aristotle's happy citizen will encounter moral conflicts and dilemmas like the rest of us. How do we resolve these dilemmas? What guides our choices?

Aristotle did not believe that human beings confront each choice as though it were the first they ever made. Rather, he thought, we develop habits that guide our choices. There are good habits and bad habits. Good habits contribute to our flourishing and are called *virtues* (Aristotle's word, *arête*, may also be translated as *excellence*). Bad habits diminish our happiness and are called *vices*. And happily, for Aristotle, the thinking person will see that there is a practical method for sorting between virtues and vices built into the nature of human beings. Aristotle insisted that human beings are animals, like any other warm-blooded creature on the earth; just as a tiger can act in ways that cause it to flourish or fail, so human beings have a natural guide to their betterment. This has come to be called Aristotle's "golden mean": the notion that our good lies between the extremes of the deficiency of an activity and its excess. Healthy virtue lies in moderation.

An example will help. Suppose you are sitting in the classroom with your professor and fellow students when a wild buffalo storms into the room. The buffalo is enraged and ready to gore all comers. What do you do? An excessive action would be to attack the buffalo with your bare hands: this would, for Aristotle, show the vice of rashness. A deficient action would be to cower behind your desk and shriek for help: this would show the vice of cowardice. But a moderate action would be to make a loud noise to frighten the buffalo, or perhaps to distract it so that others could make for the door, or to do whatever might reasonably reduce the danger to others and yourself. This moderate course of action exemplifies the virtue of courage. Notice, however, that the courageous course of action would change if an enraged tomcat came spitting into the room. Then the moderate and virtuous choice might be to trap the feline with a handy trash basket.

Aristotle's list of virtues includes courage, temperance, justice, liberality or generosity, magnificence (living well), pride, high-mindedness, aspiration, gentleness, truthfulness, friendliness, modesty, righteous indignation, and wittiness. But one could write many such lists, depending on one's own society and way of life. Aristotle would doubtless argue that at least some of these virtues are virtuous for any human being in any place or time, but a strength of his theory is that others' virtues depend on the when, where, and how of differing human practices and communities. One appeal of virtue ethics is that it insists on the context of our moral deliberations.

But is human goodness fully expressed by moderation? Or by being a good citizen? And what about people who lack Aristotle's material requirements of health, friends, and a little property? Aristotle is committed to the idea that such people cannot live fully moral lives, but can that be right? As powerful as it is, one weakness of Aristotle's virtue ethics is that it seems to overemphasize the importance of "fitting" into one's society. The rebel, the outcast, or the romantic chasing an iconoclastic ideal has no place. And Aristotle's theory may sanction some gross moral injustices—such as slavery—if they contribute to the flourishing of society as a whole. Aristotle himself would have had no problem with this: his theory was explicitly designed for the aristocratic way of life. But today we would insist that the good life, if it is to be truly *good* for any of us, must at least in principle be available to every member of our society.

Feminism and the Ethics of Care

Psychologist and philosopher **Carol Gilligan** discovered that moral concepts develop differently in young children. Boys tend to emphasize reasons, rules, and justifications; girls tend to emphasize relationships, the good of the group, and mutual nurturing. From these empirical studies Gilligan developed what came to be called the "ethics of care": the idea that morality might be better grounded on the kind of mutual nurturing and love that takes place in close friendships and family groups. The ethical ideal, according to Gilligan, is a good mother.

Gilligan's ethics of care is compelling because it seems to reflect how many of us make our daily moral decisions. Consider the moral decisions you face in a typical day: telling the truth or lying to a parent or sibling, skipping a party to take care of a heartsick friend or going to see that cute guy, keeping a promise to another student to copy your notes or saying "oops, I forgot." We often confront the moral difficulties of being a good son, sister, friend, or colleague. Generally speaking, we do not settle these moral issues on the basis of impersonal moral principles—we wonder whether it would even be appropriate to do so, given that we are personally involved in these decisions. Should you treat your best friend in precisely the same way you treat a stranger on the street? Some moralists would say, "Of course!" Yet, many of us would consider such behavior odd or psychologically impossible.

The feminist attack on traditional ethics does not accuse one Western morality or another, but indicts its whole history. Western morality has insisted on rationality at the expense of emotions, on impartiality at the expense of relationships, on punishment at the expense of forgiveness, and on "universal principles" at the expense of real, concrete moral problems. In a phrase, morality has been male at the expense of the female. Thus, the feminist argues, a radical rethinking of the entire history of morality is necessary.

As a negative attack on traditional morality, it is hard to disagree with feminism. Our moral tradition does have a suspiciously masculine cast; it is not surprising that virtually every philosopher mentioned in this appendix was a man. But feminism has struggled to develop a positive ethics of its own. Many consider Gilligan's ethics of care to be the best attempt so far, and it works well in family contexts. But when we try to extend the ethics of care into larger spheres, we run into trouble. Gilligan insists on the moral urgency of partiality (as a mother is partial to her children, and even among children). But you would object if you were a defendant in a lawsuit and saw the plaintiff enter, wave genially to the judge, and say, "Hi Mom!" The point, of course, is that in many situations we insist on *impartiality*, and for good reasons. And we all agree that people we have never met may still exercise moral demands upon us. We believe that a man rotting in prison on the other side of the world ought not be tortured, and maybe that we should do something about it if he is (if only by donating money to Amnesty International). Everyone deserves protection from torture for reasons that apply equally to all of us.

PLURALISM

When the German philosopher Nietzsche famously proclaimed that "God is dead," he was not proposing that the nature of the universe had changed. Rather, he was proposing that a change had taken place in the way we view ourselves in the universe. He meant that the Judeo-Christian tradition that has informed all of our values in the West can no longer do the job for us that it used to do. Part of that tradition, Nietzsche thought, was the unfortunate Platonic idea that there is an answer to the question "What is the good?" There is no more one "good" than there is one "God" or one "truth": there are, Nietzsche insisted, many goods, like there are many truths. Nietzsche argued the moral position that we now call *pluralism*.

Pluralism is the idea that there are many goods and many sources of value. Pluralism is explicitly opposed to Plato's insistence that all good things and actions must share some quality that makes them all good. But does this make the pluralist a relativist? No, because the pluralist argues for the moral significance of two ideas that the relativist rejects: (1) that some aspects of human nature are transcultural and transhistorical, and (2) that some methods of inquiry reveal transcultural and transhistorical human values.

When we look at human history, we see goods that repeatedly contribute to human flourishing and evils that interfere with it. War is almost always viewed as an evil in history that has consistently interfered with human flourishing; health, on the other hand, is almost always viewed as a good (with the exception of aberrant religious practices like asceticism). "Avoid war and seek health" is not a moral code—although it might go further than we think—but it does provide an example of what a pluralist is looking for. The pluralist wants concrete goods and practices that actually enrich human life. For the pluralist, the choice between Plato's absolutism and moral relativism is a false dichotomy. Just because there is no absolute "good" does not mean that all goods or values are relative to the time, place, and culture in which we find them. Some things and practices are usually bad for humans, others are usually good, and the discovery and encouragement of the good things and practices is the game the smart ethicist plays.

For this reason, pluralists emphasize the importance of investigating and questioning. Is our present culture enhancing or diminishing us as human beings? Is the American attitude toward sexuality, say, improving the human condition or interfering with it? (And before we can answer that question, what *is* the American attitude toward sexuality? Or are there many attitudes?) The ethical contribution to the history of philosophy made by the fascinating 20th-century movement called *existentialism* is its insistence on this kind of vigorous, ruthlessly honest interrogation of oneself and one's culture. The danger of hypocrisy and self-deception, or what the leading existentialist **Jean-Paul Sartre (1912–1984)** called *bad faith*, is rampant in every culture: challenging our values is uncomfortable. It is much easier for us, like the subjects of the nude ruler in H. C. Andersen's fable *The Emperor's New Clothes*, to collectively pretend that something is good (even if we know there is really nothing there at all). Thus, the project of becoming a good person becomes not just a matter of following the rules, doing one's duty, seeking happiness, becoming virtuous, or caring for others. It is also the lifelong project of discovering *if, when, and why* the apparently good things we seek are what we ought to pursue.

SUMMARY

After reading this appendix, a reasonable student might ask: "But which of these moralities is the *right* one?" Admittedly, philosophers are better at posing problems than solving them. But the lesson was not in demonstrating that one or another morality is the one a person ought to follow. Rather, this appendix has attempted to show you how different people have struggled with the enormously difficult questions of ethics. Many people think they simply know the difference between right and wrong, or unreflectively accept the definitions of right and wrong offered by their parents, churches, communities, or societies. This appendix tried to show that there is nothing simple about ethics. To understand ethics means to think, to challenge, to question, and to reflect. Accordingly, being a good person might mean attempting your own struggle with, and attempting to find your own answer to, what we called Plato's knotty question of goodness.

APPENDIX **B**

ASSOCIATION FOR COMPUTING MACHINERY (ACM) CODE OF ETHICS AND PROFESSIONAL CONDUCT

Adopted by ACM Council 10/16/92

Preamble

Commitment to ethical professional conduct is expected of every member (voting members, associate members, and student members) of the Association for Computing Machinery (ACM).

This Code, consisting of 24 imperatives formulated as statements of personal responsibility, identifies the elements of such a commitment. It contains many, but not all, issues professionals are likely to face. Section 1 outlines fundamental ethical considerations, while Section 2 addresses additional, more specific considerations of professional conduct. Statements in Section 3 pertain more specifically to individuals who have a leadership role, whether in the workplace or in a volunteer capacity, such as with organizations like ACM. Principles involving compliance with this Code are given in Section 4.

The Code shall be supplemented by a set of Guidelines, which provide explanation to assist members in dealing with the various issues contained in the Code. It is expected that the Guidelines will be changed more frequently than the Code.

The Code and its supplemented Guidelines are intended to serve as a basis for ethical decision making in the conduct of professional work. Secondarily, they may serve as a basis for judging the merit of a formal complaint pertaining to violation of professional ethical standards.

It should be noted that although computing is not mentioned in the imperatives of Section 1, the Code is concerned with how these fundamental imperatives apply to one's conduct as a computing professional. These imperatives are expressed in a general form to emphasize that ethical principles which apply to computer ethics are derived from more general ethical principles.

It is understood that some words and phrases in a code of ethics are subject to varying interpretations, and that any ethical principle may conflict with other ethical principles in specific situations. Questions related to ethical conflicts can best be answered by thoughtful consideration of fundamental principles, rather than reliance on detailed regulations.

1.GENERAL MORAL IMPERATIVES

As an ACM member I will ....

1.1 Contribute to Society and Human Well-Being

This principle concerning the quality of life of all people affirms an obligation to protect fundamental human rights and to respect the diversity of all cultures. An essential aim of computing professionals is to minimize negative consequences of computing systems, including threats to health and safety. When designing or implementing systems, computing professionals must attempt to ensure that the products of their efforts will be used in socially responsible ways, will meet social needs, and will avoid harmful effects to health and welfare.

In addition to a safe social environment, human well-being includes a safe natural environment. Therefore, computing professionals who design and develop systems must be alert to, and make others aware of, any potential damage to the local or global environment.

1.2 Avoid Harm to Others

"Harm" means injury or negative consequences, such as undesirable loss of information, loss of property, property damage, or unwanted environmental impacts. This principle prohibits use of computing technology in ways that result in harm to any of the following: users, the general public, employees, and employers. Harmful actions include intentional destruction or modification of files and programs, leading to serious loss of resources or unnecessary expenditure of human resources, such as the time and effort required to purge systems of "computer viruses."

Well-intended actions, including those that accomplish assigned duties, may lead to harm unexpectedly. In such an event the responsible person or persons are obligated to undo or mitigate the negative consequences as much as possible. One way to avoid unintentional harm is to carefully consider potential impacts on all those affected by decisions made during design and implementation.

To minimize the possibility of indirectly harming others, computing professionals must minimize malfunctions by following generally accepted standards for system design and testing. Furthermore, it is often necessary to assess the social consequences of systems to project the likelihood of any serious harm to others. If system features are misrepresented to users, coworkers, or supervisors, the individual computing professional is responsible for any resulting injury.

In the work environment the computing professional has the additional obligation to report any signs of system dangers that might result in serious personal or social damage. If one's superiors do not act to curtail or mitigate such dangers, it may be necessary to "blow the whistle" to help correct the problem or reduce the risk. However, capricious or misguided reporting of violations can, itself, be harmful. Before reporting violations, all relevant aspects of the incident must be thoroughly assessed. In particular, the assessment of risk and responsibility must be credible. It is suggested that advice be sought from other computing professionals. See principle 2.5 regarding thorough evaluations.

1.3 Be Honest and Trustworthy

Honesty is an essential component of trust. Without trust an organization cannot function effectively. The honest computing professional will not make deliberately false or deceptive claims about a system or system design, but will instead provide full disclosure of all pertinent system limitations and problems.

A computer professional has a duty to be honest about his or her own qualifications, and about any circumstances that might lead to conflicts of interest.

Membership in volunteer organizations such as ACM may at times place individuals in situations where their statements or actions could be interpreted as carrying the "weight" of a larger group of professionals. An ACM member will exercise care to not misrepresent ACM or positions and policies of ACM or any ACM units.

1.4 Be Fair and Take Action Not to Discriminate

The values of equality, tolerance, respect for others, and the principles of equal justice govern this imperative. Discrimination on the basis of race, sex, religion, age, disability, national origin, or other such factors is an explicit violation of ACM policy and will not be tolerated.

Inequities between different groups of people may result from the use or misuse of information and technology. In a fair society, all individuals would have equal opportunity to participate in, or benefit from, the use of computer resources regardless of race, sex, religion, age, disability, national origin or other such similar factors. However, these ideals do not justify unauthorized use of computer resources nor do they provide an adequate basis for violation of any other ethical imperatives of this code.

1.5 Honor Property Rights Including Copyrights and Patents

Violation of copyrights, patents, trade secrets and the terms of license agreements is prohibited by law in most circumstances. Even when software is not so protected, such violations are contrary to professional behavior. Copies of software should be made only with proper authorization. Unauthorized duplication of materials must not be condoned.

1.6 Give Proper Credit for Intellectual Property

Computing professionals are obligated to protect the integrity of intellectual property. Specifically, one must not take credit for other's ideas or work, even in cases where the work has not been explicitly protected by copyright, patent, etc.

1.7 Respect the Privacy of Others

Computing and communication technology enables the collection and exchange of personal information on a scale unprecedented in the history of civilization. Thus there is increased potential for violating the privacy of individuals and groups. It is the responsibility of professionals to maintain the privacy and integrity of data describing individuals. This includes taking precautions to ensure the accuracy of data, as well as protecting it from unauthorized access or accidental disclosure to inappropriate individuals. Furthermore, procedures must be established to allow individuals to review their records and correct inaccuracies.

This imperative implies that only the necessary amount of personal information be collected in a system, that retention and disposal periods for that information be clearly defined and enforced, and that personal information gathered for a specific purpose not be used for other purposes without consent of the individual(s). These principles apply to electronic communications, including electronic mail, and prohibit procedures that capture or monitor electronic user data, including messages, without the permission of users or bona fide authorization related to system operation and maintenance. User data observed during the normal duties of system operation and maintenance must be treated with strictest confidentiality, except in cases where it is evidence for the violation of law, organizational regulations, or this Code. In these cases, the nature or contents of that information must be disclosed only to proper authorities.

1.8 Honor Confidentiality

The principle of honesty extends to issues of confidentiality of information whenever one has made an explicit promise to honor confidentiality or, implicitly, when private information not directly related to the performance of one's duties becomes available. The ethical concern is to respect all obligations of confidentiality to employers, clients, and users unless discharged from such obligations by requirements of the law or other principles of this Code.

2. MORE SPECIFIC PROFESSIONAL RESPONSIBILITIES

As an ACM computing professional I will ....

2.1 Strive to Achieve the Highest Quality, Effectiveness, and Dignity in Both the Process and Products of Professional Work

Excellence is perhaps the most important obligation of a professional. The computing professional must strive to achieve quality and to be cognizant of the serious negative consequences that may result from poor quality in a system.

2.2 Acquire and Maintain Professional Competence

Excellence depends on individuals who take responsibility for acquiring and maintaining professional competence. A professional must participate in setting standards for appropriate levels of competence, and strive to achieve those standards. Upgrading technical knowledge and competence can be achieved in several ways: doing independent study; attending seminars, conferences, or courses; and being involved in professional organizations.

2.3 Know and Respect Existing Laws Pertaining to Professional Work

ACM members must obey existing local, state, province, national, and international laws unless there is a compelling ethical basis not to do so. Policies and procedures of the organizations in which one participates must also be obeyed. But compliance must be balanced with the recognition that sometimes existing laws and rules may be immoral or inappropriate and, therefore, must be challenged. Violation of a law or regulation may be ethical when that law or rule has inadequate moral basis or when it conflicts with another law judged to be more important. If one decides to violate a law or rule because it is viewed as unethical, or for any other reason, one must fully accept responsibility for one's actions and for the consequences.

2.4 Accept and Provide Appropriate Professional Review

Quality professional work, especially in the computing profession, depends on professional reviewing and critiquing. Whenever appropriate, individual members should seek and utilize peer review as well as provide critical review of the work of others.

2.5 Give Comprehensive and Thorough Evaluations of Computer Systems and Their Impacts, Including Analysis of Possible Risks

Computer professionals must strive to be perceptive, thorough, and objective when evaluating, recommending, and presenting system descriptions and alternatives. Computer professionals are in a position of special trust, and therefore have a special responsibility to provide objective, credible evaluations to employers, clients, users, and the public. When providing evaluations the professional must also identify any relevant conflicts of interest, as stated in imperative 1.3.

As noted in the discussion of principle 1.2 on avoiding harm, any signs of danger from systems must be reported to those who have opportunity and/or responsibility to resolve them. See the guidelines for imperative 1.2 for more details concerning harm, including the reporting of professional violations.

2.6 Honor Contracts, Agreements, and Assigned Responsibilities

Honoring one's commitments is a matter of integrity and honesty. For the computer professional this includes ensuring that system elements perform as intended. Also, when one contracts for work with another party, one has an obligation to keep that party properly informed about progress toward completing that work.

A computing professional has a responsibility to request a change in any assignment that he or she feels cannot be completed as defined. Only after serious consideration and with full disclosure of risks and concerns to the employer or client, should one accept the assignment. The major underlying principle here is the obligation to accept personal accountability for professional work. On some occasions other ethical principles may take greater priority.

A judgment that a specific assignment should not be performed may not be accepted. Having clearly identified one's concerns and reasons for that judgment, but failing to procure a change in that assignment, one may yet be obligated, by contract or by law, to proceed as directed. The computing professional's ethical judgment should be the final guide in deciding whether or not to proceed. Regardless of the decision, one must accept the responsibility for the consequences. However, performing assignments "against one's own judgment" does not relieve the professional of responsibility for any negative consequences.

2.7 Improve Public Understanding of Computing and Its Consequences

Computing professionals have a responsibility to share technical knowledge with the public by encouraging understanding of computing, including the impacts of computer systems and their limitations. This imperative implies an obligation to counter any false views related to computing.

2.8 Access Computing and Communication Resources Only When Authorized to Do So

Theft or destruction of tangible and electronic property is prohibited by imperative 1.2, Avoid Harm to Others. Trespassing and unauthorized use of a computer or communication system is addressed by this imperative. Trespassing includes accessing communication networks and computer systems, or accounts and/or files associated with those systems, without explicit authorization to do so. Individuals and organizations have the right to restrict access to their systems so long as they do not violate the discrimination principle (see 1.4). No one should enter or use another's computer system, software, or data files without permission. One must always have appropriate approval before using system resources, including communication ports, file space, other system peripherals, and computer time.

3. ORGANIZATIONAL LEADERSHIP IMPERATIVES

As an ACM member and an organizational leader, I will ....

BACKGROUND NOTE: This section draws extensively from the draft IFIP Code of Ethics, especially its sections on organizational ethics and international concerns. The ethical obligations of organizations tend to be neglected in most codes of professional conduct, perhaps because these codes are written from the perspective of the individual member. This dilemma is addressed by stating these imperatives from the perspective of the organizational leader. In this context "leader" is viewed as any organizational member who has leadership or educational responsibilities. These imperatives generally may apply to organizations as well as their leaders. In this context "organizations" are corporations, government agencies, and other "employers," as well as volunteer professional organizations.

3.1 Articulate Social Responsibilities of Members of an Organizational Unit and Encourage Full Acceptance of Those Responsibilities

Because organizations of all kinds have impacts on the public, they must accept responsibilities to society. Organizational procedures and attitudes oriented toward quality and the welfare of society will reduce harm to members of the public, thereby serving public interest and fulfilling social responsibility. Therefore, organizational leaders must encourage full participation in meeting social responsibilities as well as quality performance.

3.2 Manage Personnel and Resources to Design and Build Information Systems That Enhance the Quality of Working Life

Organizational leaders are responsible for ensuring that computer systems enhance, not degrade, the quality of working life. When implementing a computer system, organizations must consider the personal and professional development, physical safety, and human dignity of all workers. Appropriate human-computer ergonomic standards should be considered in system design and in the workplace.

3.3 Acknowledge and Support Proper and Authorized Uses of an Organization's Computing and Communication Resources

Because computer systems can become tools to harm as well as to benefit an organization, the leadership has the responsibility to clearly define appropriate and inappropriate uses of organizational computing resources. While the number and scope of such rules should be minimal, they should be fully enforced when established.

3.4 Ensure that Users and Those Who Will Be Affected by a System Have Their Needs Clearly Articulated During the Assessment and Design of Requirements; Later the System Must Be Validated to Meet Requirements

Current system users, potential users and other persons whose lives may be affected by a system must have their needs assessed and incorporated in the statement of requirements. System validation should ensure compliance with those requirements.

3.5 Articulate and Support Policies That Protect the Dignity of Users and Others Affected by a Computing System

Designing or implementing systems that deliberately or inadvertently demean individuals or groups is ethically unacceptable. Computer professionals who are in decision making positions should verify that systems are designed and implemented to protect personal privacy and enhance personal dignity.

3.6 Create Opportunities for Members of the Organization to Learn the Principles and Limitations of Computer Systems

This complements the imperative on public understanding (2.7). Educational opportunities are essential to facilitate optimal participation of all organizational members. Opportunities must be available to all members to help them improve their knowledge and skills in computing, including courses that familiarize them with the consequences and limitations of particular types of systems. In particular, professionals must be made aware of the dangers of building systems around oversimplified models, the improbability of anticipating and designing for every possible operating condition, and other issues related to the complexity of this profession.

4. COMPLIANCE WITH THE CODE

As an ACM member I will ....

4.1 Uphold and Promote the Principles of This Code

The future of the computing profession depends on both technical and ethical excellence. Not only is it important for ACM computing professionals to adhere to the principles expressed in this Code, each member should encourage and support adherence by other members.

4.2 Treat Violations of This Code As Inconsistent with Membership in the ACM

Adherence of professionals to a code of ethics is largely a voluntary matter. However, if a member does not follow this code by engaging in gross misconduct, membership in ACM may be terminated.

This Code and the supplemental Guidelines were developed by the Task Force for the Revision of the ACM Code of Ethics and Professional Conduct: Ronald E. Anderson, Chair, Gerald Engel, Donald Gotterbarn, Grace C. Hertlein, Alex Hoffman, Bruce Jawer, Deborah G. Johnson, Doris K. Lidtke, Joyce Currie Little, Dianne Martin, Donn B. Parker, Judith A. Perrolle, and Richard S. Rosenberg. The Task Force was organized by ACM/SIGCAS and funding was provided by the ACM SIG Discretionary Fund. This Code and the supplemental Guidelines were adopted by the ACM Council on October 16, 1992.

Source: ACM Web site at *www.acm.org*. Copyright © 2006 ACM. Used with permission.

ASSOCIATION OF INFORMATION TECHNOLOGY PROFESSIONALS (AITP) CODE OF ETHICS

I acknowledge:

That I have an obligation to management, therefore, I shall promote the understanding of information processing methods and procedures to management using every resource at my command.

That I have an obligation to my fellow members, therefore, I shall uphold the high ideals of AITP as outlined in the Association Bylaws. Further, I shall cooperate with my fellow members and shall treat them with honesty and respect at all times.

That I have an obligation to society and will participate to the best of my ability in the dissemination of knowledge pertaining to the general development and understanding of information processing. Further, I shall not use knowledge of a confidential nature to further my personal interest, nor shall I violate the privacy and confidentiality of information entrusted to me or to which I may gain access.

That I have an obligation to my College or University, therefore, I shall uphold its ethical and moral principles.

That I have an obligation to my employer whose trust I hold, therefore, I shall endeavor to discharge this obligation to the best of my ability, to guard my employer's interests, and to advise him or her wisely and honestly.

That I have an obligation to my country, therefore, in my personal, business, and social contacts, I shall uphold my nation and shall honor the chosen way of life of my fellow citizens.

I accept these obligations as a personal responsibility and as a member of this Association. I shall actively discharge these obligations and I dedicate myself to that end.

AITP Standards of Conduct

These standards expand on the Code of Ethics by providing specific statements of behavior in support of each element of the Code. They are not objectives to be strived for, they are rules that no true professional will violate. It is first of all expected that an information processing professional will abide by the appropriate laws of their country and community. The following standards address tenets that apply to the profession.

In recognition of my obligation to management I shall:

- Keep my personal knowledge up-to-date and insure that proper expertise is available when needed.
- Share my knowledge with others and present factual and objective information to management to the best of my ability.
- Accept full responsibility for work that I perform.
- Not misuse the authority entrusted to me.
- Not misrepresent or withhold information concerning the capabilities of equipment, software or systems.
- Not take advantage of the lack of knowledge or inexperience on the part of others.

In recognition of my obligation to my fellow members and the profession I shall:

- Be honest in all my professional relationships.
- Take appropriate action in regard to any illegal or unethical practices that come to my attention. However, I will bring charges against any person only when I have reasonable basis for believing in the truth of the allegations and without any regard to personal interest.
- Endeavor to share my special knowledge.
- Cooperate with others in achieving understanding and in identifying problems.
- Not use or take credit for the work of others without specific acknowledgement and authorization.
- Not take advantage of the lack of knowledge or inexperience on the part of others for personal gain.

In recognition of my obligation to society I shall:

- Protect the privacy and confidentiality of all information entrusted to me.
- Use my skill and knowledge to inform the public in all areas of my expertise.
- To the best of my ability, insure that the products of my work are used in a socially responsible way.
- Support, respect, and abide by the appropriate local, state, provincial, and federal laws.
- Never misrepresent or withhold information that is germane to a problem or situation of public concern, nor will I allow any such known information to remain unchallenged.
- Not use knowledge of a confidential or personal nature in any unauthorized manner or to achieve personal gain.

In recognition of my obligation to my employer I shall:

- Make every effort to ensure that I have the most current knowledge and that the proper expertise is available when needed.
- Avoid conflict of interest and insure that my employer is aware of any potential conflicts.
- Present a fair, honest, and objective viewpoint.
- Protect the proper interests of my employer at all times.
- Protect the privacy and confidentiality of all information entrusted to me.
- Not misrepresent or withhold information that is germane to the situation.

- Not attempt to use the resources of my employer for personal gain or for any purpose without proper approval.
- Not exploit the weakness of a computer system for personal gain or personal satisfaction.

Source: Courtesy of AITP (*www.aitp.org*)

SOFTWARE ENGINEERING CODE OF ETHICS AND PROFESSIONAL PRACTICE

(Version 5.2) as recommended by the IEEE-CS/ACM Joint Task Force on Software Engineering Ethics and Professional Practices and Jointly approved by the ACM and the IEEE-CS as the standard for teaching and practicing software engineering.

Preamble

Computers have a central and growing role in commerce, industry, government, medicine, education, entertainment and society at large. Software engineers are those who contribute by direct participation or by teaching, to the analysis, specification, design, development, certification, maintenance and testing of software systems. Because of their roles in developing software systems, software engineers have significant opportunities to do good or cause harm, to enable others to do good or cause harm, or to influence others to do good or cause harm. To ensure, as much as possible, that their efforts will be used for good, software engineers must commit themselves to making software engineering a beneficial and respected profession. In accordance with that commitment, software engineers shall adhere to the following Code of Ethics and Professional Practice.

The Code contains eight Principles related to the behavior of and decisions made by professional software engineers, including practitioners, educators, managers, supervisors and policy makers, as well as trainees and students of the profession. The Principles identify the ethically responsible relationships in which individuals, groups, and organizations participate and the primary obligations within these relationships. The Clauses of each Principle are illustrations of some of the obligations included in these relationships. These obligations are founded in the software engineers humanity, in special care owed to people affected by the work of software engineers, and the unique elements of the practice of software engineering. The Code prescribes these as obligations of anyone claiming to be or aspiring to be a software engineer.

It is not intended that the individual parts of the Code be used in isolation to justify errors of omission or commission. The list of Principles and Clauses is not exhaustive. The Clauses should not be read as separating the acceptable from the unacceptable in professional conduct in all practical situations. The Code is not a simple ethical algorithm that generates ethical decisions. In some situations standards may be in tension with each other or with standards from other sources. These situations require the software engineer to use ethical judgment to act in a manner which is most consistent with the spirit of the Code of Ethics and Professional Practice, given the circumstances.

Ethical tensions can best be addressed by thoughtful consideration of fundamental principles, rather than blind reliance on detailed regulations. These Principles should influence software engineers to consider broadly who is affected by their work; to examine if they and their colleagues are treating other human beings with due respect; to consider how the public, if reasonably well informed, would view their decisions; to analyze how the least empowered will be affected by their decisions; and to consider whether their acts would be judged worthy of the ideal professional working as a software engineer. In all these judgments, concern for the health, safety and welfare of the public is primary; that is, the "Public Interest" is central to this Code.

The dynamic and demanding context of software engineering requires a code that is adaptable and relevant to new situations as they occur. However, even in this generality, the Code provides support for software engineers and managers of software engineers who need to take positive action in a specific case by documenting the ethical stance of the profession. The Code provides an ethical foundation to which individuals within teams and the team as a whole can appeal. The Code helps to define those actions that are ethically improper to request of a software engineer or teams of software engineers.

The Code is not simply for adjudicating the nature of questionable acts; it also has an important educational function. As this Code expresses the consensus of the profession on ethical issues, it is a means to educate both the public and aspiring professionals about the ethical obligations of all software engineers.

Principles

Principle 1 PUBLIC

Software engineers shall act consistently with the public interest. In particular, software engineers shall, as appropriate:

1.01. Accept full responsibility for their own work.

1.02. Moderate the interests of the software engineer, the employer, the client and the users with the public good.

1.03. Approve software only if they have a well-founded belief that it is safe, meets specifications, passes appropriate tests, and does not diminish quality of life, diminish privacy or harm the environment. The ultimate effect of the work should be to the public good.

1.04. Disclose to appropriate persons or authorities any actual or potential danger to the user, the public, or the environment, that they reasonably believe to be associated with software or related documents.

1.05. Cooperate in efforts to address matters of grave public concern caused by software, its installation, maintenance, support or documentation.

1.06. Be fair and avoid deception in all statements, particularly public ones, concerning software or related documents, methods and tools.

1.07. Consider issues of physical disabilities, allocation of resources, economic disadvantage and other factors that can diminish access to the benefits of software.

1.08. Be encouraged to volunteer professional skills to good causes and contribute to public education concerning the discipline.

Principle 2 CLIENT AND EMPLOYER

Software engineers shall act in a manner that is in the best interests of their client and employer, consistent with the public interest. In particular, software engineers shall, as appropriate:

2.01. Provide service in their areas of competence, being honest and forthright about any limitations of their experience and education.

2.02. Not knowingly use software that is obtained or retained either illegally or unethically.

2.03. Use the property of a client or employer only in ways properly authorized, and with the client's or employer's knowledge and consent.

2.04. Ensure that any document upon which they rely has been approved, when required, by someone authorized to approve it.

2.05. Keep private any confidential information gained in their professional work, where such confidentiality is consistent with the public interest and consistent with the law.

2.06. Identify, document, collect evidence and report to the client or the employer promptly if, in their opinion, a project is likely to fail, to prove too expensive, to violate intellectual property law, or otherwise to be problematic.

2.07. Identify, document, and report significant issues of social concern, of which they are aware, in software or related documents, to the employer or the client.

2.08. Accept no outside work detrimental to the work they perform for their primary employer.

2.09. Promote no interest adverse to their employer or client, unless a higher ethical concern is being compromised; in that case, inform the employer or another appropriate authority of the ethical concern.

Principle 3 PRODUCT

Software engineers shall ensure that their products and related modifications meet the highest professional standards possible. In particular, software engineers shall, as appropriate:

3.01. Strive for high quality, acceptable cost and a reasonable schedule, ensuring significant tradeoffs are clear to and accepted by the employer and the client, and are available for consideration by the user and the public.

3.02. Ensure proper and achievable goals and objectives for any project on which they work or propose.

3.03. Identify, define and address ethical, economic, cultural, legal and environmental issues related to work projects.

3.04. Ensure that they are qualified for any project on which they work or propose to work by an appropriate combination of education and training, and experience.

3.05. Ensure an appropriate method is used for any project on which they work or propose to work.

3.06. Work to follow professional standards, when available, that are most appropriate for the task at hand, departing from these only when ethically or technically justified.

3.07. Strive to fully understand the specifications for software on which they work.

3.08. Ensure that specifications for software on which they work have been well documented, satisfy the users' requirements and have the appropriate approvals.

3.09. Ensure realistic quantitative estimates of cost, scheduling, personnel, quality and outcomes on any project on which they work or propose to work and provide an uncertainty assessment of these estimates.

3.10. Ensure adequate testing, debugging, and review of software and related documents on which they work.

3.11. Ensure adequate documentation, including significant problems discovered and solutions adopted, for any project on which they work.

3.12. Work to develop software and related documents that respect the privacy of those who will be affected by that software.

3.13. Be careful to use only accurate data derived by ethical and lawful means, and use it only in ways properly authorized.

3.14. Maintain the integrity of data, being sensitive to outdated or flawed occurrences.

3.15 Treat all forms of software maintenance with the same professionalism as new development.

Principle 4 JUDGMENT

Software engineers shall maintain integrity and independence in their professional judgment. In particular, software engineers shall, as appropriate:

4.01. Temper all technical judgments by the need to support and maintain human values.

4.02. Only endorse documents either prepared under their supervision or within their areas of competence and with which they are in agreement.

4.03. Maintain professional objectivity with respect to any software or related documents they are asked to evaluate.

4.04. Not engage in deceptive financial practices such as bribery, double billing, or other improper financial practices.

4.05. Disclose to all concerned parties those conflicts of interest that cannot reasonably be avoided or escaped.

4.06. Refuse to participate, as members or advisors, in a private, governmental or professional body concerned with software related issues, in which they, their employers or their clients have undisclosed potential conflicts of interest.

Principle 5 MANAGEMENT

Software engineering managers and leaders shall subscribe to and promote an ethical approach to the management of software development and maintenance. In particular, those managing or leading software engineers shall, as appropriate:

5.01. Ensure good management for any project on which they work, including effective procedures for promotion of quality and reduction of risk.

5.02. Ensure that software engineers are informed of standards before being held to them.

5.03. Ensure that software engineers know the employer's policies and procedures for protecting passwords, files and information that is confidential to the employer or confidential to others.

5.04. Assign work only after taking into account appropriate contributions of education and experience tempered with a desire to further that education and experience.

5.05. Ensure realistic quantitative estimates of cost, scheduling, personnel, quality and outcomes on any project on which they work or propose to work, and provide an uncertainty assessment of these estimates.

5.06. Attract potential software engineers only by full and accurate description of the conditions of employment.

5.07. Offer fair and just remuneration.

5.08. Not unjustly prevent someone from taking a position for which that person is suitably qualified.

5.09. Ensure that there is a fair agreement concerning ownership of any software, processes, research, writing, or other intellectual property to which a software engineer has contributed.

5.10. Provide for due process in hearing charges of violation of an employer's policy or of this Code.

5.11. Not ask a software engineer to do anything inconsistent with this Code.

5.12. Not punish anyone for expressing ethical concerns about a project.

Principle 6 PROFESSION
Software engineers shall advance the integrity and reputation of the profession consistent with the public interest. In particular, software engineers shall, as appropriate:

6.01. Help develop an organizational environment favorable to acting ethically.

6.02. Promote public knowledge of software engineering.

6.03. Extend software engineering knowledge by appropriate participation in professional organizations, meetings and publications.

6.04. Support, as members of a profession, other software engineers striving to follow this Code.

6.05. Not promote their own interest at the expense of the profession, client or employer.

6.06. Obey all laws governing their work, unless, in exceptional circumstances, such compliance is inconsistent with the public interest.

6.07. Be accurate in stating the characteristics of software on which they work, avoiding not only false claims but also claims that might reasonably be supposed to be speculative, vacuous, deceptive, misleading, or doubtful.

6.08. Take responsibility for detecting, correcting, and reporting errors in software and associated documents on which they work.

6.09. Ensure that clients, employers, and supervisors know of the software engineer's commitment to this Code of ethics, and the subsequent ramifications of such commitment.

6.10. Avoid associations with businesses and organizations which are in conflict with this code.

6.11. Recognize that violations of this Code are inconsistent with being a professional software engineer.

6.12. Express concerns to the people involved when significant violations of this Code are detected unless this is impossible, counter-productive, or dangerous.

6.13. Report significant violations of this Code to appropriate authorities when it is clear that consultation with people involved in these significant violations is impossible, counter-productive or dangerous.

Principle 7 COLLEAGUES

Software engineers shall be fair to and supportive of their colleagues. In particular, software engineers shall, as appropriate:

7.01. Encourage colleagues to adhere to this Code.

7.02. Assist colleagues in professional development.

7.03. Credit fully the work of others and refrain from taking undue credit.

7.04. Review the work of others in an objective, candid, and properly documented way.

7.05. Give a fair hearing to the opinions, concerns, or complaints of a colleague.

7.06. Assist colleagues in being fully aware of current standard work practices, including policies and procedures for protecting passwords, files and other confidential information, and security measures in general.

7.07. Not unfairly intervene in the career of any colleague; however, concern for the employer, the client or public interest may compel software engineers, in good faith, to question the competence of a colleague.

7.08. In situations outside of their own areas of competence, call upon the opinions of other professionals who have competence in that area.

Principle 8 SELF

Software engineers shall participate in lifelong learning regarding the practice of their profession and shall promote an ethical approach to the practice of the profession. In particular, software engineers shall continually endeavor to:

8.01. Further their knowledge of developments in the analysis, specification, design, development, maintenance and testing of software and related documents, together with the management of the development process.

8.02. Improve their ability to create safe, reliable, and useful quality software at reasonable cost and within a reasonable time.

8.03. Improve their ability to produce accurate, informative, and well-written documentation.

8.04. Improve their understanding of the software and related documents on which they work and of the environment in which they will be used.

8.05. Improve their knowledge of relevant standards and the law governing the software and related documents on which they work.

8.06. Improve their knowledge of this Code, its interpretation, and its application to their work.

8.07. Not give unfair treatment to anyone because of any irrelevant prejudices.

8.08. Not influence others to undertake any action that involves a breach of this Code.

8.09. Recognize that personal violations of this Code are inconsistent with being a professional software engineer.

This Code was developed by the IEEE-CS/ACM Joint Task Force on Software Engineering Ethics and Professional Practices (SEEPP):

Executive Committee: Donald Gotterbarn (Chair),

Keith Miller and Simon Rogerson;

Members: Steve Barber, Peter Barnes, Ilene Burnstein, Michael Davis, Amr El-Kadi, N. Ben Fairweather, Milton Fulghum, N. Jayaram, Tom Jewett, Mark Kanko, Ernie Kallman,

Duncan Langford, Joyce Currie Little, Ed Mechler, Manuel J. Norman, Douglas Phillips, Peter Ron Prinzivalli, Patrick Sullivan, John Weckert, Vivian Weil, S. Weisband and Laurie Honour Werth.

This Code may be published without permission as long as it is not changed in any way and it carries the copyright notice.

Source: www.computer.org/portal/site/ieeecs/menuitem.c5efb9b8ade9096b8a9ca0108-bcd45f3/index.jsp?&pName=ieeecs_level1&path=ieeecs/education/certification&file=ethics.xml&xsl=generic.xsl&

PMI MEMBER ETHICAL STANDARDS AND MEMBER CODE OF ETHICS

The Project Management Institute (PMI) is a professional organization dedicated to the development and promotion of the field of project management. The purpose of the PMI Member Code of Ethics is to define and clarify the ethical responsibilities for present and future PMI members.

Preamble

In the pursuit of the project management profession, it is vital that PMI members conduct their work in an ethical manner in order to earn and maintain the confidence of team members, colleagues, employees, employers, customers/clients, the public, and the global community.

Member Code of Ethics

As a profession in the field of project management, PMI members pledge to uphold and abide by the following:

- will maintain high standards of integrity and professional conduct
- will accept responsibility for my actions
- will continually seek to enhance my professional capabilities
- will practice with fairness and honesty
- will encourage others in the profession to act in an ethical and professional manner

Source: PMI Code of Ethics, www.oliverlehmann-training.de/code-of-conduct/ PMI-Members-Ethic-Standard.pdf.

ANSWERS TO SELF-ASSESSMENT QUESTIONS

Chapter 1 answers: 1. virtues; 2. True; 3. True; 4. Intel; 5. True; 6. True; 7. code of ethics; 8. social audit

Chapter 2 answers: 1. d; 2. True; 3. c; 4. c; 5. True; 6. False; 7. d; 8. True; 9. b; 10. False; 11. d; 12. b

Chapter 3 answers: 1. False; 2. exploit; 3. c; 4. hacker; 5. False; 6. False; 7. risk assessment; 8. b; 9. virtual private network; 10. False; 11. True; 12. d

Chapter 4 answers: 1. b; 2. True; 3. Freedom of Information; 4. True; 5. European, United States; 6. False; 7. a; 8. Cryptography; 9. d; 10. personalization software; 11. True; 12. True

Chapter 5 answers: 1. c; 2. True; 3. Miller v. California; 4. b; 5. True; 6. a; 7. False; 8. John Doe lawsuit; 9. True; 10. USA Patriot Act; 11. False; 12. a

Chapter 6 answers: 1. d; 2. True; 3. Digital Millennium Copyright Act; 4. patent; 5. True; 6. c. trade secret; 7. False; 8. True; 9. b; 10. True; 11. True

Chapter 7 answers: 1. True; 2. software quality; 3. d; 4. software development methodology; 5. True; 6. a; 7. c; 8. safety critical system; 9. a; 10. FMEA

Chapter 8 answers: 1. False; 2. coemployment; 3. d; 4. False; 5. U.S. Congress and U.S. Citizenship and Immigration Services; 6. b; 7. False; 8. c; 9. b; 10. True

Chapter 9 answers: 1. b; 2. c; 3. False; 4. True; 5. b; 6. d; 7. Education Rate (E-Rate) Program; 8. optimization technologies; 9. d; 10. shielding; 11. True; 12. protecting patient privacy and finding a consistent way to identify each patient

affiliated Web sites   A collection of Web sites served by a single advertising network.

anonymous expression   The ability to state your opinions without revealing your identity.

anonymous remailer   An Internet service that allows an e-mail sender to remain anonymous by stripping the originating address from an e-mail message and then forwarding it to its intended recipient.

black-box testing   Unit testing of software that has expected input and output behaviors, with the inner workings unknown.

breach of the duty of care   The failure to act as a reasonable person would act.

breach of warranty   The failure of a product to meet its warranty.

business information system   A set of interrelated hardware, software, databases, networks, people, and procedures that collect data, process it, and disseminate the output.

Capability Maturity Model Integration   A model that defines five levels of software development maturity and identifies the issues that are most critical to software quality and process improvement.

chargebacks   Disputed credit card transactions.

chief privacy officer (CPO)   A senior manager responsible for training employees about privacy, checking the company's privacy policies for potential risks and then figuring out how to fill gaps, and managing customer privacy disputes.

click-stream data   Data that tracks each point and click of a visitor to a Web site.

code of ethics   A statement that highlights an organization's key ethical issues and identifies the overarching values and principles that are important to the organization and its decision making.

coemployment relationship   A relationship between two or more employers in which each has actual or potential legal rights and duties with respect to the same employee or group of employees.

collaborative filtering   A form of personalization software that offers consumer recommendations based on the types of products purchased by other people with similar buying habits.

collusion   A conspiracy to commit fraud between a company employee and someone outside the company.

common good approach   An approach to ethical decision making that is based on a vision of society as a community whose members work together to achieve a common set of values and goals.

Communications Assistance for Law Enforcement Act (CALEA)   A 1994 law that required the telecommunications industry to build tools into its products that federal investigators could use—after getting court approval—to eavesdrop on conversations.

competitive intelligence   The legal gathering of publicly available information that helps a company gain an advantage over its rivals.

compiler   A language translator that converts computer program statements from a source language into binary machine language that the computer can execute.

contextual commerce   A form of personalization software that associates product promotions and other e-commerce offerings with specific content a user may be receiving in a news story online.

contingent workforce   Independent contractors, employees brought in through employment agencies, on-call or day laborers, and on-site workers whose services are provided by contract firms.

contributory negligence   A defense in a negligence case in which the defendant proves that the plaintiffs' own actions contributed to their injuries.

cookie   A unique identifier that passes information back to a marketer's computer when a person surfs the Web.

copyright   A form of protection that grants authors the exclusive right to distribute, display, perform, or reproduce their work in copies or to prepare derivative works based on their work.

corporate ethics officer   A senior-level manager responsible for establishing policies and training to improve the ethical behavior of an organization.

cracker   A hacker who performs illegal acts such as breaking into other people's networks and systems, defacing Web pages, crashing computers, spreading harmful programs or hateful messages, or writing scripts and automatic programs that let other people do the same things.

cryptography   The science of encoding messages so that only the sender and intended receiver can understand them.

cybercriminal   A person who hacks into corporate computers for personal gain.

cybersquatter   Someone who registers a Web site domain that contains a famous trademark or company name to which the person has no connection, with the hope that the trademark's legitimate owners will pay to gain ownership of the domain.

cyberterrorist   Someone who intimidates or coerces a government to further a political or social objective by launching attacks or threatening attacks against computers, networks, and the information stored on them.

decision support system   A type of business information system used to improve decision making.

decompiler   A software tool that can read machine language and produce the source code.

defamation   The publication of a statement of alleged fact that is false and that harms another person.

definitions   Virus detection information used to update antivirus software.

deliverables   Products created during the various stages of software development.

demographic filtering   A form of personalization software that employs an Internet user's demographic information to provide targeted Web advertising.

denial-of-service attack   An attack in which a malicious hacker takes over computers on the Internet and causes them to flood a target site with demands for data and other small tasks.

digital divide   A term used to describe the gulf between people who do and don't have access to modern information and communications technology.

duty of care   The obligation people have to each other not to cause any unreasonable harm or risk of harm.

dynamic testing   The testing of software by entering test data and then comparing the actual results to expected results.

egress filtering A process by which corporations ensure that packets with false return addresses do not leave their corporate network.

Electronic Communications Privacy Act of 1986 (ECPA) A law that set standards for access to stored e-mail and other electronic communications and records.

electronic health record A summary of health information generated by a patient encounter in a healthcare delivery setting.

employee leasing An arrangement in which a company outsources all or part of its workforce to a professional employer organization.

encryption The process of converting an original message into a form that can be understood only by the intended recipients.

ethics A set of beliefs about right and wrong behavior.

Executive Order 12333 The legal authority for electronic surveillance outside the United States. The order was issued by President Reagan in 1982.

exploit An attack on an information system that takes advantage of a particular system vulnerability.

failure mode and effects analysis (FMEA) A reliability evaluation technique used to determine the effect of system and equipment failures.

fair use doctrine A set of criteria that courts employ to determine the fair use of copyrighted property and whether it can be allowed without penalty.

fairness approach An approach to ethical decision making that focuses on how fairly actions and policies distribute benefits and burdens among those affected by the decision.

False Claims Act A federal law enacted during the U.S. Civil War to combat fraud by companies that sold supplies to the Union Army. It was designed to entice whistle-blowers to come forward by offering them a share of the money recovered from a fraud lawsuit.

Federal Wiretap Act A 1968 law that outlines processes to obtain court authorization for surveillance of all kinds of electronic communications, including e-mail, fax, Internet, and voice, in criminal investigations.

firewall A hardware or software device that serves as a barrier between a company and the outside world, limiting access through the company's network based on its IT usage policy.

Foreign Intelligence Surveillance Act of 1978 (FISA) A law that allows wiretapping of aliens and citizens in the United States, based on a finding of probable cause that the target is a member of a foreign terrorist group or an agent of a foreign power.

GET data During a Web browsing session, data gathered to identify visited sites and requested information.

government licensing A process in which a professional must demonstrate the ability to operate ethically and safely. In the United States, licensing is generally administered at the state level.

H-1B worker A foreign person with specialized skills who works in the United States under the terms of an H-1B visa.

hacker A person who tests the limitations of computer systems out of intellectual curiosity, not necessarily to cause trouble.

honeypot A decoy computer server that gives hackers fake information about a network in order to confuse them and log their activities.

industrial espionage The use of illegal means to obtain business information not available to the general public.

industrial spies Hackers who use illegal means to obtain trade secrets about competitors of the firm that hired them.

ingress filtering   A process by which Internet service providers prevent incoming packets with false IP addresses from being passed on.

integration testing   Software testing that follows successful unit testing, where the software units are combined into an integrated subsystem that undergoes rigorous testing.

integrity   Adherence to a moral or ethical code.

intellectual property   Distinct works such as art, books, films, formulae, inventions, music, and processes that are "owned" or created by a single entity.

intentional misrepresentation   Fraud that occurs when a seller or lessor knowingly either misrepresents the quality of a product or conceals a defect in it.

Internet filter   Software that can be installed with a Web browser to block access to certain Web sites that contain inappropriate or offensive material.

intrusion detection system   A system that monitors a computer network and then notifies authorities when it identifies possible intrusions from outside the organization or misuse from within the organization.

ISO 9000   A series of standards that require organizations to develop formal quality management systems that focus on identifying and meeting customer needs and expectations.

IT user   A person for whom computer hardware or software is designed.

John Doe lawsuit   A lawsuit in which the true identity of the defendant is temporarily unknown.

(encryption) key   A variable value that is applied using an algorithm to produce encrypted text or to decrypt encrypted text.

lamer   A technically inept hacker.

libel   A written statement of alleged fact that is false and that harms another person.

live telemedicine   Medical treatment enhanced by live links to audio and video communications, which enables physicians at remote sites to assist in patient care.

logic bomb   A type of Trojan horse that executes under specific conditions.

moral code   Rules that establish the boundaries of generally accepted behavior.

morality   Social conventions about human conduct that are widely shared by society.

National Security Letter (NSL)   A government notification that requires financial institutions to turn over electronic records about the finances, telephone calls, e-mail, and other personal information of suspected terrorists or spies.

negligence   Failure to do what a reasonable person would do, or doing something that a reasonable person would not.

noncompete agreement   An agreement that requires employees not to work for a competitor for a period of time after leaving an employer.

nondisclosure clause   A clause in an employment contract that prohibits employees from revealing company trade secrets.

N-version programming   A form of redundancy in which two computer systems simultaneously execute a series of program instructions.

offshore outsourcing   A variation of outsourcing in which work is done by an organization whose employees are in a forieign country.

open source code   A program whose source code is available for use or modification by other developers.

optimization technologies   New information technologies that require little capital and can reduce the digital divide to make computing and communication better, cheaper, faster, and more available around the world.

opt-in An approach to data collection that requires permission from consumers before their data is collected.

opt-out An information gathering method that requires consumers to inform companies not to collect data about them; otherwise, companies assume they can collect the consumers' personal information.

outsourcing A company's reliance on services from an outside organization that has expertise in providing a specific function.

patent Legal protection of an invention that prevents its unauthorized use.

patent farming An unethical strategy of influencing a standards organization to make use of a patented item without revealing the existence of the patent; later, the patent holder might demand royalties from all implementers of the standard.

pen register Technology that reads electronic or other impulses that identify the numbers dialed for outgoing calls.

personalization software Software that marketers use to optimize the number, frequency, and mixture of their ad placements and to test how visitors react to new ads.

phishing An attempt to steal private data by tricking users into entering the information on a counterfeit Web site.

plagiarism Stealing or passing off someone else's ideas or words as one's own.

Platform for Privacy Preferences (P3P) Screening technology that can shield users from Web sites that don't provide the level of privacy protection they desire.

POST data Data entered into blank fields on a Web site when a consumer signs up for a service.

principle-based decision making Decision making based on principles in a corporate code of ethics.

prior art The body of knowledge available to a person of ordinary skill in an art. A patent cannot be issued for an invention whose professed improvements already exist in the prior art.

private key encryption system A system that uses only one key to both encode and decode messages.

product liability The liability of manufacturers, vendors, and others for injuries caused by defective products.

productivity The amount of output produced per unit of input.

profession A calling that requires specialized knowledge and often long and intensive academic preparation.

professional code of ethics A statement of essential principles and core values for a particular occupation.

professional malpractice The breach of a professional duty of care, which makes professionals liable for any injury caused by their negligence.

public key encryption system A system that uses a public key to encode messages and a private key to decode messages.

quality management The definition, measurement, and refinement of product quality during the development of information systems and other products.

qui tam A provision of the Federal False Claims Act that allows private citizens to sue government contractors for fraud in the name of the U.S. Government. These citizens may also share in money awarded in the suit.

reasonable assurance The recognition that managers must use their judgment to ensure that the cost of system control does not exceed a system's benefits or the risks involved.

reasonable person standard How an objective, careful, and conscientious person would act under certain circumstances.

reasonable professional standard　How a trained and experienced professional would act under certain circumstances.

redundancy　The use of interchangeable hardware components to perform a single function in order to cope with failures and errors.

reliability　The probability that a component or system will perform dependably well over a certain length of time.

résumé inflation　On a résumé, the false claim of competency in an IT skill that is in high demand.

reverse engineering　The process of analyzing finished software to create a new representation of it in a different form or at a higher level of abstraction.

risk　The likelihood of a negative event occurring multiplied by the impact of such an occurrence.

risk assessment　An organization's review of potential threats to its computers and network and the probability of those threats occurring.

rules-based software　Personalization software that uses customer-provided preferences or online behavior to determine the most appropriate Web page views and product information to display.

safety-critical system　A system whose failure may cause injury or death to human beings.

Sarbanes-Oxley Act Section 404　Legislation that requires corporate annual reports to contain signed assurances by CEOs and CFOs that the information in an SEC filing is accurate. Section 404 also requires companies to submit to an audit to prove that it has controls in place to ensure accurate information.

script kiddie　A technically inept hacker.

security policy　A document that defines an organization's security requirements and describes the controls and sanctions used to meet those requirements.

slander　An oral statement of alleged fact that is false and that harms another person.

smart card　A credit card containing a memory chip that is updated with encrypted data every time the card is read.

social audits　Studies that identify past corporate ethical lapses and set directives for avoiding similar missteps in the future.

software defect　An error that, if not removed, would cause a system to fail to meet the needs of its users.

software development methodology　A standard, proven work process that enables systems analysts, programmers, project managers, and others to develop high-quality software.

software piracy　The act of illegally making copies of software or enabling others to access software to which they are not entitled.

software quality　The degree to which the attributes of a software product enable it to meet the needs of its users.

software quality assurance　Methods within a software development process that are designed to guarantee reliable operation of the software.

spamming　The transmission of a commercial e-mail message to a large number of people.

spoofing　The use of false return addresses on computer packets so that the source of a computer attack is obscured and cannot be identified and turned off.

spyware　Keystroke-logging software that is downloaded to users' computers without adequate notice, consent, or control for the user.

stakeholder　Someone who stands to gain or lose from a company's performance or business decision.

static testing　The use of special software to look for suspicious patterns in programs that might indicate a defect.

store-and-forward telemedicine The acquisition of data, sound, images, and video from a patient and subsequent transmission to a medical specialist for assessment.

strict liability A version of product liability in which a defendant is held responsible for injuring another person, regardless of negligence or intent.

submarine patent A patent that is not known to be part of a standard until it "surfaces" after the standard is broadly accepted.

system testing A form of software testing that follows successful integration testing, in which various subsystems are combined and tested as a complete entity.

telemedicine Modern telecommunications and information technologies that provide medical care to people who live far away from healthcare providers.

trade secret Information that a company tries to keep confidential, that represents something of economic value, that required effort or cost to develop, and that has some degree of uniqueness or novelty.

trademark A logo, package design, phrase, sound, or word that helps a consumer distinguish one company's products from another's.

trap and trace Technology that can identify the originating number for incoming calls.

TRIPS Agreement The Agreement on Trade-Related Aspects of Intellectual Property Rights, which was established by the World Trade Organization (WTO) to set minimum levels of protection that each government must provide to the intellectual property of all WTO members.

Trojan horse A secretly installed computer program that plants a harmful payload and allows a hacker to steal passwords or spy on users by recording keystrokes and transmitting them to a third party.

USA Patriot Act of 2001 A law passed after the terrorist attacks of September 11, 2001, that gave sweeping new surveillance powers to domestic law enforcement and international intelligence agencies.

user acceptance testing An independent test performed by trained users to ensure that a system operates as expected from their viewpoint.

utilitarianism approach An ethical approach to decision making in which people choose the course of action that has the best overall consequences for everyone involved, either directly or indirectly.

value system The complex scheme of moral values that a person elects to live by.

vice A moral habit that inclines a person to do what is generally unacceptable to society.

virtual private network (VPN) A network that uses the Internet to relay communications, but maintains privacy through security procedures and tunneling protocols that encrypt data at the sending end and decrypt it at the receiving end.

virtue A moral habit that inclines people to do what is generally acceptable to society.

virtue ethics approach An ethical approach to decision making in which people who are faced with an ethical dilemma do what they are most comfortable doing, or do what they think a person they admire would do.

virus A computer program that attaches to a file and replicates itself repeatedly, typically without user knowledge or permission.

virus signature A specific sequence of bytes that identifies a virus.

warranty An assurance to buyers or lessees that a product meets certain standards of quality.

whistle-blowing An effort by an employee to attract attention to a negligent, illegal, unethical, abusive, or dangerous act by a company that threatens the public interest.

white-box testing   Unit testing of software that has expected input and output behaviors and whose internal workings are known. It involves the testing of all possible logic paths through the software unit and is done with thorough knowledge of the software's logic.

worms   Harmful computer programs that differ from viruses because they can self-propagate without human intervention.

zombies   Computers that are taken over by a hacker during a denial-of-service attack and directed to send repeated requests for access to a single target site.

zero-day attack   An attack that occurs before the security community or a software developer knows about the vulnerability or has been able to repair it.

INDEX

A

Accolade, 187
accounting practices
    Computer Associates (CA), 26–27
    Enron's, 5
    SAS (Statement on Auditing Standards), 250
ACFE (Certified Fraud Examiners), 77
ACM (Association for Computing Machinery), 45, 46
ACP (Associate Computing Professional), 48
ACPA (Anti-cybersquatting Consumer Protection Act), 193
activity logs, 90
Adelphia Communications Corp., 6
Aegis radar, 221
affiliated Web sites, 124
AFHCAN (Alaska Federal Health Care Access Network), 288
Airbus, 262–264
AITP (Association of Information Technology Professionals), 45
al Qaeda, 81
Alaska Federal Health Care Access Network (AFHCAN), 288
Altria Group, 1, 2
altruism vs. egoism, 297–298
Amazon.com, shopping cart patent, 178
America Online, 154
American Society for Quality Control (ASQC), 48
animal rights, 302
anonymity
    anonymous expression, remailers, 153–156
    and privacy issues, 113–118
answers to Self-Assessment questions, 331
Anti-cybersquatting Consumer Protection Act (ACPA), 193
AOL (America Online) customer lists, 182–183
Apple
    software defects, 209
    and trade secrets, 171–172
appraisals, employee, 14
Aristotle, 1, 304, 305
Arms Export Control Act, 121
Artzt, Russel, 27
Associate Computing Professional (ACP), 48
Association for Computing Machinery (ACM)
    Code of Ethics and Professional Conduct, 309–316
    organization described, 45, 46
Association of Information Technology Professionals (AITP), 45
attack. *See specific attack*

audits
    IT security, 87
    SAS (Statement on Auditing Standards), 250
    social, 13–14
Avaya, 60

B

backup processes, 87
Barailer, Dr. David J., 277
Barnes & Noble, 178
Barry, Dave, 207
behavior
    code of ethics' effects on, 44
    effect of management on employees' ethical, 15–16
    ethical, 1–4
    moral. *See* morality
    and safety-critical systems, 221
Bentham, Jeremy, 301
Bergonzi, Albert, 29
Berners-Lee, Tim, 179
Better Business Bureau (BBB) Online, 112
Bill of Rights, 147, 154
black-box testing, 216
Black Box Voting: Ballot Tampering in the 21st Century (Harris), 251
Black Hat, 99
Blaster worm, 70
blocking Web sites, 150–151
blogging policies, 165
BLS (Bureau of Labor Statistics), 239
boards of directors, ethical standards set by, 11–12
Boeing, 262, 263
Bonzi Software Inc., 111–112
Borland, Lotus v., 202–203
Brandeis, Justice Louis, 108
breach of contract, 40–41
breach of the duty of care, 50
breach of warranty, 214
bribery, 41–42, 60, 296
Brown & Williamson Tobacco Company, 259–260
Brynjolfsson, Erik, 287
Bureau of Labor Statistics (BLS), 239
Bush, President George W., 117, 277

business ethics
    codes. *See* codes of ethics
    creating ethical work environment, 14–16
    ethical decision making, 17–20
    importance of, 8–10
    improving corporate ethics, 10–14
business information systems, 211
Business Software Alliance (BSA), 36
businesses. *See* corporations, *or specific business*

C

CALEA (Communications Assistance for Law Enforcement Act), 116
camera surveillance, 130
Camilleri, Louis G., 1
CAN-SPAM (Controlling the Assault of Non-Solicited Pornography and Marketing), 129, 159
Capability Maturity Model Integration (CMMI) for software, 216–217
CardSystems Solutions, 76
Catalytic Software, 260–261
CCIE (Cisco Certified Internetworking Engineer), 47
CCP certification, 48
CDA (Communications Decency Act), 149
CDDL (Common Development and Distribution License), 189
Centers for Medicare and Medicaid Services, 267–268, 278
Central Intelligence Agency (CIA), 109, 137
CERT/CC (Computer Emergency Response Team Coordinating Center), 69, 85
certification of IT professionals, 46–48, 59–60
Certified Computing Professional (CCP) certification, 48
Certified Fraud Examiners (ACFE), 77
chargebacks, 78–79
Check Clearing for the 21st Century Act (Check 21), 79–80
checklists
    consumer data responsibilities (table), 127
    contingent employees' use (table), 243
    for establishing ethical work environment (table), 16
    for establishing IT usage policy (table), 52
    ethical competitive intelligence operation (table), 191
    handling freedom of expression in workplace (table), 160
    improving software quality (table), 223
    organization's Internet security incident response (table), 92–93
chief information officers (CIOs), 35
chief privacy officers (CPOs), 126–127
Child Online Protection Act (COPA), 149–150
children
    freedom of expression issues, 149–150
    pornography, 159–160
Children's Internet Protection Act (CIPA), 151–153
Children's Online Protection Act (COPA), 111
China
    and freedom of expression, 145–146
    and outsourcing, 237–238

Chinook helicopters, 218
CIA (Central Intelligence Agency), 109, 137
Ciarelli, Nicholas, 171–172
CIOs (chief information officers), 35
CIPA (Children's Internet Protection Act), 151–153
CipherTrust, 129
Cisco, 100, 166
Cisco Certified Internetworking Engineer (CCIE), 47
CISSP examination, 59
Cleveland State University, 33–34
click-stream data, 124
Clinton, President Bill, 182
CMMI (Capability Maturity Model Integration) for software, 216–217
codes of ethics
    ACM Code of Ethics and Professional Conduct, 309–316
    Association of Information Technology Professionals (AITP) Code of Ethics, 317–319
    establishing, 12–13
    PMI Member Ethical Standards and Member Code of Ethics, 329
    professional, 44–46
    Software Engineering Code of Ethics and Professional Practice, 321–327
coemployment relationships, 241
Cohen, Gerald, 261
collusion, malicious insiders, 77
commerce, contextual, 125
Common Development and Distribution License (CDDL), 189
common good approach to ethical decision making, 18–19
common sense, and relativism, 295–298
Common Sense (Paine), 153
Communications Act of 1934, 109
Communications Assistance for Law Enforcement Act (CALEA), 116
Communications Decency Act (CDA), 149
community good will and good business ethics, 8
companies. *See* corporations
competitive intelligence, 189–191, 195
compilers, and reverse engineering, 186–187
Computer Associates (CA), 6, 26–27, 29
Computer Emergency Response Team Coordinating Center (CERT/CC), 69, 85
computer science majors, 239
computer viruses, 72
computers and optimization technologies, 274
computing resources, inappropriate use, 51
Compuware, 181
consequentialism, 302
Constitution, U.S.
    First Amendment rights, 146–148, 158
    Fourth Amendment rights, 128
    privacy rights, 107–108
consumer profiling, 123–126, 133, 182–183
contextual commerce, 125
contingent workforce, 240–243, 257

Controlling the Assault of Non-Solicited Pornography and Marketing (CAN-SPAM), 129, 159
Cooke, Alistair, 33
cookies, 123
COPA (Children's Online Protection Act), 111, 149–150
copying
    copyright law and patent protection, 172–176, 194
    and intellectual property. *See* intellectual property
    piracy, software. *See* software piracy
    plagiarism, 184–186, 194
corporate ethics
    See also business ethics
    improving, 10–14
    officers, 10–11
corporations
    See also organizations
    ethics programs in, 9–10
    installing firewall, 51, 84–85
    noncompete agreements, 181
    nondisclosure clauses, 180
    satisfied with code of ethics, 12–13
    and unethical behavior, 5–7
    whistle-blowing, 38–39, 251–256
countries
    digital divide across, 272–274
    healthcare spending (table), 275
    offshore outsourcing leaders, 246–247
    regulation of IT professionals, 49
Cox, Rep. Christopher, 166
CPOs (chief privacy officers), 126–127
crackers, hackers, 76
credit card information, stolen, 78–79, 105–106, 139
crime, computer
    See also specific attack
    federal sentencing guidelines, 9–10
    perpetrators of, 75–81
cross-licensing agreements, 178
cryptography described, 118–119
customer lists, 182–183
cybercriminals, 78–79
cybersquatting, 192–193, 195
cyberterrorists, 80–81

D

data
    click-stream, 124
    encryption, 118–121, 132
Data General, 29
databases, and reverse engineering, 186
Db4objects, 188
decision making
    and corporate ethics officers, 11
    ethical, 17–20
    principle-based, 14
decision support systems (DSSs), 211
decompilers, and reverse engineering, 186–187
defamation, 148, 157–158
definitions of terms (glossary), 333–340
demographic filtering, 125
denial-of-service (DoS) attacks, 73–74, 75

deontology, 298–300
Department of Homeland Security, 110
DES (Digital Encryption Standard), 120
Deutch, John, 182
Dewey, Kathleen, 25, 26
Diamond v. Diehr, 177
Diebold, 251
digital divide, 272–274, 285–286
Digital Encryption Standard (DES), 120
Digital Millennium Copyright Act (DMCA), 175–176
DMCA (Digital Millennium Copyright Act), 175–176
Dodd, Cliff, 267
DoS attacks, 73–74, 75
DoubleClick, 124, 126
DSSs (decision support systems), 211
dumpster diving, 191
duty of care, breach of the duty, 50
DVD movies, and copyrights, 176

E

e-mail and spamming, 128–129, 133
Ebbers, Bernard, 5
Echelon eavesdropping system, 139–141
Economic Espionage Act (EEA), 77, 180
ECPA (Electronic Communications Privacy Act of 1986), 114
Ed-Tech program, 273–274
Education Rate (E-Rate) Program, 273
EEA (Economic Espionage Act) of 1996, 77, 180
EFF (Electronic Frontier Foundation), 156, 167
egoism vs. altruism, 297–298
egress filtering, 74
EHRs (electronic health records), 276–277, 281
Eldred, Eric, 173
Electronic Communications Privacy Act of 1986 (ECPA), 114
Electronic Frontier Foundation (EFF), 150, 156, 167
electronic health records, 276–277, 281
employee leasing, 241
employees
    See also workers
    and ethics training, 14
    recruitment and applicant tracking systems (RATS), 264
    reporting violations, 15
    security education of, 83–84
    and trade secrets, 180
    use of nontraditional workers, 239–240
    whistle-blowing, 38–39, 251–257, 258
    workplace monitoring, 127–128, 133
employers and IT professionals, 36–39
encryption, data, 118–121, 132
Energy Reorganization Act, 253
Enron, 5, 6
enterprise resource planning systems (ERPs), 33–34, 48
environmental protection, and whistle-blowing, 252
Eolas Technologies, 179
Epicurus, 300

ERPs (enterprise resource planning systems), 33–34, 48
ethical decision making, 17–20, 24
ethics
    See also business ethics, morality
    in the business world, 5–20
    of care, and feminism, 305–306
    codes of, 12–13, 309–319, 321–327, 329
    issues for IT users, 51–53
    overview of, 1–4
Ethics Officer Association (EOA), 11
European Community Directive 95/46/EC, 111–112
European Union privacy protections, 111–112
Executive Order 12333 (electronic surveillance), 116
exploits described, 70
expression
    freedom issues, 151–153
    legislation regarding freedom of, 153–160

F

facial recognition software, 130
failure mode and effects analysis (FMEA), 222
Fair Credit Billing Act, 78
Fair Credit Reporting Act of 1970, 109
fair use doctrine, 174
fairness approach to ethical decision making, 18–19
False Claims Act, 252–253
Fannie Mae (Federal National Mortgage Association), 10
farming, patent, 179–180
FCC (Federal Communications Commission), 273
FCPA (U.S. Foreign Corrupt Practices Act), 41–42, 60–61
Federal Communications Commission (FCC), 273
Federal Computer Security Report Card, 87–88
Federal National Mortgage Association (Fannie Mae), 10
Federal Wiretap Act, 113–114
feminism, and ethics of care, 305–306
filtering
    demographic, 125
    ingress, egress, 74
    Internet, 150–151
firewall
    and censorship, 145–146
    corporate, 51, 84–85
First Amendment rights, 146–148, 158
FISA (Foreign Intelligence Surveillance Act of 1978), 115
FMEA (failure mode and effects analysis), 222
FOIA (Freedom of Information Act), 109
Ford Motor Company, 286–287
Foreign Corrupt Practices Act (FCPA), 60–61
Foreign Intelligence Surveillance Act of 1978 (FISA), 115
foreign workers, 243–251
Franklin, Benjamin, 105
fraud
    described, 40
    online credit card, 78–79
freedom of expression
    China, and dissent, 145–146
    First Amendment rights, 146–148
    key issues, legislation, 148–158
    pornography and, 158–160

Freedom of Information Act (FOIA), 109
Frye, Robert W., 60

G

Gates, Bill, 261
GCE (Global Connections and Exchange), 285–286
General Motors, 275
Geometric Software Solutions Ltd. (GSSL), 78
GET data, 124
gifts and bribes, 42
Gilligan, Carol, 305–306
Global Connections and Exchange (GCE), 285–286
Global Crossing, 6
Global Internet Freedom Act, 166
Global Positioning System chips, 131
glossary, 333–340
GNU General Public License (GPL), 189
goodness, question of, 294–295
Google, 165, 210
government licensing of IT professionals, 48–50
governmental electronic surveillance, 113–118, 139–141
GPS (Global Positioning System) chips, 131
Gramm-Leach-Bliley Act, 113
Greek virtue, and the good life, 304–306
GSSL (Geometric Software Solutions Ltd.), 78
Gupta, Gautam, 26

H

H-1B workers, 243–246, 257, 261–262
hackers, crackers, 76
Harris, Bev, 251
hate speech, 157–158
HBO (Huff, Barrington, and Owens) & Company, 28–29
Health Buddy program, 268
healthcare
    digital divide in, 267–268
    impact of IT on costs, 275–280, 281
hedonism, 300
Helix Player, 209
Hobbes, Thomas, 303
Homer's *Illiad*, 294
honeypots, 88, 89
HTML (Hypertext Markup Language), 91
Huff, Barrington, and Owens & Company (HBOC), 28–29

I

IBM
    and Lotus, 203
    and noncompete agreements, 181
    patents, 176, 178
    vs. SCO Group, 200–201
iCANN (Internet Corporation for Assigned Names and Numbers), 193

ICCP (Institute for Certification of Computing Professionals), 47
ICRA (Internet Content Rating Association), 150
identity theft, 121–123, 133
Identity Theft and Assumption Deterrence Act of 1998, 122
IEEE (Institute of Electrical and Electronics Engineers), 46
Iliad (Homer), 294
industrial espionage, 189–190
industrial spies, 77–78
industry association certifications, 47–48
information
    controlling Internet access, 148
    credit card theft, 105–106, 139
    customer lists, 182–183
    healthcare (table), 279
    sharing, inappropriate, 43, 51
    technology. *See* IT (information technology)
Information Builders, 261
Information Security Professionals Working Group, 60
information technology. *See* IT
infringement, copyright, 174
ingress filtering, 74
insiders, malicious, 77, 85–86
Institute for Certification of Computing Professionals (ICCP), 47
Institute of Electrical and Electronics Engineers (IEEE), 46
integration testing, 216
integrity
    importance of, 4
    and Philip Morris, 1–2
Intel Corporation, 13–14
intellectual property
    copyright law, 172–176
    cybersquatting, 192–193, 195
    generally, 171–172, 194
    open source code, competitive intelligence, 188–191, 195
    patents, 176–180, 194
    plagiarism, 184–186, 194
    reverse engineering, 186–188, 195
    trade secret laws, 180–184
Intellectual Ventures, 179
intelligence, competitive, 189–191
intentional misrepresentation, 214
International Information Systems Security Certification (ISC2), 59
International Organization for Standardization (ISO) quality management standards, 221–222
Internet
    censorship in China, 145–146
    controlling access to information on, 148–150
    filtering, 150–151
    service providers. *See* ISPs
Internet Content Rating Association (ICRA), 150
Internet Corporation for Assigned Names and Numbers (iCANN), 193
Internet Explorer, 71, 179, 209
Internet Relay Chat (IRC), 72
Internet Security Systems (ISS), 99
intrusion detection systems, 88

intrusion preventing systems (IPAs), 88–89
IRC (Internet Relay Chat), 72
ISO quality management standards, 221–222
ISPs (Internet Service Providers)
    defamation and hate speech, 157–158
    Internet filtering, 150–151
IT (information technology)
    departments, internal controls, 7
    ethics in, 20–21
    impact on healthcare costs, 275–280
    impact on standard of living, worker productivity, 269–271, 281
    professionals. *See* IT professionals
    project fraud, misrepresentation, breach of contract, 40–41
IT companies
    EOA members (table), 11
    outsourcing and, 237–238
IT professionals
    certification, 46–48, 59–60
    government licensing of, 48–50
    malpractice, 50
    relations with clients, employers, 35–43
    résumé inflation, 42–43
    security issues for, 68–71
    and users, 43, 51–53
iTunes, 73

J

Jaschen, Sven, 67–68
John Doe lawsuits, 155–156

K

Kant, Immanuel, 298, 299, 300
Kapor, Mitch, 202
Kaspersky Labs, 209
Katz v. United States, 128
kickbacks, 41–42
Klaudia Consulting Group, 33–34
Kolb, Ben, 229
Koppel, Ross, 230
Kumar, Sanjay, 27

L

Labor Condition Attestations (LCAs), 245
lamers, 76
LAN administrators, 35
laws
    See also specific law
    copyright, 172–176, 194
    intellectual property. *See* intellectual property
lawsuits, John Doe, 155–156
legal departments and ethical work environments, 15

legal issues
    See also ethics
    fraud, misrepresentation, breach of contract, 40–41
    John Doe lawsuits, 155–156
    privacy. *See* privacy
    software product liability, 212–214
legislation
    See also specific law
    about freedom of expression, 148–160
    privacy, 107–113, 132
Lexar, 38
Lexmark, 174–175
liability, software product, 212–214
libel, 148
licensing of IT professionals, 49
"Lincoln Law" (False Claims Act), 252–253
Linux, 200–201, 209
live telemedicine, 278–279
local area network (LAN) administrators, 35
Lockheed Martin, 217
logic bombs, 73
LonghornSingles.com, 129
Lotus v. Borland, 201
Lucent Technologies, 60–61
Lynn, Mike, 99, 100

M

M.A. Mortenson Company, 214
MAA (Machine Accountants Association), 45
Mac computers, Apple trade secrets case, 171–172
Machine Accountants Association (MAA), 45
macro viruses, 72
malicious insiders, 77, 85–86
malpractice, IT professional, 50
management, effect on employees' ethical behavior, 15–16
manager's checklists. *See* checklists
Mangan, Joseph, 263
marketing and spamming, 128–129, 133
Marrero, Maikel, 26
Martin, Clancy, 293
Matrix (Multistate Anti-Terrorism Information
    Exchange), 137
McGinn, Richard, 60, 61
McKesson HBOC, Inc., 28–29
MCSE (Microsoft Certified System Engineer), 47
Media Lab Asia, 274
medical, medicine
    See also healthcare
    information Web sites, 279–280
Medicare, Medicaid, 267–268
MEDPRO, and McKesson HBOC, 28–29
methodology, software development, 215, 224–225
Microsoft
    collection of personal information by, 125
    cross-licensing agreements, 178
    and cybercrime, 100–101
    and H-1B workers, 261
    and outsourcing, 237–238
    software defects, 211
    use of temporary workers by, 260
    Vizcaino v. Microsoft, 242
Microsoft Certified System Engineer (MCSE), 47, 59
Mill, John Stuart, 301
Miller, Dennis, 87
Miller, Erika, 26
Miller, Rick, 100
Miller v. California, 147–148
misrepresentation
    intentional, 214
    offense described, 40
mobile, wireless technology in healthcare, 278
monitoring the workplace, 127–128
Moore, Demi, 149
moral codes and ethics, 3, 293
morality
    defined, 3
    deontology, 298–300
    goodness, 294
    Greek virtue, and the good life, 304–306
    introduction to, 293
    pluralism, 306–307
    relativism, 295–298
    utilitarianism, 300–304
Moss, Jeff, 99
Multistate Anti-Terrorism Information Exchange
    (Matrix), 137
MyDoom e-mail worm, 75

N

n-version programming, 220
National Group for Communications and Computers, 60
National Health Information Network (NHIN), 277
National Infrastructure Security Co-ordination Centre
    (NISCC), 98
National Security Agency (NSA), 139
National Security Letters (NSLs), 156–157
National Traffic Safety Administration (NHTSA), 230–231
NATO and cyberterrorists, 81
NCLB (No Child Left Behind Act), 273–274
negligence suits, 213
NHIN (National Health Information Network), 277
NHTSA (National Traffic Safety Administration), 230–231
Nietzsche, Friedrich, 302, 306
No Child Left Behind Act (NCLB), 273–274
Nolan, Joseph, 34
noncompete, nondisclosure agreements, 180, 181
Norton, Eleanor Holmes, 145
NSA (U.S. National Security Agency), 139

O

obscene speech, 147–148
Odysseus (Homer), 296
OECD (Organization for Economic Cooperation and
    Development), 110–111

Office of Federal Housing Enterprise Oversight
  (OFHEO), 10
officer, corporate ethics, 10–11
offshore outsourcing, 246–251
OFHEO (Office of Federal Housing Enterprise
  Oversight), 10
Omnibus Crime Control and Safe Streets Act of 1968, 114
online crime
    See also specific attack
    identify theft, 121–122
    perpetrator profiles, 75–81
Opanki worm, 73
open source code, 188–189
Open Source Initiative (OSI), 189
OpenSolaris, 189
optimization technologies, 274
Oracle Corporation, 39
Organization for Economic Cooperation and Development
  (OECD), 110–111
organizations
    See also corporations
    ethics programs in, 9–10
    Internet security incident response readiness
      (table), 92–93
    professional, IT, 45–46
    risk assessments, security policies, 82–83
Orms, Dr. Michael, 287–288
OSI (Open Source Initiative), 189
outsourcing, 237–238, 246–251

P

P3P (Platform for Privacy Preferences), 126
Paine, Thomas, 153
patents, 176–180, 194
Patriot Act, 116–118, 156
Patriot missile failure, 231–233
payoffs, 41–42
PayPal, 122
PCAOB (Public Company Accounting Oversight Board), 7
pen register, 114
PeopleSoft, 33–34
perpetrators of online crime, 75–81
personal negligence, 50
personalization software, 125
Peters, Tom, 237
Peterson, Donald, 60
Phelps, Elizabeth Stuart, 171
Philip Morris, 1
phishing, 122, 123
piracy, software, 3, 36–37, 51
plagiarism, 184–186, 194
Platform for Privacy Preferences (P3P), 126
Plato, 294–295
pluralism, 306–307
PMI (Project Management Institute), 46
PMI Member Ethical Standards and Member Code of Ethics, 329

policies
    blogging, 165
    IT security, 87
    regulating freedom of expression, 147
    security, establishing, 82–83
    University of Cincinnati's, on IT, 61–64
political bribes, 41–42
pornography, 147–148, 158–160
POST data, 124
Pre-Paid Legal v. Sturz et al., 155–156
press, freedom of, 153
prevention
    of cookie deposits, 124
    of security threats, 84–88
principle-based decision making, 14
prior art, copyright protection, 176, 179
Prius hybrid, 230–231
privacy
    advanced surveillance technology, 130–131
    consumer data responsibilities, 123–127
    data encryption, 118–121
    governmental electronic surveillance, 113–118
    identify theft, 121–123
    legal protections, 107–113
    spamming, 128–130
    workplace monitoring, 127–128
Privacy Act of 1974, 109–110
private key encryption, 120
Proctor & Gamble Company (P&G), 191–192
product liability, 213
productivity, and IT impact, 269–271, 281
professional codes of ethics, 12–13, 44, 309–319, 321–327
professional malpractice, 50
professional organizations, 45–46
professions, and IT, 35
Project Management Institute (PMI), 46
psychological egoism, 297–298
Public Company Accounting Oversight Board (PCAOB), 7
public key encryption systems, 119–120

Q

QA (quality assurance), 215
QAZ worm, 101
quality management, 210
Quattro spreadsheets, 202–203
Qwest Communications International Inc., 5, 6

R

radio frequency identification (RFID) tags, 138
RAID (redundant array of independent disks) systems, 219
Rambus, 179–180
rating system, ICRA, 150
RATS (recruitment and applicant tracking systems), 264
Rawls, John, 303

RCADS (Remote Cyber Attack Detector Sensor), 89
RealNetworks Inc, 71
RealPlayer Music Store, 198–199
RealPlayer software defects, 209
reasonable assurance, risk assessments, 82
reasonable person, professional standards, 50
Reasoning (company), 188–189
Recording Industry Association of America (RIAA), 261
recruitment and applicant tracking systems (RATS), 264
redundant array of independent disks (RAID) systems, 219
relativism, 295–298
reliability, system, 220
remailers, anonymous, 154
Remote Cyber Attack Detector Sensor (RCADS), 89
Reno v. ACLU, 149
reporting
    incident notification, 89–90
    responsibilities for, 39
    whistle-blowing. *See* whistle-blowing
    violations by employees, 15
résumé inflation, 42
reverse engineering, 186–188, 195
RFID tags, 138
RIAA (Recording Industry Association of America), 261
Richter, Scott, 130
Rigas, John and Timothy, 6
right and wrong, 295–298
rights
    First Amendment, 146–148
    Fourth Amendment, 128
    freedom of expression issues, 148–160
    privacy. *See* privacy
risks
    assessments of, 82
    in the business world, 5–7
    growing IT security, 70–71
RSA (public key encryption), 120
rules-based personalization, 125

S

Sachs, Jonathan, 202
safety-critical systems, 218–221
SANS (System, Administration, Networking, and Security) Institute, 86
Sapient Corporation, 59
Sarbanes-Oxley Act (SOX), 6, 12, 27, 250
Sartre, Jean-Paul, 307
SAS (Statement on Auditing Standards), 250
Sasser worm, 67–68
satellites, intelligence, 140
schools and CIPA regulations, 151–153
Schrage, Michael, 25
Schuetze, Walter, 27
SCO Group vs. IBM, 200–201
script kiddies, 76
Seagate Technology, 121
SEC (U.S. Security and Exchange Commission), 5, 12, 29, 60
secret, trade, 38

Secure Flight airline safety program, 110
security
    audits, 87
    incident notification, 89–90
    issues for IT industry, 68–71
    policies, establishing, 82–83
    reducing vulnerabilities, 81–93
security threats
    detecting, 88–89
    preventing, 84–88
    response to, 89–93
Sega, 187
Self-Assessment questions, answers to, 331
sexual orientation, 154
Singer, Pete, 302
Slammer worm, 91
slander, defamation, 148
slush funds, 41–42
smart cards, 79
Smith v. Maryland, 114
social audits, 13–14
Society of Competitive Intelligence Professionals, 190
Socrates, 294
software
    commercial, with known vulnerabilities, 70–71
    company, guidelines for using, 51
    copyright protection, 175–176
    facial recognition, 130
    improving quality (checklist), 223
    license agreements, and reverse engineering, 187–188
    patents, 177–179
    personalization, 125
    piracy, 3
    quality assurance (QA), 215
    testing, 216
software development
    Capability Maturity Model Integration (CMMI) for software, 216–217
    methodology, 215, 224–225
    process of, 215–216
    quality management standards, 221–223
    safety-critical systems, 218–221
    strategies for, 209–214
Software Engineering Code of Ethics and Professional Practice, 321–327
software piracy, 3, 36–37, 51
Sonny Bony Copyright Term Extension Act, 173
source code, open, 188–189
SOX (Sarbanes-Oxley Act), 6, 12, 27, 250
spamming, 128–130, 133, 154
speech
    See also freedom of expression
    defamation, 148
    obscene, 147–148
spoofing, 74–75
spying, industrial, 77–78
spyware, 73, 122
SQL Slammer worm, 70
stakeholders described, 16–17
standard of living, IT impact on, 269–272, 281

standards
    ISO (International Organization for Standardization), 221–222
    and patents, 179
    and Plato's idea of the good, 295
    quality software management, 221–223
    SAS (Statement on Auditing Standards), 250
Statement on Auditing Standards (SAS), 250
Static Control Components (SCC), 174–175
static testing, 216
Strassmann, Paul, 287
student plagiarism, 185–186
Subimo, 280
submarine patents, 179–180
Sun Microsystems, 71, 165
SunTrust Banks Inc., 7
surveillance
    advanced technology, 130–131
    governmental electronic, 113–118
    privacy issues. See privacy
Sutter Health, 277
swiping (identify theft), 105–106
Symantec Corporation, 210
system testing, 216
Szekely, Dr. Daniel, 288

T

Tao, Shi, 166
Tata Consultancy Services, 237–238
technology, information. See IT
Tekiela, Robert, 59
teleological theories, 302
temporary workers, 240–251
Tennant, Don, 59
testing software, 216
theft, identify, 121–123, 133
Therac-25 radiation machine, 218
Thus Spoke Zarathustra (Nietzsche), 302
Timberline, 214
Toshiba Corporation, 38
Toyota's Prius hybrid, 230–231
Toysmart.com, 126–127
Trade-Related Aspects of Intellectual Property Rights (TRIPs) Agreement, 183–184
trade secrets, 38, 77, 91, 180, 180–184, 200–201
trademarks, 192
trainings in ethics, 14
trap and trace orders, 114
TRIPs (Trade-Related Aspects of Intellectual Property Rights) Agreement, 183–184
Trojan horses, 73, 98–99, 100–101
TRUSTe, 112, 126–127
TTTech Computertechnik, 263
Tyco, 6

U

UK government and Trojan horse attack, 98–99
Uniform Trade Secrets Act (UTSA), 180
Unilever, 191–192
United Auto Workers, 275
University of Cincinnati, 61–62
UNIX, SCO Group vs. IBM, 200–201
U.S. Code of Federal Regulations' definition of profession, 35
U.S. Constitution
    Bill of Rights, 154
    First Amendment rights, 146–148, 158
    Fourth Amendment rights, 128
U.S. Foreign Corrupt Practices Act (FCPA), 41–42
U.S. Justice Department's sentencing guidelines for ethical violations, 9–10
U.S. National Security Agency (NSA), 139
U.S. Patent and Trademark Office (USPTO), 176, 177, 192
U.S. Security and Exchange Commission (SEC), 5, 12, 29, 60
USA Patriot Act of 2001, 116–118, 156
USB flash drives, 38
Usenet, 128–129
user acceptance testing, 216
USPTO (U.S. Patent and Trademark Office), 176, 177, 192
utilitarian approach to ethical decision making, 18–19
utilitarianism, 300–304
UTSA (Uniform Trade Secrets Act), 180
Uzbekistan, digital divide in, 285–286

V

value systems of organizations, 9
Van Gogh, Vincent, 302
Van Umersen, Claire, 33
Vanity Fair, 149
vendor certification, 47
Verizon, 5, 91
Verma Shekhar, 78
Veterans Administration's use of telemedicine, 288
virtual private networks (VPNs), 83
virtue ethics approach to ethical decision making, 18–19
viruses
    and cyberterrorists, 80–81
    described, 72
    Sasser worm, 67–68
    signatures, 85
Visa Reform Act of 2004, 246
Visual Basic, and worms, 72–73
Viscaino v. Microsoft, 242
VoIP technology, 116
VPNs (virtual private networks), 83

W

Wang, Charles, 27
warranties, software, 213–214
Web sites
    affiliated, 124
    blocking, 150–151
    Business Software Alliance (BSA), 36
    defaced, 91
    health information (table), 279
whistle-blowing, 38–39, 99, 251–257, 258
white-box testing, 216
White Buffalo Ventures, 129
white hats, 99
Whitehead, Alfred North, 294–295
Wigand, Dr. Jeffrey, 259–260
Williams, Bernard, 302
Windows Genuine Advantage program, 76
WIPO (World Intellectual Property Organization) Copy-
    right Treaty, 175
wireless
    Ford's use of, 286–287
    and GPS chips, 131
    technology in healthcare, 278
wiretapping, 113–118
work environment, creating ethical, 14–16

workers
    See also employees
    contingent workforce, 240–243, 257
    H-1B workers, 243–246, 257
    productivity, and IT impact, 269–271, 281
    use of nontraditional, 239–240, 257
workplace monitoring, 127–128, 133
World Intellectual Property Organization (WIPO) Copy-
    right Treaty, 175
World Trade Organization (WTO), TRIPs Agreement,
    183–184
WorldCom, 5, 6
worms, 67–68, 72–73, 75, 80–81, 86

Y

Yahoo!, 166

Z

Zenger, John, 153
zero-day attacks, 70–71
zombie attacks, 74
Zotob computer worm, 70